THE
NEW EUROPEAN
ECONOMY

The Politics and Economics
of Integration

Second Revised Edition

LOUKAS TSOUKALIS

D0109430

OXFORD UNIVERSITY PRESS

Oxford University Press, Walton Street, Oxford OX2 6DP
Oxford New York Toronto
Delhi Bombay Calcutta Madras Karachi
Kuala Lumpur Singapore Hong Kong Tokyo
Nairobi Dar es Salaam Cape Town
Melbourne Auckland Madrid
and associated companies in
Berlin Ibadan

Oxford is a trade mark of Oxford University Press

Published in the United States
by Oxford University Press Inc., New York

British Library Cataloguing in Publication Data
data available

Library of Congress Cataloging in Publication Data
Tsoukalis, Loukas.
The new European economy: the politics and economics of
integration/Loukas Tsoukalis.—2. rev. ed.
Includes bibliographical references.
1. European Economic Community. 2. Europe 1992. 3. European
Economic Community countries—Economic policy.
HC241.2.T74 1993   337.1'42—dc20   92–32479
ISBN 0–19–828795–X

3 5 7 9 10 8 6 4 2

Printed in Great Britain
on acid-free paper by
Biddles Ltd., Guildford and King's Lynn

To Marta

Preface to the Second Edition

When the writing for the first edition of this book was completed in January 1991, accompanied by a strong sigh of relief, I had not expected that I would be going back to it a year later in order to prepare the second, revised edition. The book was well received, public interest in European integration was as high as ever, and the pace of events indeed breathtaking. It is for those reasons, I presume, that Oxford University Press came up with the idea of an early second edition of the book in paperback form. Some updating would be desirable, it was suggested; but this could be, perhaps, limited to a few interventions, especially in the concluding chapter. I welcomed the idea, even though I realized at the time that revision would entail more than the updating of figures here and there and incorporating recent events in the conclusions. I had not, however, anticipated the scale of the enterprise involved, which became increasingly clear once I started working on individual chapters.

There is so much that has happened in a very short period of time: the Maastricht treaty; important new legislation in the context of the internal market programme; new far-reaching agreements between the Twelve and several other European countries, which have not, however, eliminated the issue of further enlargement from the top of the agenda; the continuing effects of German unification and the breakdown of the old political and economic order in Eastern Europe; not to mention autonomous developments in the market-place, which have had a significant impact on the nature and shape of the new European economy. This second edition contains a substantial revision of several chapters, and most notably those on social policy, economic and monetary union, regional and external policies. Most of the data used in the first edition have been brought up-to-date, and this is also true of all tables and figures. The main arguments and the ideas developed in the book have hardly changed, although minor adaptations have been made in response to new developments and theoretical insights.

Writing on regional integration and the European economic

system in general is becoming an almost impossible task, and this was strongly felt by the author in the preparation of the second edition. Gone are those 'happy' days when a book on this subject was quite manageable; and how comfortable for the author when the book remained more or less up-to-date for several years. Nowadays, superficiality is a very high risk for anybody who still resists the temptation of narrow specialization. On the other hand, writing in times of rapid change and trying to identify the main line of direction and the links between apparently unconnected events and policies may be very challenging and exciting, but also extremely risky. It is again up to the reader to pronounce judgement on the final product.

I should like to express my thanks to all those who have contributed to this second edition with their comments and suggestions. I am particularly grateful to my assistant at the College of Europe, Outi Jääskeläinen, for her research assistance and for the excellent work she has done with respect to the preparation of tables and graphs. Thanks are also due to Niall Bohan for his research assistance, and to John Watson for compiling the index. As always, all remaining errors are entirely mine.

L.T.

June 1992

Preface to the First Edition

The urge to write a general book on the European economy has grown slowly over the years, nourished by many lectures delivered to different audiences, research and writing on a variety of issues in the field of regional integration, and periodic entries into the real world of decision-making.

Having worked in the past in several areas of European political economy, a term which has become once again fashionable after having been banished for many years from the vocabulary of respectable economics, I decided to embark upon a much more ambitious exercise, namely a book which would try to capture the shape of the emerging regional economic system in Europe. This meant going against the general trend of specialization largely forced upon any social scientist by the growing complexity of issues and the sometimes embarrassing wealth of data available. The continuous expansion and 'deepening' of regional integration in Western Europe already justifies the use of the term 'European economy' or 'European economic system' which is qualitatively different from the set of highly interdependent national economies which we have known for some time.

Trying to figure the shape of this new economic animal has been a challenging task which required a certain degree of familiarization with many different sectors and areas of economic policy, not to mention the often conflicting interests and perceptions of the major actors involved. It also required a dangerous kind of navigation between the Scylla of overgeneralization and grand theorizing on the one hand, and the Charybdis of narrow empiricism on the other. It has been, indeed, a challenging task and sometimes also highly frustrating because of the mountains of material which had to be digested and the obstinate refusal of the real world to stay still. I have learned a great deal in the process. It is now up to the reader to decide whether this exercise has been at all worthwhile.

I would not have dared to embark on this project had I not lectured on various aspects of the subject both inside and outside

Europe. My lecture course at the College of Europe in Bruges formed the basis from which I started. Successive *Promotions* of students from all over Europe provided me in Bruges with the inspiration and the stimulus to develop my ideas and to think 'European'. Oxford offered a different kind of testing-ground for those ideas: a very international outlook, a critical and slightly detached approach to European integration, and, most importantly, its very high standards. Both places have had a strong influence on my ideas and method of work.

This book started being written in Bruges and Oxford, and it has been completed in Athens where I have returned recently, dare I say a more mature person, after many years of absence, to live and work at home. Europe is a relatively small place, intra-European frontiers are rapidly disappearing, and national loyalties are no longer incompatible with a strong sense of belonging to this new Europe. Having lived and worked in different countries has helped me to acquire a different perspective on issues European. This will be, perhaps, considered as an asset by the reader of a book which does not treat economics as only a positive science and in which different interests and perceptions are deemed to play a decisive role in the economic process. I have made a deliberate effort not to view the various issues discussed in this book from a particular national angle. Needless to say, the views expressed in it are not necessarily those of any institution with which I have been or I am presently associated; although I hope that this disclaimer should be absolutely superfluous for anybody who cares to read more than a few pages of this book. This should not, however, be interpreted as a rejection of value judgements. I am sure there are plenty of them in this book and I have certainly not tried to hide them behind a thick cloak of positivism. To be fair to the reader, I have tried as much as possible to back statements with empirical evidence while also presenting, as dispassionately as humanly possible, the alternative points of view.

This book is based on the experience and knowledge I have accumulated during the many years I have been studying and working in this field. I have learned from many people and sometimes I may not have done sufficient justice to their ideas. I am particularly grateful to my students who have taught me a great deal, while also helping me to clarify my own thinking. I shall name only a few people whose assistance has been much appreciated in the preparation of this book. Most of all, my assistant at the College of Europe, Niall Bohan, who has done much of the statistical work and has also helped a great deal with documentation. I should also like to thank Declan Costello, Mads Kieler, Jean-Yves Muylle, and Kristin Schreiber who have provided me with valuable research assistance; and Eric de Souza who offered a helping hand whenever I got lost in the modern world of computers. Several people have read drafts of individual chapters. I am especially grateful to Helen Wallace, and also to Andrea Boltho, Anthony Kastrissianakis, and John Morley. I should like to acknowledge the financial support I have received from the Nuffield Foundation which helped to pay for some of the costs of research assistance. Last but not least, I am grateful to my former colleagues at Oxford for the mental challenge with which they provided me for years and also for their tolerance during my frequent periods of absence in other parts of the world, to my colleagues in Bruges for an exciting and truly European environment, and to my new colleagues in Athens for the warmth of their welcome; my new environment still waits to be discovered.

This book is addressed to the interested observer and active participant of the European economy in the making. It is, of course, also addressed to the student of European and international economics who hopefully does not believe that economics takes place in a political vacuum. It assumes regular reading of the quality press and some prior knowledge of economic theory, although special effort has been made to avoid the use of technical jargon whenever this was not absolutely necessary. It has a strong policy orientation, although it does not always refrain from raising some more 'academic' issues (part of the inevitable *déformation professionelle*). It is meant as a contribution, small though it may be, to the debate about the nature and future development of the European economy, without shying away from some of the wider political ramifications of the economic issues involved.

L.T.

January 1991

Contents

List of Figures

Map

List of Tables

Note: The symbol — means that figures are not available.

List of Abbreviations

ACP	Africa-Caribbean-Pacific
AFNOR	Association Française de Normalisation
BRITE	Basic Research in Industrial Technologies for Europe
BSI	British Standards Institution
CAP	Common Agricultural Policy
CCP	Common Commercial Policy
CEN	Centre Européen de Normalisation
CENELEC	Centre Européen de Normalisation Électrotechnique
CET	Common External Tariff
CIS	Commonwealth of Independent States
CMEA	Council of Mutual Economic Assistance
CSCE	Conference on Security and Cooperation in Europe
DIN	Deutsches Institut für Normung
DM	Deutschmark
EAEC	European Atomic Energy Community
EAGGF	European Agricultural Guidance and Guarantee Fund
EBRD	European Bank for Reconstruction and Development
EC	European Community
ECOFIN	Council of Economic and Finance Ministers
ECSC	European Coal and Steel Community
ECB	European Central Bank
ECU	European Currency Unit
EDC	European Defence Community
EDF	European Development Fund
EEA	European Economic Area
EEC	European Economic Community
EFTA	European Free Trade Association
EIB	European Investment Bank
EMCF	European Monetary Cooperation Fund
EMI	European Monetary Institute

EMS	European Monetary System
EMU	Economic and Monetary Union
EOTC	European Organisation of Testing and Certification
EPC	European Political Cooperation
EPU	European Payments Union
ERDF	European Regional Development Fund
ERM	Exchange Rate Mechanism
ESC	Economic and Social Committee
ESCB	European System of Central Banks
ESF	European Social Fund
ESPRIT	European Strategic Programme for Research and Development in Information Technology
ETSI	European Telecommunications Standards Institute
ETUC	European Trade Union Confederation
EUREKA	European Research Co-ordinating Agency
FDI	foreign direct investment
GATT	General Agreement on Tariffs and Trade
GCC	Gulf Co-operation Council
GDP	Gross Domestic Product
GFCF	gross fixed capital formation
GNP	Gross National Product
GSP	Generalized Scheme of Preferences
HDTV	high-definition television
ILO	International Labour Organisation
IMF	International Monetary Fund
IMP	Integrated Mediterranean Programme
JESSI	Joint European Submicron Silicon Initiative
LTA	Long-Term Arrangement
MCA	monetary compensatory amount
MEP	Member of the European Parliament
METS	minimum efficient technical scale
MFA	Multifibre Arrangement
MFN	Most Favoured Nation
NAFTA	North Atlantic Free Trade Agreement
NATO	North Atlantic Treaty Organisation
NCI	New Community Instrument

NIC	newly industrializing country
NIEO	New International Economic Order
NTB	non-tariff barrier
OECD	Organisation for Economic Co-operation and Development
OEEC	Organisation for European Economic Co-operation
OJ	Official Journal
OPEC	Organisation of Petroleum Exporting Countries
PHARE	Pologne, Hongrie assistance pour la restructuration économique
PPS	purchasing power standard
PSE	Producer Subsidy Equivalent
RACE	Research in Advanced Communications for Europe
SAD	Single Administrative Document
SDI	Strategic Defence Initiative
SEA	Single European Act
SME	small and medium enterprises
UNICE	Union des Confédérations de l'Industrie et des Employeurs d'Europe
VAT	value added tax
VER	voluntary export restraint
WEU	Western European Union

1. Introduction

This book is about the process of integration in Western Europe and the gradual emergence of a new regional economic entity. It concentrates mainly, although not exclusively, on the latest phase of this process of integration which started around the mid-1980s and has been closely associated with the establishment of the European internal market. It attempts to identify some of the distinguishing features of the new European economy and the main outstanding issues as governments prepare for the next phase of integration leading to economic and monetary union (EMU).

After a long period of stagnation, during which early federalist and functionalist hopes built around the three European Communities — the European Coal and Steel Community (ECSC), the European Economic Community (EEC), and the European Atomic Energy Community (EAEC) — seemed to have been buried under the resurgence of nationalism, economic recession, and the inertia of common institutions, things started moving again in the second half of the 1980s; and at a very rapid pace indeed. Thus, in the course of only a few years, the economic and political climate in Western Europe was radically transformed.

In the early and mid-1980s, terms such as 'Euro-pessimism' and 'Euro-sclerosis'[1] were meant to descibe the functioning of European economies and their prospects in the context of international competition. And then the picture suddenly became brighter: regional integration picked up a much faster pace, the European Community (EC)[2] occupied again the centre of the stage, while most macroeconomic variables registered a dramatic and sustained improvement. Thus, the economic

---

[1] This term was first introduced by Herbert Giersch, a German economist. It referred to the apparent rigidities in European labour and product markets.
[2] This is the term used for the three above-mentioned Communities.

climate — the variable which has consistently defied the awkward attempts made by economists to tame it with their econometric models — had changed once again, to an extent that would have been unimaginable only a few years earlier.

The process of integration has always been characterized by fits and starts; and the latest phase may be considered as yet another, but still very powerful, push on the accelerator, to be followed by the inevitable shift to a smaller gear and a lower speed. Whether we are already experiencing such a shift is not yet clear, and the signs are rather contradictory. On the one hand, the agreement reached at Maastricht to proceed towards a complete EMU suggests that integration is likely continue at a high speed in the post-1992 period. On the other, the worsening of the economic environment and the doubts expressed about the feasibility of the Maastricht agreement may point to the end, even if temporary, of 'Euro–optimism'. Future developments will obviously depend on a host of domestic and international factors which are still unknown variables; the sequence of political and economic events in recent years has been, indeed, breathtaking.

The early stages of European integration were mainly about the elimination of trade barriers, and this eventually brought about a, still incomplete, customs union in goods and the rapid growth of trade interdependence. Developments since the mid-1980s constitute a qualitative shift in this historical process. What is now at stake is the creation of a European economic system as a superstructure built on top of already densely knit national economies. And much of it seems to be irreversible, to the extent that this term can be used at all with respect to the affairs of human beings. The main issues now revolve around the regulation of markets, the development of common policies at both the micro- and macroeconomic levels, and the restructuring of European industry and the services sector, involving increased cross-border co-operation among firms.

European integration has been and continues to be largely about economics; but economics which has wider political ramifications. The distinction made by traditional theories of international relations between high politics, referring to matters of national security and prestige, and low politics, reserved for more mundane issues such as trade and money (thus relegating economic issues to the second division of international affairs) has always looked somewhat suspect; although, perhaps, understandable when seen through the eyes of superpowers at the peak of the Cold War. Such a distinction can be positively misleading when applied to the contemporary European reality.

Therefore, the study of economic issues will be placed in its proper political context. It will also be set against a historical background in order to facilitate a better understanding of the evolution of the attitudes and policies of the parties concerned. In other words, a political economy approach will be adopted in an attempt to obtain a deeper insight into the ongoing process of economic integration in Europe and its possible outcomes. Since so much revolves around questions of regulation and distribution, we shall try to explore the likely effects of the completion of the internal market and the associated policies on the prevailing economic order, thus shedding light on the kind of interaction between state and markets in the emerging European economic system. Is 1992 simply a policy of deregulation aimed at the gradual dismantling of the mixed economy and the welfare state established in post-war Western Europe or is it an attempt to introduce new forms of regulation at a higher level? What is the importance of accompanying policies and what will be the implications of integration in the macroeconomic sphere leading to the adoption of a single currency? Last but not least, how does Europe define its role in the global economy?

The efficient allocation of resources and the maximization of global welfare have been almost the exclusive concern of neo-classical economics. However, politics in the real world is not only about efficiency but also about distribution. Economic policies will be discussed in this book with a close eye on their distributional aspects among countries, regions, and social classes, in full awareness of the strong limitations in this respect of the available data and our tools of analysis. Attention will also be devoted to a very different, albeit highly sensitive, dimension of distribution, namely the distribution of economic power between different levels of decision-making. As integration approaches the very heart of economic sovereignty, the resistance of the European nation-state is bound to grow; and this may be only half concealed by the European rhetoric of many national politicians. Economic integration is also political integration; the search for greater efficiency does not take place in a power vacuum.

This book does not follow the usual structure found in most

other books on European integration, with chapters devoted to individual EC policies. For example, the reader may be surprised not to find a separate chapter on agriculture; this omission may even be considered as a sacrilege. Sectoral issues are, however, treated in different chapters of the book, and this is certainly true of agriculture. Furthermore, this is a subject which has been written to death in many other publications and which, although still important, no longer occupies the same central position in the politics of economic integration. There is no separate chapter on industrial policy either, although in this respect there may be less need for an apology in view of the rather debatable existence of this policy at the European level. Questions of market regulation and policies aimed at influencing the allocation of resources are, however, dealt with extensively in this book.

European integration has evolved within the context of growing international economic interdependence. The period since the end of the Second World War has been characterized by a continuous and rapid growth of international economic exchange. Trade in goods has almost invariably expanded faster than world output. Services have become increasingly internationalized, following the explosion of transnational flows of capital and finance. National frontiers are now only one among many different factors influencing business decisions about the location of production. Time lags in the transfer of information and technology have been drastically reduced. And people have been moving from one country to the other in ever-increasing numbers, changing their residence and place of work or simply for visits of variable duration, for work or pleasure.

What is, perhaps, striking is that this process of economic internationalization has continued almost irrespective of the significant changes which have intervened both in terms of the world economy and the political balance of forces. True, the process of economic internationalization has not been entirely smooth. However, reversals which may have occurred until now have proved to be shortlived, and, adjusting for 'seasonal' fluctuations, almost all available indicators have been in the ascendant.

The period of rapid economic growth of the 1950s and the 1960s was succeeded by a prolonged world recession. Yet the dire predictions about the imminent end of international economic interdependence were not confirmed by trade, investment, and finance figures for the 1970s and the first half of the 1980s. True, there was a deceleration of the process, particularly during the early 1980s. Yet it was a deceleration and not a reversal of the earlier rising trend, unlike, for example, the experience of the inter-war period. With the return to high rates of economic growth in the second half of the last decade, the process of economic internationalization gathered new momentum; and this became more pronounced in the area of finance.

Similarly, the period of now almost fifty years since the end of the Second World War has been characterized by significant changes in international relations and the world balance of power. Many Western observers, especially American, have taken pains to demonstrate the decline of the hegemonic power of the United States, on which the post-war international economic order had been founded.[3] Although the causes and the nature of this decline can be a matter of dispute, such a decline, in relative terms, has, undoubtedly, occurred; and the internationalization of economic exchange may have been influenced as a result. But again, to the extent that there has been a negative effect, itself a debatable proposition, there has been no reversal of the above process. Once more, contrary to academic doomsday theories.

Intra-European developments have been, of course, strongly influenced by changes in the world economy. European countries, because of their size and history, have been active participants in the process of economic internationalization. Yet European integration cannot be seen as simply another manifestation of the phenomenon of economic interdependence. The difference between the European and the world level is both quantitative and qualitative. It lies in the different degree and nature of economic interaction between national units and the relatively advanced form of joint rule-setting and management. Autonomous economic processes and policy co-operation have been mutually reinforcing. This is particularly true of more recent developments. Thus, there has been and still remains a clear

---

[3] There was for years a rapidly growing literature on this subject which started from the thesis, itself based on limited historical evidence, that international economic order depends largely on the existence of a hegemonic power which, in post-war years, was identified with the United States (Calleo, 1982; Keohane, 1984; Gilpin, 1987; Kennedy, 1988). After much breast-beating about the decline of American power and its alleged negative effects on the international economic order, US observers have started rediscovering, with glee, those ingredients of power that have remained untouched (Nye, 1990). And this has been strengthened by the collapse of the Soviet empire. We are unlikely to witness, as yet, the end of the American superpower.

discontinuity, in terms of the intensity of economic interaction and joint policy-making, between the European economic system and the rest of the world.

Any talk about discontinuity implies the existence of boundaries. And in this case boundaries have never been clearly marked. Although it is perfectly true that the EC, with its ever-increasing membership, has always been at the centre, acting more and more as the main focus of attention and the motor for regional integration, the latter, as measured by trade flows and movements of factors of production, has not been limited to member countries of the EC.

Economic transactions have a life of their own and are not entirely determined, although still strongly influenced, by institutional arrangements. The wider Western European area, which includes at least the present members of the European Free Trade Association (EFTA), is characterized by a high degree of economic interdependence. Autonomous economic flows have been strengthened by various forms of policy co-operation between the EC and individual EFTA countries. In fact, in certain areas of the economy, countries outside the EC have been more closely integrated into the European economic system than some of the existing members. This applies, for example, to monetary policy and exchange rates and also to co-operation in the field of high technology.

The recent acceleration of the process of integration inside the EC has had a strong impact on the Community's other Western European neighbours. These countries have been closely following EC legislation and trying to adjust unilaterally their domestic regulations and policies to those of the EC in order to avoid their exclusion from the large European market.

A discussion of the emerging European economic system cannot be confined to the present members of the EC. Such an attempt would fly at the face of economic reality, not to mention the new agreement signed between the EC and EFTA and also the strong possibility of further enlargement in the foreseeable future. This, undoubtedly, adds a further complicating factor to an already complex exercise. In examining the various aspects of regional integration, this book will take the EC as its main point of reference, while extending the analysis, whenever possible, to include the other economies of Western Europe.

Until the extraordinary events of 1989, the *annus mirabilis* for

Eastern Europe, regional integration was meant by virtually everybody to refer to the European Community, and sometimes to Western Europe more generally. Since the political and geographical definitions of Western Europe were not identical (itself a remnant of the post-war political settlement which brought a country like Greece to the western side of the dividing line), the concept of integration was more specifically applied to countries with pluralist political systems and market economies. Countries to the east of the Iron Curtain were outside this process and were involved in their own, and not particularly successful, form of integration. But the rapid and very unexpected dismantling of the Soviet empire in Eastern Europe and the collapse of communist regimes in those countries have put an end to this division.

Developments in Eastern Europe, combined with the acceleration of regional integration in the West and the strengthening of the EC, have radically transformed the economic and political map of the old continent, turning, perhaps, the last decade of this turbulent century into a decade of Europe. They are also likely to interact closely with each other. The further development of the EC will be influenced in many different ways by the unfolding of events in the East. German reunification and the transition of other Eastern European countries to pluralist political systems and market economies, a difficult and painful transition, will certainly have a significant impact on the integration process inside and around the EC. For the first time, there is also a concrete possibility of the European economic system extending to the whole of the continent. But this is bound to be a long process.

Having tried to define the parameters for this book, the author is still left with an enormous task. With the years, European integration has extended to a very large number of areas. In view of the complexity of the issues involved and also the fact that patterns of decision-making often vary considerably from one area of activity to the other, depending on historical circumstances, the nature of the activity, and the internal as well as external balance of forces, most observers of the European scene have abandoned any attempt to analyse the phenomenon of integration as a whole, opting instead for narrow specialization. The relative failure of earlier attempts at general theorizing (be it from the political or the economic angle) to contribute much to our understanding of the whole process, not to mention their predictive capacity, has acted as a further disincentive against the search for the overall view.

In my early period of initiation to European afairs, I remember reading an article in which the author likened the process of integration to an elephant and researchers to blind men who by touching different parts of the body try to draw conclusions about the shape of the whole animal (Puchala, 1972). I have embarked on this project painfully aware of the limitations and difficulties of the exercise. This book is an attempt to provide a general picture of economic integration and the emerging European economic system, by concentrating especially on the links between different areas of policy and the real economy. Writing on recent and contemporary issues also involves an additional risk in view of the time-lag between writing and publication. In fact, in spite of the tremendous improvement in information flows, we are still forced to make predictions about the past, extending to several months or even years in the case of certain economic variables. One may sometimes feel jealous of the historian who writes about more distant events; but there is also some excitement associated with uncertainty. The aim of this book is not towards grand theorizing nor is it to engage in an exercise in futurology. On the contrary, there will be a strong policy orientation and as much reliance as possible on 'hard facts'. Hopefully, the search for the overall shape of the elephant will not lead to too much of a fuzzy picture.

The book starts with a historical section which is intended to provide a *tour d'horizon* of European economic integration and its different phases. Chapter 2 covers the period between the end of the Second World War and the 'dark years' of the post-1973 prolonged economic recession, when the whole European edifice looked in danger of collapsing. It traces the evolution of the integration model from the period of the economic boom to the years of the recession, its links with national and international developments, and the effects of regional integration in terms of trade and welfare. It also examines the apparent contradiction between the trade liberalization of early years and the development of the mixed economy and the welfare state in individual Western European countries; a contradiction which proved manageable only in the years of rapid economic growth. Changing economic circumstances and successive enlargements of the

EC have had a noticeable effect on the overall package deal which has sustained the process of integration, thus also having a direct impact on common policies and institutions.

The next chapter concentrates on the events of the 1980s and early 1990s in an attempt to understand the dramatic transformation of the political and economic climate in Western Europe during this period and the reasons behind it; a transformation which has left an indelible mark on the process of integration. The 1992 phenomenon is about the successful interaction between government policies and autonomous market developments, and in the process the European political agenda has changed beyond recognition. The transition from the 'Europessimism' of earlier years to the virtuous circle of higher growth and closer integration in the second half of the decade offers a fascinating case-study of political leadership, excellent marketing skills, and, perhaps, also luck, this necessary ingredient of success. The last section of this chapter is devoted to the treaty revision agreed at Maastricht, which sets the framework for economic integration for the rest of the decade.

The nature of the remaining barriers and the large gap still separating the EC in the mid-1980s from a true internal market are discussed in Chapter 4. In the context of mixed economies, integration is a long process and the old distinction between the different stages loses much of its meaning. The emphasis gradually shifts from border controls, mainly in the form of tariffs, to domestic government intervention and the multitude of manifestations of the mixed economy. It also slowly shifts from goods to services and factors of production. The internal market programme is about stronger competition and the freer interplay of market forces, which is in turn consistent with the general shift to liberal economic ideas during this period. Its effects will be felt over several years, and there is little reason to expect that those effects will be evenly distributed between different countries, regions, and social classes. The 1992 programme is also an attempt to influence expectations and the orientation of European businesses; in this respect, the effect of the programme is already pronounced.

The next chapter goes into the heart of the 1992 process by examining specific attempts to eliminate remaining barriers in the intra-EC movement of goods, services, and factors of production. Since integration does not take place in a *laisser-faire* environment, the elimination of those barriers raises delicate questions about the appropriate mixture of liberalization and regulation in the context of mixed economies; in other words, about the interaction between the State and the market. What is also at stake in this respect is the distribution of power between national and supranational levels of authority. This chapter contains four case-studies: the first one examines the recent wave of cross-border mergers and acquisitions and its influence on the development of a new EC regulatory framework as part of the common competition policy; the second deals with the liberalization of financial services and the new supervision rules adopted at the Community level; the third one examines the difficulties encountered in terms of tax harmonization and the implications for public finances; and the fourth one tries to make sense of the highly technical world of standards and regulations as potent instruments of industrial policy and external protection. The conclusions drawn from each case-study in terms of the mixture of liberalization and regulation are not identical; and why should they be so? There is no single pattern of integration followed in each and every sector of economic activity. Those case-studies offer some insight into the complexity of the integration process and the changing economic order in Europe.

Given the special characteristics of labour-markets, the attempts to agree on a common regulatory framework at the European level are discussed separately in Chapter 6. As with social policy more generally, regulation is intimately linked with the objective of redistribution. This is an area where national differences have largely survived the impact of economic integration; and this in turn explains why the role of EC institutions had remained relatively marginal until recently. Attempts to add a social dimension to the internal market and the whole process of 1992 have come with a considerable time-lag, almost as an afterthought. In the debate which has ensued and which has been largely an attempt to win the hearts of trade unionists, ideology and political symbols seem to have taken precedence over concrete measures. Social dumping has been the term coined by the better-off to give expression to their fears about excessive deregulation of labour-markets and unfair competition. Little progress has been achieved until now towards the creation of a European 'social space'; and it remains to be seen whether the highly complex agreement reached at Maastricht on this subject will succeed in breaking the political deadlock.

The discussion about new forms of regulation in product and labour-markets is followed by an examination of stabilization policies at the European level. The Community had started from virtually zero and gradually developed over the years mechanisms for economic policy co-ordination centred on a system of fixed but adjustable exchange rates. The snake was the precursor to a much more successful European Monetary System (EMS), and this is in turn intended to provide the basis for the creation of a complete EMU by the end of the century. In view of its importance, this subject is likely to dominate the European political agenda for the rest of the decade. After all, a complete EMU will radically transform the European economic system and may also bring political union through the back door; hence sometimes the tendency for economic fundamentals to be sacrificed at the altar of political expediency. This chapter attempts to draw a balance sheet of the snake and the EMS, and it then concentrates on some of the main issues associated with the transition to EMU: the importance of the exchange rate instrument for the open economies of the EC; the likely costs for some of the weaker members; the issue of interregional transfers; the link between monetary and economic union; and last but not least, the link between monetary and political union. The strict criteria adopted at Maastricht for the admission of countries to the final stage of EMU also raises the question of different tiers or speeds inside the Community.

The discussion of EMU is followed by an examination of redistributive policies at the Community level. Redistribution is one of the central elements of the European mixed economy; it can also be considered as an indicator of the degree of internal cohesion of a political system. Chapter 8 discusses the nature and size of regional disparities inside the EC, and the link between those disparities and the process of European integration. Since it is virtually impossible to generalize on such very broad issues, the introductory comments on the evolution of regional disparities are followed by a discussion of the experience of three less developed and geographically peripheral countries, namely Ireland, Greece, and Spain, as members of the EC. This chapter also examines the development of EC regional policy instruments, and the link between effectiveness and central co-ordination and control. Regional policy is only one, albeit still the most important, form of redistribution in the EC budget. The size of the latter and the extent of its redistributive function have become major issues in the European debate. The whole process of integration in the post-1992 era will largely depend on the agreements reached in the area of public finance.

The next chapter is devoted to the Community's role in the world economy and the external dimension of EC policies. The economic size and importance of the regional bloc make the Community a powerful actor on the international scene, albeit still a rather untypical actor in view of the uncertain division of powers between central and national institutions. This chapter starts by examining trade policies as a means of influencing the EC position in the international division of labour. They still remain the principal instrument in relations with the rest of the world, despite the fact that the area of EC competence has expanded considerably over the years. The difficulties in reconciling the common commercial policy with the proliferation of policy instruments at the national level and the resurgence of protectionist pressures are discussed in more detail in the first section of this chapter. For many years, there were some noticeable gaps in the common policy. The internal market programme will fill some of those gaps, while the Community's participation in international trade negotiations will have a considerable impact on both internal and external policies. Is there any justification in the fears of many outsiders about the 'Fortress Europe'? Protectionism is one issue; regionalism and trade preferences another. The use of such preferences by the EC as an instrument of both low and high policy is also discussed in this chapter. The last section is devoted to the Community's changing relations with other countries on the old continent and the growing pressures to develop a proper European policy which extends beyond strictly economic issues. The policy on further enlargement will have to be an integral part of this European policy.

The concluding chapter, like so many other concluding chapters, attempts to pull together the loose threads of the argument. The economic map of Western Europe has undergone a radical transformation over the years. This chapter examines the overall shape of this map and some of the main features of the new European geography. The main emphasis is on the changing relationship between private and public power and the effects of integration on the allocation, stabilization, and redistribution functions of the State. The chapter points to the discrepancy between economic and political integration and its wide consequences. It ends up by raising some general questions about further integration and the challenges which lie ahead for Europe.

2. The Ups and Downs of European Integration

The Foundations of European Regionalism

European economies emerged from the Second World War faced with enormous physical destruction. For many countries, 1936 production levels were only recovered in 1949, a year which is considered by most historians as marking the end of the reconstruction phase. The latter was relatively short and successful. It was characterized by high economic growth and rapid expansion of intra-European trade, with both trends continuing, without any major interruption, for almost thirty years.

The foundations of regional economic co-operation were laid during the reconstruction period. The initiative and the money came from the Americans, in the form of the Marshall plan. American aid to the dollar-hungry economies of Western Europe provided the finance for the large payments deficits, which were in turn the inevitable outcome of the ambitious growth strategies pursued by European governments. At the same time, US aid was conditional on the effective co-operation among European governments and the progressive liberalization of intra-European trade and payments. Both the Organisation for European Economic Co-operation (OEEC), which was later transformed into the Organisation for Economic Co-operation and Development (OECD), with the entry of non-European, Western industrialized countries, and the European Payments Union (EPU) were the result of American pressure on the European recipients of Marshall aid. The aim was to achieve a more effective use of the money given and also promote the economic and political integration of the old continent.

Through the Marshall plan, the United States offered Europe what it had adamantly refused to do at Bretton Woods only three years earlier. The refusal to become 'the milch cow of the world in general' (MacBean and Snowden, 1981: 38), by agreeing to the large amounts of unconditional liquidity envisaged by the Keynes plan, was replaced by a US initiative for large bilateral aid for Europe. True, it was now bilateral aid and not drawing rights with an international clearing bank. Furthermore, the granting of Marshall aid and the creation of the OEEC and the EPU meant that the American Administration appeared ready to accept, albeit reluctantly, that the principle of multilateralism would have to be put in cold storage. Multilateralism had been one of the main planks of US policy with respect to the new post-war economic order.

European governments continued with their extensive system of trade controls, and the slow process of liberalization, mainly through the OEEC and the EPU, was carried at the regional level, thus implying a discrimination against American exporters. Economic necessity and long-term political objectives were behind the pragmatism and the generosity of the new US policy towards Europe.

The European Recovery Programme and the creation of the first regional organizations also marked the institutionalization of the political division of Europe, established by the armies of the allied powers, as they converged into Germany from the east and the west. The refusal, or more precisely the inability, of the European countries which had come under Soviet control to accept US aid and participate in the new regional organizations led to the first institutionalized form of a Western European system. The early divisions among members of the OEEC and its political counterpart, the Council of Europe, mainly along supranational-intergovernmental lines and the nature and extent of economic co-operation-integration, quickly created the basis for a further division. And this time, the line cut across the Western European region.

The OEEC and the EPU contributed significantly towards the rapid expansion of intra-European trade and economic recovery in general. They also provided the initiation stage for economic co-operation at the regional level, which was later taken further through other organizations. The war had followed the pro-tectionism of the 1930s, thus further severing the economic links across national frontiers. Gunnar Myrdal (1956) aptly described the 1930s as the period of national economic integration and international disintegration, thus referring not only to autarchic national policies but also to the expanding economic role of governments during that period. The first efforts at regional co-operation in the early years of reconstruction marked a conscious attempt to reverse an earlier trend which had, obviously, acquired a much larger dimension during the war. The disastrous experience of the past and the considerable disillusionment with the nation-state provided a fertile ground for new ideas on European co-operation, and even federalism, to grow.

However, external liberalization, conducted with much caution and conditioned by the fragile state of European economies, did not signify the adoption of more general *laisser-faire* attitudes by the governments concerned. On the contrary, the early post-war years were characterized by a much wider acceptance of Keynesian ideas regarding the role of governments in manipulating aggregate demand and managing the national economy more generally. It was exactly those ideas, backed by popular expectations, and the weak state of their economies which explained the strong reluctance of European governments to go along with the liberal vision of the post-war economic order held by the Americans. Interestingly enough, American power and money were largely unsuccessful in shaping the European economic order according to Washington's preferences (Milward, 1984).

The next step was taken with the Schuman plan of May 1950, which led to the signing of the Treaty of Paris one year later and the establishment of the European Coal and Steel Community (ECSC). This also marked the further division of Europe, with the Federal Republic of Germany, the Benelux countries, and Italy following the French initiative, while Britain, still the most powerful country on the western side of Europe's dividing line, decided to stay out of this new organization. Different interests and priorities were coupled with a complete misjudgement by the government in London of the intentions of the Six (Camps, 1964; Monnet, 1976). This was to be repeated several times subsequently; a reassuring sign that people rarely learn any lessons from history.

The Schuman plan was a French initiative intended to deal essentially with the 'German problem' which belonged to the sphere of high politics. It was also directly related to the emergence of the Cold War, and it was for long perceived as such by the Soviet Union. The Schuman plan laid the foundations of Franco-German reconciliation which later developed into close co-operation, thus providing the cornerstone and the main driving force of regional integration. Wider political considerations were behind the decision of the other four countries to join. The compatibility of economic interests was not always obvious. Hence, for example, the special provisions made in the treaty for Belgian coal and Italian steel. This also explains the complicated nature of many of the legal provisions intended to safeguard different national interests in a very complex package deal.

The ECSC was about the integration of two sectors with central importance for both economic and defence reasons. Two decades later, they would have been among the least likely candidates to be chosen in the first step towards the integration of industrialized economies. However, their strategic importance was hardly in doubt at the time. Although the main objective of the Treaty of Paris was the elimination of barriers and the encouragement of competition in the sectors of coal and steel, many specific provisions were hardly compatible with economic liberalism. This was virtually inevitable in view of the long history of direct government intervention and restrictive business practices in both sectors.

The High Authority, the supranational executive organ of the ECSC, was given extensive powers including the right to levy taxes, influence investment decisions, and also impose minimum prices and production quotas in times of 'imminent' and 'manifest crisis' respectively. Walter Hallstein referred to the economic system set up by the Paris Treaty as one of 'regulated competition' (Haas, 1958: 247), and this was clearly very different from the liberal order dreamt of by post-war US administrations. On the other hand, the development of an international organization, and for that matter any political system, cannot be determined solely by the specific provisions contained in a treaty or a constitution. In fact, Alan Milward (1984: 420) has referred to the ECSC as a protoplasmic organization which could take virtually any shape the High Authority and the member countries would eventually wish to give it. Flexibility and suppleness are after all the main weapons of legal documents in the battle against time. Milward's observation is supported by the actual deveopment of the ECSC.

There was a clear discrepancy in the treaty between the institutional framework envisaged and the specific economic tasks which those institutions were expected to perform. The discrepancy was intentional, since the integration of the two sectors was only partially an end in itself; it was also a means towards the achievement of wider and more long-term political objectives. Subsequent attempts to explain and analyse the Schuman plan relied heavily on functionalist ideas about the central role of economic issues in Western pluralist systems and their potential use in promoting international co-operation and world peace (Haas, 1958; Lindberg, 1963). And this often led to exaggerated expectations about the influence and the scope of functionalist strategies which were allegedly behind the whole process of integration. Many academics appeared to be only too ready to write off the old nation-states of Europe as the undisputed protagonists of the political scene.

The failure of an even more ambitious French plan, also put forward in 1950 by the Defence Minister, Mr Pleven, for the creation of a European Defence Community (EDC), should have served as an early warning sign for those who hoped to find European political union around the corner. The failure of the EDC had another consequence, namely to place defence co-operation firmly within an Atlantic framework (Fursdon, 1980).

The establishment of the ECSC came in the aftermath of the economic reconstruction period. Output continued to grow fast, and external trade, especially between European countries, grew even faster. The OEEC and the EPU helped to liberalize European trade and also multilateralize the system of payments. Meanwhile, following the 1949 devaluation of several European currencies and the economic boom associated with the Korean war, domestic prices began to stabilize. Soon, new plans were put forward for the integration of other sectors, and especially agriculture. European governments were slowly regaining their confidence and were looking for new ways of extending their co-operation in the economic field.

With the signing of the Treaty of Rome in 1957, two new Communities were created: the European Economic Community (EEC) and the European Atomic Energy Community (EAEC — better known as Euratom). The latter was the favourite child of the French who continued their search for strategically placed sectors in the economy, following the earlier example of coal and steel. Interestingly enough, France was later instrumental in placing Euratom in a permanent state of hibernation.

The EEC is by far the most important and far reaching in terms of scope and instruments among the three Communities which constitute what has been for years generally referred to as the European Community (EC).[1] Through its contents and, perhaps, even more sharply through its omissions, the EEC treaty reveals the combination of economic preferences and political objectives as well as the balance of power which prevailed during the negotiations leading to its signing in 1957.

The central pillar of the new construction was to be the creation of a common market, meaning an economic area within which there would be free movement of goods, services, persons, and capital. The main emphasis was on the progressive elimination of tariffs and quantitative restrictions and the suppression of any form of discrimination based on nationality. The stress on the abolition of state barriers, the provisions made for an EEC competition policy, along the lines of the US anti-trust model, and the negative attitude expressed on the subject of state aids, despite the qualifications included in the relevant article (92), all pointed towards a more liberal economic approach than the one found in the Treaty of Paris. This could be explained in terms of the much wider scope of the new treaty. It was also a reflection of the greater confidence of European governments in the efficacy of market mechanisms; a confidence which had been gradually gained through the successful economic performance of previous years.

There were, however, important exceptions, and the most notable one was agriculture which at the time represented more than one-fifth of the total labour force in the six member countries. Given the special characteristics of this sector and the long history of extensive government intervention, the inclusion of agriculture in the common market could not be achieved through the simple elimination of existing barriers, but only through the adoption of a common policy at the EEC level. On the other hand, the incorporation of agriculture was an essential part of the overall package deal on which the Treaty of

---

[1] Following the treaty revision signed at Maastricht in February 1992, the term 'European Community' (EC) will substitute for the old term EEC, while the other two Communities retain their old terms.

Rome was based. Hence the provisions made for a common policy, which were, however, kept at a very general level; thus, basically, registering a joint commitment and providing a framework for the negotiations which were to follow. A similar comment could be made about transport policy, also mentioned in the treaty. As regards the common commercial policy, it was essentially the inevitable outcome of the establishment of a customs union and the adoption of a common external tariff.

Industrial policy and regional policy were most notable by the absence of any specific provisions in the treaty. This absence was striking, especially in view of the importance attached to those policies by European governments during the 1960s. But this reflected the inability of the six signatories to agree on anything specific, because of political and ideological differences. Although Article 2 of the treaty referred to the 'harmonious development of economic activities' and to 'a continuous and balanced expansion', there were hardly any instruments created to ensure the achievement of those general objectives.

The European Investment Bank (EIB) and the European Social Fund (ESF), together with provisions made for the free movement of labour, were designed to work mainly in favour of the Mezzogiorno, the least developed area of the EEC of Six. But there was virtually nothing else to ensure a 'harmonious development' and a 'balanced expansion'. There was a lack of correspondence between objectives and instruments. It could be argued that the signatories may have hoped or believed that those objectives could be achieved through the working of the market mechanism; but this must have, surely, required strong faith and little regard to historical experience. A more plausible explanation would be that the complicated package deal, expressed through the various provisions made in the treaty, was expected to bring about a fairly equal distribution of gains and losses among participants, and that this was the main thing that counted; or perhaps that this was the maximum that could be agreed by the founding members at the time. The fundamental point is that the treaty made virtually no provision for redistributive instruments of policy.

On the other hand, despite various pious wishes about the co-ordination of economic policies and the consideration of the exchange rate as a problem of common interest, there were virtually no specific provisions regarding co-operation in the macroeconomic field. Furthermore, the objective of the common market was accompanied by a very clear caution regarding the liberalization of capital movements which was to take place only 'to the extent necessary to ensure the proper functioning of the common market' (Article 67).

National differences were one explanation behind the legal void in the macroeconomic area. Considerations of economic and political feasibility were another. The creation of a common market was in itself an extremely ambitious objective. Trying to go beyond that might stretch dangerously the limits of the politically possible, and thus jeopardize the success of the whole enterprise. On the other hand, the predominance of the US currency in the international monetary system and the remaining perception of the 'dollar gap' among European policy-makers made the creation of a regional monetary system in Western Europe virtually unthinkable at the time of the signing of the treaty (Tsoukalis, 1977*a*).

There was still another important reason. The influence of Keynes's ideas meant that governments were keen on retaining direct control over fiscal and monetary policies to be used for the attainment of the objective of full employment at home. Therefore, there should be no readiness for the transfer of real powers to the European or international level. Greater mobility of capital across national frontiers would undermine the effectiveness of national monetary instruments. Hence the caution regarding the liberalization of capital movements in the Treaty of Rome, which could also be found in the Articles of Agreement of the International Monetary Fund (IMF).

The emphasis in the treaty was clearly on the creation of a common market, and more precisely on the establishment of a customs union for goods, with the addition of a limited number of sectoral policies. In view of its much wider scope, the Treaty of Rome was generally less specific in its provisions, when compared with the Treaty of Paris. This explains the distinction sometimes made between *traité-cadre* and *traité-règle* (Lagrange, 1971). The heavy institutional framework provided for in both treaties, when seen in relation to the concrete economic tasks undertaken, can be largely explained in terms of the more long-term political ambitions of the authors. It was evident that some federalist ideas had infiltrated the higher echelons of government in at least a number of Western European countries.

The division created by the ECSC inside Western Europe, resulting from the refusal by a group of countries, led by the UK, to accept the supranational aspects of the Treaty of Paris, and which was further strengthened by the establishment of two new Communities in 1958, was taken to its logical conclusion with the creation of the European Free Trade Association (EFTA) in 1960. As the name implies, this was a much less ambitious initiative than the EEC. The seven members of EFTA (Austria, Denmark, Norway, Portugal, Sweden, Switzerland, and the UK) retained their independence in the conduct of external trade policies, hence the problem of trade deflection and the need to introduce rules of origin in intra-EFTA trade. Agricultural goods were virtually excluded from the free trade area, and again, as the name implies, there was no attempt to extend co-operation beyond trade. This was also reflected in the institutional provisions which were kept to a bare minimum.

EFTA should be seen more as a defensive response by the European countries which had decided, essentially for political reasons, to stay away from the more ambitious integration projects of the Six. However, with the exception of Portugal which in economic and political terms was clearly an oddity inside EFTA, all the other members of the two rival groups constituted the core of Western Europe. By core, we mean here the most advanced industrialized countries of the region, with pluralist, democratic regimes. Members of the European periphery, as defined at the time, were excluded from those integration efforts, their participation being limited, at best, to the OEEC and the Council of Europe.

Economic Growth, Trade Liberalization, and the Mixed Economy

The establishment of the EEC in 1958 coincided with an important new phase in the international economic system. It was heralded, perhaps unconsciously, by the decision of all member countries, together with the UK and Japan, to restore the external convertibility of their currencies. This also meant that the Bretton Woods agreement could at last be taken off the shelf. However, the international monetary system was in practice to evolve in a very different way from what had been originally envisaged by those gathered in the small resort of New Hampshire, and especially by its American architects.

The US current account progressively deteriorated, and the small surpluses were no longer sufficient to compensate for the large outflows of capital. Meanwhile, European and Japanese payments surpluses grew. This led to the weakening of the dollar in foreign exchange markets and to successive speculative crises. Sterling shared the fate of the dollar, if only in more misery and despair. It usually acted as the soft underbelly of the American currency until the UK government gave way to market pressure by devaluing sterling in 1967. On the other hand, the Six enjoyed healthy surpluses and mounting foreign reserves (Strange, 1976). The 1960s was a decade marked by a significant shift of the economic balance of power from the United States and Britain to continental Europe and Japan, and this was bound to have a positive effect on confidence among the Six and on their integration effort.

The twelve-year transitional period envisaged by the treaty also coincided with what appears with the benefit of hindsight as the last phase of the post-war economic miracle. It was a period of high and steady growth and unprecedented levels of employment. Although high growth and low unemployment also characterized the large majority of Western European countries, the performance of the Six was significantly better than that of the United States and the UK. The average rate of growth of the six countries was slightly higher than the average for OECD Europe. The only exception were the southern European NICs (newly industrializing countries) which approached the growth rates registered by the emerging economic giant in East Asia (Tables 2.1 and 2.2). There was only one dark spot, namely the accelerating rate of inflation which was already noticeable towards the end of the 1960s (Table 2.3). But inflation had not yet reached dangerous levels. Tariffs and quantitative restrictions in intra-EC trade were abolished eighteen months ahead of schedule, and the common external tariff (CET) was adopted by all six member countries. In fact, the CET had already been considerably reduced as a result of the Kennedy Round negotiations within GATT (General Agreement of Tariffs and Trade), in which the Six had participated as a single unit.

External negotiations were closely linked to internal develop-

Table 2.1. Growth Rates, 1960–1990

(Average annual % change of real GDP at constant prices)[a]

	1960–7	1968–73	1974–9	1980–5	1986–90
Belgium	4.6	5.3	2.2	1.3	3.2
France	5.5	5.2	2.8	1.5	2.9
Germany	3.8	5.0	2.3	1.1	3.1
Italy	5.6	4.9	3.7	1.9	3.0
Luxembourg	2.8	5.5	1.3	2.2	4.3
Netherlands	4.6	5.1	2.7	1.0	2.7
Denmark	4.6	4.0	1.9	2.1	1.5
Ireland	3.6	5.2	4.9	2.6	4.4
United Kingdom	2.9	3.5	1.5	1.2	3.2
Greece	7.4	8.0	3.7	1.4	1.7
Portugal	6.2	7.6	2.9	1.5	4.6
Spain	7.7	6.7	2.2	1.4	4.5
EC-12	4.5	4.9	2.5	1.4	3.1
Austria	4.2	5.7	2.9	1.6	3.1
Sweden	4.5	3.7	1.8	1.8	2.1
Switzerland	4.5	4.3	−0.4	1.9	2.8
OECD-Europe	4.5	4.8	2.5	1.5	3.1
United States	4.5	3.3	2.4	2.4	3.0
Japan	9.8	9.3	3.6	3.7	4.6
Total OECD	5.0	4.7	2.7	2.8	2.7

[a] Calculated using 1985 price levels and exchange rates.
Source: OECD.

ments, as an integral part of the intra-EC package. This is a pattern which was to be regularly repeated in subsequent years. Thus, the mandate given to the Commission to negotiate tariff reductions in the Kennedy Round was dependent on an agreement between member countries to proceed with the customs union and the CAP. The Kennedy Round was also the first manifestation of the role and power of the EEC as an international actor, and this had profound effects on the self-perception of the Six and the balance of power on the world economic scene. The EEC successfully challenged the dominant position of the United States which was finally forced to modify substantially its own expectations about the outcome of the trade negotiations (Shonfield, 1976).

Table 2.2. Unemployment Rates, 1960–1990

(As % of the labour force; average of annual unemployment rates)

	1960–7	1968–73	1974–9	1980–5	1986–90
Belgium	2.1	2.3	5.7	11.5	10.3
France	1.5	—	4.5	8.3	9.8
Germany	0.8	0.8	3.5	6.3	7.2
Italy	4.9	5.7	6.6	8.8	11.4
Luxembourg[a]	0.0	0.0	0.6	2.8	2.2
Netherlands	0.7	1.5	4.9	10.1	9.0
Denmark	1.6	1.0	—	9.2	6.4[b]
Ireland	4.9	5.6	7.6	12.6	16.2
United Kingdom	1.5	2.4	4.2	9.8	8.5
Greece	5.2	3.7	1.9	6.1	7.4
Portugal	2.4	2.5	6.0	7.9	6.2
Spain	2.3	2.7	5.3	16.4	18.6
EC-12	2.2	2.7	4.8	9.2	9.9
Austria	2.0	1.4	—	3.2	3.4
Sweden	1.6	2.2	1.9	2.8	1.8
Switzerland	—	—	—	0.6	0.6
OECD-Europe	2.8	3.4	5.1	8.8	9.2
United States	5.0	4.6	6.7	8.0	5.8
Japan	1.3	1.2	1.9	2.4	2.5
Total OECD	3.1	3.4	5.1	7.5	7.1

[a]*Source: Eurostat 1985–9.*

Source: OECD.

At the same time, the Kennedy Round strengthened and confirmed the shift in British policy towards Brussels, a shift which had already led to the first British application for membership in 1961. The EEC was a concrete economic reality, and this was now manifested through a strong presence on the international economic scene. The British who were, according to the architect of European integration, Jean Monnet, unable to understand an idea but supremely good at grasping hard facts, had decided to join. However, two successive applications for membership in 1961 and 1967 were blocked by General de Gaulle's veto.

During the transitional period, trade grew much faster than output, and this was almost entirely due to the growth of

Table 2.3. Consumer Price Indices, 1960–1990

(Average annual % change)

	1961–8	1969–73	1974–9	1980–5	1986–90
Belgium	2.5	4.5	8.5	6.9	2.1
France	3.5	5.8	10.7	10.2	3.1
Germany	2.5	4.3	4.7	4.2	1.4
Italy	4.1	5.1	16.1	15.0	5.7
Luxembourg	2.0	4.3	7.4	7.0	1.7
Netherlands	3.8	6.4	7.2	4.6	0.7
Denmark	5.3	6.7	10.8	8.6	3.9
Ireland	4.0	8.2	15.0	13.3	3.3
United Kingdom	3.1	7.0	15.6	8.9	5.9
Greece	2.1	4.7	16.1	21.4	17.4
Portugal	3.0	7.8	23.7	22.1	11.3
Spain	6.2	6.7	18.3	12.8	6.5
EC-12	3.7[a]	5.8	12.3	9.8	4.4
Austria	3.5	4.8	6.3	5.1	2.2
Sweden	4.1	5.3	9.8	9.8	6.2
Switzerland	3.3	5.1	4.0	4.2	2.5
OECD-Europe	3.8[a]	6.0	12.8	11.5	6.8
United States	1.7	4.9	8.5	6.7	4.0
Japan	5.5	6.8	9.9	3.6	1.3
Total OECD	2.8	5.6	10.5	8.3	4.8

[a] 1961–7.

Source: OECD.

intra-EC trade (Fig. 2.1). In other words, the dependence of the six economies on each other grew steadily in relation to their dependence on the rest of the world. Virtually all empirical studies point to a close link between EC integration and the growth of trade; and by far the largest component of the latter is attributed to trade creation rather than trade diversion (Robson, 1987; Mayes, 1989).[1] The exact magnitude of the integration

---

[1] Trade creation and trade diversion are the two main concepts developed by customs union theory. The former arises when domestic production is substituted by cheaper imports from a partner country, and the latter when cheaper imports from a third country are substituted by dearer imports from a partner country. Trade creation represents a shift to a more efficient producer as a result of the establishment of a customs union, while the opposite is true of trade diversion.

% of total trade

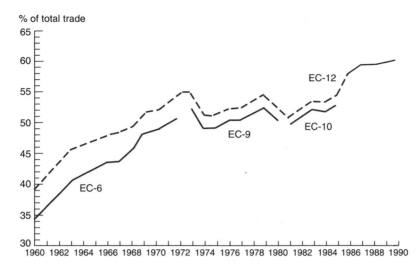

Fig. 2.1. Intra-EC Trade in Goods, 1960–1990
Source: Eurostat.

effect is, of course, impossible to determine, since the rapid growth of intra-EC trade during the transitional period was bound to be influenced by many different factors. The results of empirical studies vary widely, and this should have been expected. Müller and Owen (1989) refer to a 'consensus view' that trade creation in 1970 accounted for 30 per cent of intra-EC trade in manufactures, which may be considered as a rough indicator of the orders of magnitude involved.

There is, however, hardly a 'consensus view' among economists regarding the contribution of increased trade to welfare and growth. On the basis of the static theory of customs union, the only developed part of economic integration theory for many years, the results obtained were meagre: usually, less than 1 per cent of the Gross Domestic Product (GDP) of member countries (Robson, 1987). This was hardly the result to match the time and efforts spent, not to mention the expectations associated with European integration.

One possible and perhaps obvious conclusion to draw from this empirical finding would therefore be that European integration was taking place for essentially political reasons; the customs

union was an economic means to a political end, and this was
spectacularly manifested in its marginal economic effects. This
does not, however, exclude another possible explanation, namely
that the economic theory, on which those econometric studies
were based, was in itself highly inadequate and thus unable to
capture the effects of integration.

Customs union theory suffers from all the limitations of traditional theories of international trade based on comparative
advantage. It concentrates on static effects. It relies on totally
unrealistic assumptions such as perfect competition, homogeneous
products, full employment of resources and no adjustment costs.
It pretends that the role of the State is only limited to border
interventions, thus denying the existence of mixed economies
(Pelkmans, 1984). It also ignores, like most neo-classical economics, the fundamental question of distribution. On the other
hand, its own inherent logic is totally suspect, since, on the
basis of the standard assumptions, the conclusion drawn is that
unilateral tariff disarmament is preferable to participation in a
customs union. This presumably means that there is something
totally irrational in the whole process. Last but not least, customs
union theory is unable to explain an important fact which has
emerged from empirical studies, namely that the largest part of
Community trade is intra-industry and not inter-industry trade, as
should have been expected from the theory.

Some economists have introduced economies of scale, as part
of the dynamic effects. The results obtained are much more
significant. Growth in trade has made possible a substantial
increase in plant size, and this in turn has led to cost reductions.
According to Owen (1983), this could have amounted to some 3–
6 per cent of the combined GDP of the Six for 1980. There were
also signs of a positive investment effect of integration, both from
internal and external sources; US foreign investment in Western
Europe being partly attributed to the creation of the EEC
(Davenport, 1982; Pelkmans, 1984). Thus economists gradually
switched their attention to the more important variables, but
their tools of analysis and measurement, despite frequent claims
to the contrary, have remained woefully deficient.

The causality between trade and growth is arguably not only
one-way, as it is usually implied. The rapid elimination of intra-
EEC tariffs and quantitative restrictions between 1958 and 1968
was made possible largely because of the favourable macro-

Table 2.4. Total Outlays of Government, 1960–1990

(% of GDP)

	1960	1974	1980	1985	1990
Belgium	30.3	39.4	59.0	62.4	54.7
France	34.6	39.3	46.1	52.2	50.4
Germany	32.4	44.6	48.3	47.5	46.0
Italy	30.1	37.9	41.7	50.8	53.2
Luxembourg	30.5	35.6	54.8	51.7	51.3[a]
Netherlands	33.7	47.9	57.5	59.7	56.3
Denmark	24.8	45.9	56.2	59.3	57.5
Ireland	28.0	43.0	50.8	55.1	44.5
United Kingdom	32.2	44.8	44.7	46.1	42.9
Greece[b]	17.4	25.0	30.5	43.7	50.4
Portugal	17.0	24.7	25.9	43.4	41.7[c]
Spain	—	23.1	32.9	42.2	41.9
EC-12	31.8	40.7	45.9	49.7	47.8[c]
Austria	35.7	41.9	48.9	51.7	48.6
Sweden	31.0	48.1	61.6	64.7	61.5
Switzerland[b]	17.2	25.5	29.3	31.0	29.9[c]
OECD-Europe	31.3	40.4	45.9	49.5	48.6
United States	27.0	32.2	33.7	36.7	37.0
Japan	17.5	24.5	32.6	32.3	30.7
Total OECD	28.0	34.8	39.4	40.7	40.7

[a] *Source*: Eurostat.
[b] Current disbursements only (e.g. gross capital formation excluded).
[c] 1989.
Source: OECD.

economic environment, characterized by high rates of growth and low unemployment. Increased exposure to international trade brings with it adjustment costs for both labour and capital. They are much more easily absorbed in times of rapid growth, thus minimizing the resistance from potential losers. This points to a possible virtuous circle: the favourable macroeconomic environment of the late 1950s and the 1960s, attributable to a combination of different factors, created the conditions which permitted the signing of the Treaty of Rome and the successful implementation of its trade provisions. Liberalization then led to more trade and this, in turn, contributed to the remarkable growth rates of this period (Boltho, 1982: chs. 1 and 8; Maddison, 1982).

There is, however, another major feature of Western European economies during this period, which may appear, at least superficially, in contradiction to the trend towards liberalization. These years were characterized by the continuous strengthening of the mixed economy and the welfare state. At both the micro- and the macroeconomic levels, the role of the State became increasingly pronounced. One clear indication are the figures on government expenditure, which show a steep rise in terms of GDP (Table 2.4). The figures in this table also show a significant difference between the large majority of Western European countries, including the Six, on the one hand, and the United States and Japan on the other. The Western European experience of the time suggests that high economic growth is not necessarily incompatible with an active economic role of the State.

Fiscal policy and, increasingly from the mid-1960s onwards, also monetary policy were actively used by political authorities as a means of influencing aggregate demand. On the other hand, intervention was not limited to the macroeconomic level. Governments also resorted to regional, industrial, and incomes policies, albeit in varying degrees of enthusiasm and with different instruments, in order to influence the domestic allocation of resources and the distribution of income (Boltho, 1982: chs. 10–14).

The promotion of 'national champions',[2] especially in the fields of high technology, through active fiscal and credit policies, discriminatory public procurement, and only half-concealed measures of external protection, was one but very important manifestation of the political answer, at the national level, to growing international competition. These were the high days of optimism about the ability of political institutions to guide the 'invisible hand'. The contrast with the economic ideas prevailing in the 1980s is, indeed, striking.

Meanwhile, the growing power of trade unions, combined and not unrelated with welfare state provisions and job security legislation, led to the creation of highly regulated labour-markets.

---

[2] This is a term first used by Raymond Vernon of Harvard University to refer to large firms, actively supported by governments through various means, which were usually seen as national representatives in international oligopolistic competition (Vernon, 1971). The promotion of 'national champions' and the need to face the 'American challenge' (Servan-Schreiber, 1967) also explains the rather relaxed view taken by most European governments during this period to the ever-increasing concentration which characterized many industrial sectors.

More generally, it could be argued that the 1960s were characterized by a constant attempt made by Western European governments to reduce the incidence of micro- and macroeconomic risk on both individuals and firms, thus complementing, if not partly negating, the operation of the market mechanism. This was, in turn, a function of social pressures and the existing political balance of power; in other words, the prevalence of social democratic forces and values.

There is, certainly, a strong element of contradiction between the increasing openness of national economies to world competition, closely related to both international and EEC liberalization processes, and the conscious attempt to regulate the market domestically. It was yet another manifestation of 'Keynes at home and Smith abroad' (Mayall quoted in Gilpin, 1987: 355). Both happened simultaneously, and the coincidence may be a surprise only to the economic purists. In fact, it could be argued that both the mixed economy and the welfare state facilitated the process of internationalization and European integration by enabling governments to smooth out and, sometimes, even slow down the necessary adjustment arising from the increased openness of national economies.

It is true that domestic intervention sometimes led to the introduction of non-tariff barriers (NTBs), thus replacing border controls which had been previously negotiated away in the context of the EEC or GATT. But this represents only part of the overall picture. The mixed economy and the welfare state also helped to buy social acceptance of greater international competition by alleviating its effects on potential losers. It may, therefore, have contributed significantly to the process of European integration in the late 1950s and 1960s. The two developments appear to have been closely linked and mutually reinforcing, although the relationship was based on a very delicate equilibrium which was soon to break down. This delicate equilibrium sustained the process of economic integration in Western Europe for many years, based on a symbiosis between external liberalization and the strengthening of the economic role of the State at the domestic level. But this also meant that for years European integration was essentially limited to trade.

The establishment of the customs union in the EEC was coupled with the creation of a CAP relying on three main principles: the unity of the market, based on the instrument of common prices,

Community preference, and financial solidarity (Tracy, 1989). This, in turn, led to the progressive liberalization and also rapid expansion of intra-EC trade in agricultural goods, more the result of trade diversion than trade creation. The adoption of a common policy for agricultural products meant the implementation of the original agreement expressed in the Treaty of Rome and the Stresa conference of 1958; it was also an absolutely integral part of a package deal which included both the intra-EC liberalization of trade in industrial goods and the significant reduction of the CET which resulted from the Kennedy Round negotiations.

The elimination of many border controls and the establishment of the CAP were accompanied by the first steps towards an EEC competition policy. The latter concentrated on restrictive business practices and the 'abuse of dominant position' by private firms, while state aids remained almost untouched. No progress was registered in the field of transport, despite the provisions originally made in the treaty. Economic integration also had little effect on intra-European factor movements. Cross-border labour mobility was limited by cultural and linguistic factors as well as the slow progress in the transferability of social security rights and the mutual recognition of professional qualifications. To the extent that it did take place, it was mainly unskilled labour from the south of Italy and non-member countries of the Mediterranean. As for the liberalization of capital movements, the EC had virtually no effect during the 1960s; international factors continued to play the decisive role.

No substantial progress was made in the macroeconomic field either. Although there were signs of growing interest in EC co-operation in this area, the political sensitivity of the issues involved, policy differences, and the inability to reach a common position *vis-à-vis* the United States, and the question of the dollar in particular, prevented the Six from registering any serious progress in macroeconomic policy co-operation during the 1960s (Tsoukalis, 1977a). Although growing trade interdependence made policy instruments less effective, the use of the latter remained exclusively in the domain of national discretion. The international environment, and the United States in particular, continued to have a stronger impact on national policies than any half-hearted attempt to co-ordinate at the EC level.

The external dimension has played a major role in the process of integration, closely linked to the development of internal policies. During the same period, the Six were also involved in numerous bilateral negotiations with third countries. Those negotiations concentrated mainly, but not exclusively, on trade matters, and led to the signing of preferential and non-preferential agreements which were not always reconcilable with the GATT principle of multilateralism. Thus, the EC, through its very nature as a customs union and also through the signing of many association and trade preferential agreements, helped to undermine the already fragile international trading order.

The creation of a regional bloc in Western Europe was bound to bring about attempts by neighbouring countries to reduce the trade diversion effects arising from the process of regional integration. The Six, very much flattered by the flood of requests for the negotiation of bilateral agreements and in need of international recognition, especially in the early years of the EEC, became involved, almost absent-mindedly, in those negotiations. And sometimes they signed agreements which they were later to regret (Tsoukalis, 1977*b*).

Largely as a result of the influence of General de Gaulle, who overshadowed the European scene during the 1960s, the supranationalist element of integration was seriously weakened. But this trend had already become obvious in the first years of the ECSC. The High Authority did not exercise the full powers given to it by the Treaty of Paris. Later on, the EEC treaty confirmed this trend and further reduced the powers enjoyed by the 'supranational' organ, namely the Commission. General de Gaulle then made sure that the high ambitions entertained by the first Commission, and especially its President, Mr Walter Hallstein, were completely squashed. The 'empty chair' crisis and the Luxembourg agreement of 1966, which reintroduced the national veto through the back door, were crucial turning-points on the road towards more intergovernmentalism (Camps, 1967). It took the Brussels executive a long time before it could recover from this shock.

While power was concentrated in the Council of Ministers, as a collective expression of national interests, the Court of Justice was busy building on the foundations of a federal system. And this was done in a very quiet manner, away from the political lime light. According to Weiler (1983), the establishment of the doctrine of direct effect, the doctrine of supremacy, and the doctrine of pre-emption for Community legislation during those years created a constitutional framework for a federal-type political structure. Indeed, the Community legal framework is arguably what most distinguishes the EC from traditional inter-governmental organizations; and this was bound to have long-term political effects.

Enlargement and Economic Crisis: A Question of Survival

At the end of the decade, which also coincided with the end of the twelve-year transitional period envisaged by the Treaty of Rome, the Six were again in search of a new driving force for the process of integration. This was finally provided by the Hague summit which took place in December 1969. The relaunching of European integration took the form of a new package deal. This was described by President Pompidou in terms of a triptych consisting of completion, deepening, and enlargement; and it was based on a, albeit uneasy, Franco-German bilateral agreement, something which was later to become a recurrent phenomenon.

The logic of the new package deal was rather simple: the French green light for the accession of Britain and the other applicant countries required the creation of the Community's 'own resources', a means of strengthening its financial independence and also providing a more solid revenue basis for the CAP (*l'enfant chéri* of French governments). This, in turn, would require a modest increase in the powers enjoyed by the European Parliament: something which had been vehemently resisted by General de Gaulle only a few years earlier. Deepening meant basically the creation of an intergovernmental system of foreign policy co-ordination which became known as European Political Co-operation (EPC), and the establishment of an Economic and Monetary Union (EMU). Thus the relaunching of integration would take place in both the areas of high and low politics, although low politics was clearly a misnomer, at least as far as money was concerned. This became patently obvious during the early stages of the negotiations for the establishment of EMU.

Most of the above ideas had entered the intra-European debate some years earlier. However, it was only at the Hague summit of 1969 that the political environment seemed ripe for their formal adoption. Although partly for different reasons, France and West Germany, under their new leadership, were ready to give the Community a new push forward. The positive experience of the 1960s and the fear of the EC being diluted into a free trade area, as a result of British accession, combined to produce extremely ambitious initiatives which went much closer to the heart of national sovereignty than anything which had preceded them. With the benefit of hindsight, it appears that some of those initiatives were taken in a light-hearted fashion.

The first enlargement of the Community took place in January 1973, with the accession of Britain, Denmark, and Ireland. Norway had to drop out at the last minute, when the Treaty of Accession was rejected at a national referendum. Enlargement removed a major issue of contention which had divided the Six for more than a decade. It also marked the final and undisputed victory of the group of countries, led by France, in a division which dated back to the late 1940s and which had led to the creation of two rival organizations in Western Europe, namely the EC and EFTA. The dominant position of the EC was to become even more evident a decade later, with the further strengthening of Community institutions and the process of integration, which in turn forced virtually all the remaining members of EFTA to reconsider their relations with the EC.

The financial independence of the Community was achieved through the introduction of its own sources of revenue, consisting of customs duties, agricultural levies, and a percentage of the value added tax (VAT). In so doing, member countries also adopted a common form of indirect taxation, while the budgetary powers of the European Parliament were extended. The whole process was not, however, completed before the end of the decade, because of long delays in the introduction of VAT in some member countries.

The rest of the overall package of measures agreed at the Hague summit was in the end only very partially implemented. Political resistance to any serious encroachment of national sovereignty was one explanation. Another and, perhaps, the most important explanation is associated with the dramatic change in the political and economic environment which had sustained and protected the process of regional integration until then. Additional difficulties were created by enlargement itself, as full participation in the internal decision-making process enabled the new members to challenge some important aspects of the *acquis communautaire* (the whole body of EC legislation) and thus try to shift the balance of gains and losses in their favour.

The exceptionally long period of high and stable growth, combined with unprecedented levels of employment, gradually came to an end. After 1973 it gave way to a new situation characterized by a deceleration of economic growth, declining rates of investment and productivity, galloping inflation, loss of international competitiveness, and, last but not least, a dramatic increase in unemployment. In terms of unemployment, most EFTA countries fared much better than EC members (Tables 2.1–2.3).

Numerous explanations have been put forward regarding the end of the golden age (Boltho, 1982; Emerson, 1984; Strange, 1985; Lawrence and Schultze, 1987). Some lay the emphasis on external economic shocks and the inadequate response of European governments, mainly in terms of macroeconomic policies. Such explanations obviously point to the large increase in oil prices, which happened in 1973–4 and then again in 1978–9 and led to a significant worsening of the terms of trade and the transfer of real resources to oil producers. European governments were criticized for excessively restrictive policies which exacerbated the deflationary impact of the oil shocks. Imported inflation from the United States, Washington's policies of 'benign neglect', and the advent of floating exchange rates were also considered as additional destabilizing factors.

Another set of explanations concentrated on internal structural changes and the rigidities associated with the growth of the welfare state. In addition to the loss of cheap energy, European entrepreneurs were also faced with escalating labour costs. This was, in turn, attributed to the gradual exhaustion of labour reserves, both domestic and imported, and the intensification of the political struggle for larger income shares; the latter being also a reflection of the breakdown of the political consensus which had provided the foundation of the economic miracle of the 1950s and the 1960s.

The result was a continuous increase in the share of wages and salaries at the expense of profits, with negative effects on investment. A kind of nemesis theory has been advocated by some economic observers (Giersch, 1983 and 1985; Minford, 1985) arguing that the stagflation (a new term invented to describe the hitherto unthinkable combination of recession and inflation) of the 1970s was the inevitable outcome of the full employment policies of the previous decade and the new bargaining power of European trade unions, itself closely related to the previous high levels of employment. Along similar lines, Mancur Olson (1982) has argued that economic stagnation in pluralist systems is largely attributable to societal immobilism, gradually created during long periods of peace through the proliferation and strengthening of pressure groups which try to defend their entrenched positions and thus resist any form of change.

The real increase in wages and salaries, which regularly exceeded the growth rates in productivity, was one important factor behind the deterioration of the business climate in Western Europe. Another factor was, according to those theories, the rigidities created in the labour-market through job security legislation and the growth of the welfare state. Changes in the international economic environment brought about the need for domestic adjustment. But this required a degree of flexibility and mobility on behalf of European labour and the European societies in general, which clearly did not exist. Again according to those theories, which later provided the ideological justification for supply-side measures, the inability of European economies to adjust was a major factor behind the loss of international competitiveness and the continuing high levels of unemployment.

This was a very different argument from the one advocated earlier with respect to the experience of the 1950s and the 1960s. Although the fundamental premise is clearly different, it is also true that the real world had undergone a major transformation in the meantime. The external economic shocks of the 1970s and the gradual disintegration of the internal consensus had brought about a major change in the political economy of Western Europe. The pre-conditions for economic growth were no longer there.

On the other hand, the deterioration of the international economic environment sparked off different policy responses on behalf of individual Western European countries. Broadly speaking, two groups of countries can be distinguished: the first group, consisting of West Germany, Switzerland, Austria, and, to a lesser extent, the Benelux countries, was characterized by a strong anti-inflationary policy stance; the second group, which included most of the other Western European countries, tried to ride out the storm by accommodating monetary and fiscal policies (Wegner, 1985). Interestingly enough, membership of the EC did not seem to make any difference. The post-1973 recession was accompanied by a widening of economic divergence among national economies, which was manifested both in terms of macroeconomic policies and performances (see, for example, Table 2.3). An increase in economic disparities between the more and less advanced countries and regions of the Community was another characteristic of the recession, thus reversing the trend of the 1960s when there had been a progressive narrowing of the economic gap.

National economies were affected in different ways by the crisis. Furthermore, nation-states behaved as autonomous political units, and the dialectic of economic, political, and social forces differed considerably from one to another. Under such circumstances, monetary union became a totally infeasible objective. In complete contrast to their verbal commitments, individual member countries showed hardly any interest in using the EC as a framework for an effective co-ordination of macroeconomic policies. To the extent that an interest in co-ordination did exist at all, it was more obvious in the context of international fora.

A similar observation can be made with respect to energy matters. Here, the failure of the Nine to co-ordinate their policies and adopt a common stance in the crisis diplomacy which followed the quadrupling of oil prices in 1973 was particularly striking. Caught between the conflicting pressures and demands emanating from Washington and the capitals of the Arab oil producers, Western European countries offered a sad spectacle of disunity and inaction. The more assertive stance of the United States during this period only served to exacerbate intra-EC divisions (Odell, 1986). The economic crisis and the new international environment appeared to indicate that members of the EC were not yet ready or able to extend co-operation much beyond the area of trade.

Faced with the collapse of the Bretton Woods system, characterized by the end of the dollar–gold convertibility and the abandonment of fixed exchange rates, the EC countries tried to build the foundations of a regional currency bloc in the form of the snake. In a world of generalized floating, which had already become true by March 1973, the objective was to maintain bilateral exchange rates within relatively narrow margins.

However, the attempt to preserve some of the elements of Bretton Woods at the regional level proved to be a somewhat futile exercise. The EC snake was quickly transformed into a Deutschmark-zone, after the withdrawal of sterling, the Irish punt, the lira, and the French franc (Kruse, 1980).

This development could be considered as the first clear indication of a new phase in the history of the European Community. One important feature of this new phase was the increasingly dominant economic position of the Federal Republic of Germany and the *Modell Deutschland*, especially in terms of macro-economic management (Markovits, 1982). Another feature was the gradual departure from regional arrangements applying *erga omnes*. The snake, as it gradually developed, was the first important manifestation of a two-tier Community,[3] with some countries outside the EC participating in the exchange rate arrangement. Different variations of the two-tier model were to become increasingly popular in subsequent years, largely in response to the wide economic divergence which characterized the enlarging Community.

On the other hand, the prolonged economic recession and the rapid increase in the numbers of unemployed led to a resurgence of protectionist pressures in Western Europe; a phenomenon which could also be observed in the rest of the industrialized world. This 'new protectionism' took mainly the form of non-tariff measures; it was often of a bilateral nature and directed mostly against Japan and the new Asian NICs.

Various forms of indirect protection of domestic production, coupled with 'voluntary export restraints' and 'orderly marketing arrangements' (euphemisms for essentially unilateral restrictions on imports) against third countries, were introduced in an increasing number of sectors (Page, 1981). To all intents and purposes, agriculture, fuel, and textiles had previously been taken outside the GATT framework. With the advent of the economic recession in the mid-1970s and the emergence of large surplus capacity, new sectors were added to the above list. They included steel, shipbuilding, cars, chemicals, electronics, and footwear. Meanwhile, protection in high technology sectors, mainly through

---

[3] This term was first introduced by Willy Brandt. It was later taken up by Mr Tindemans in his report on European union (Tindemans, 1975).

the promotion of national champions, continued stronger than ever.

The common commercial policy was bound to suffer as a result. It was undermined by the increasing recourse to NTBs which fell largely outside the competence of EC institutions, while being progressively diluted by tariff reductions in the framework of GATT. National measures and bilateral agreements with Japan and Eastern Europe were the most obvious holes in the external armoury of the EC. A common position in international negotiations was sometimes preserved at the expense of free intra-EC trade. Thus, the EC quantitative restrictions on imports, imposed in the context of the various multifibre arrangements (MFAs), were subdivided into national quotas, and this implied the need for controls at national frontiers and the introduction of rules of origin in intra-EC trade.

The Tokyo Round of GATT negotiations (1975–9) was as much a logical continuation and extension of the post-war process of international trade liberalization, concentrating mainly on NTBs, as it was also an attempt to resist the reversal of this process because of the new protectionist pressures. Its success in this respect was limited at best (Cline, 1983a). On the other hand, the new round confirmed the 'pyramidal style of multilateral negotiation, where issues would first be negotiated bilaterally between the larger powers [the United States and the EC] and then later multilateralized as the negotiations went on' (Winham, 1986: 371). The EC often adopted a defensive stance in view of internal problems and pressures, and this was particularly evident on two important issues, namely agriculture and safeguards against imports supposed to cause injury to domestic producers.

Meanwhile, EC agreements with third parties became increasingly compatible with GATT rules. Largely in response to international criticism, emanating mainly from Washington, the Community tried to conform to the requirements of Article XXIV of GATT, regarding the creation of free trade areas and customs unions, mainly by abandoning its demands for reciprocal trade concessions as far as developing countries were concerned. It was in this spirit that the new Mediterranean agreements were signed in the 1970s. The same principle applied to the successive Lomé conventions signed with a large and ever-growing group of developing countries, the large majority of which were former colonies of different members of the Community. On the other hand, free trade agreements were signed in 1972 with the remaining members of EFTA, thus marking an important step towards the creation of a large free trade area in Western Europe. Regionalism was thus further strengthened at the expense of multilateralism (Hine, 1985).

The rates of growth of international trade registered a significant decline (Boltho and Allsopp, 1987). However, this decline was in terms of growth rates and not in absolute figures, with the exception of only one year (1982). Thus the experience of the 1970s and the early 1980s was substantially different from that of the inter-war period when international trade had declined in absolute terms.

Protectionism spilled over into intra-Community trade. There was clear evidence of increasing recourse to state aids and technical barriers, while the number of infringements of treaty articles dealing with the free movement of goods and services multiplied during this period. Protectionism was mainly the result of strong social resistance to economic adjustment, and the emphasis was laid on declining sectors. In 1981 the EC Commissioner for Industry, Etienne Davignon, warned that 'the industrial activism of certain member states . . . has become a veritable challenge to the Community' (Noelke and Taylor, 1981: 219). The incomplete customs union of the 1960s did not come crashing down with the advent of the economic recession; and this was, perhaps, a major achievement in itself. The building did, however, suffer a certain degree of damage, and the economic relevance of the EC was considerably reduced as a consequence.

In some cases, in an attempt to preserve the common market and also strengthen the role of the Community, the Commission tried to replace national policies of intervention by common EC policies, thus following the example of agriculture. The sector where it became most active was steel where the Commission, taking advantage of the provisions of the Treaty of Paris and the crisis situation which developed in the European steel industry after 1973, gradually set up the mechanisms for a highly interventionist policy, which included production quotas, minimum prices, and severe import restrictions. It was ironic that the crisis measures in the steel industry, fully implemented in 1980, relied largely on the operation of a cartel of producers (basically a rationalization cartel) created under the guidance and supervision of the EC Commission. One of the main objectives of the Treaty of Paris had been precisely to avoid the reappearance of such a cartel. The latter brought together some prominent national champions from different countries, whose survival, thirty years after the creation of the ECSC, was a perfect illustration of the limitations of the trade model in European integration (Tsoukalis and Strauss, 1985; Messerlin, 1987).

The sun did not shine over European agriculture either; although here the problem was qualitatively different. The CAP, the most developed common policy of the Community, came increasingly under attack in the 1970s. The combination of high guaranteed prices, steady increases in productivity, and stagnant demand led to mounting surpluses which had to be stored or dumped in foreign markets at a large expense for the EC taxpayer. The growing financial cost of the CAP, which accounted for approximately two-thirds of total EC expenditure, was coupled with large and persistent income inequalities among the European farming population; inequalities to which the policy itself was contributing. The opposition from member countries with small agricultural sectors, especially the UK, who were the main losers, coupled with ever-growing pressures from third country producers, gradually succeeded in turning the Community's agricultural surpluses into a major political issue. One of the biggest successes of the 1960s was thus transformed into an albatross hanging from the neck of an embattled Community (Rosenblatt *et al.*, 1988).

The earlier trend for intra-EC trade to grow faster than total trade was reversed (Fig. 2.1). This reversal coincided with the first oil shock and the economic recession and continued until 1981. During this period there was a significant decline in relative terms of intra-EC trade for the old members of the Community, which was only partly compensated by the increase in intra-EC trade registered by the new members who joined in 1973. This phenomenon was, however, hardly attributable to protectionist measures. For this to be a valid explanatory factor for the stagnation of intra-EC trade as a percentage of total trade (and not in absolute terms, since trade continued to grow), there should be evidence to suggest that protectionist measures were directed exclusively or mainly against other members of the Community. And this was clearly not the case. The decline of intra-EC trade, which lasted for approximately ten years (a new upward trend started in 1982), has been attributed more to the gradual loss of European competitiveness, especially in strong-demand sectors, which led to growing import penetration of European markets, and the increasing integration of EC firms in the world division of labour (Jacquemin and Sapir, 1988). The oil factor should also not be ignored: the large increase in the price of imported oil led to a shift of European exports to the booming economies of OPEC (Organisation of Petroleum Exporting Countries).

While the trade integration model came under increased strain, the Community budget and the distributive impact of EC policies became, for the first time, a major political issue. This was closely related to the economic crisis and the enlargement of the EC. In times of slow growth, the struggle for larger shares of the pie usually intensifies; and this was proved true for the Community as a whole. On the other hand, the 1973 enlargement brought inside the EC two countries with serious regional problems, namely the UK and Ireland; and the former, expecting to be a big loser from its participation in the EC budget, decided to turn this into a major political issue. It was to remain so for many years to come. Last but not least, the question of redistribution was directly linked to the creation of an EMU, since the latter would remove more instruments of national economic policy and thus increase the risks of wider inter-country and regional disparities.

The developments of the 1970s and the accession of the UK in particular brought into question the original package deal which had sustained for years the process of regional integration. The result was serious internal divisions and interminable negotiations which virtually paralysed the Community for a long time. The EC budget and the so-called 'British problem' became the central issue and the focus of intra-EC negotiations (H. Wallace, 1983).

Although fighting over who pays and who gets what and when is the main stuff that politics is made of, and it would be unnatural if Community politics were very different, the amount of time and energy devoted for many years by heads of state and government down to low-ranking officials in Brussels and the various national capitals to EC budgetary disputes was totally out of proportion with the actual amounts of money involved. Disputes over sums which represented only a few decimal points of EC GDP dominated for years the Community agenda; and this was in itself a reflection of the deep political crisis inside the EC.

The Regional Fund, established in 1975, was originally conceived as a partial compensation to the UK for the budgetary loss resulting from its participation in the CAP. The progressive increase of expenditure through the Regional Fund was coupled with a similar development with respect to the Social Fund. However, regional and social policies, still representing a small percentage of total EC expenditure, remained the only two policies with an explicit redistributive bias.

The result of all the above developments was a more inward-looking and defensive Community; and also a less important Community in economic terms. The emphasis was on the preservation of the *status quo*. The creation of the Regional Fund, together with the introduction of VAT and the establishment of the Community's 'own resources', were the main exceptions to this rule. At the political level, there were three main developments: the creation of European Political Co-operation (EPC), following the initiative taken at the Hague summit of 1969; the transformation of irregular summit conferences into the European Council; and the first direct elections to the European Parliament in 1979. In fact, the first two developments were interpreted, although not very convincingly, by some observers as yet another manifestation of the trend towards more inter-governmentalism (Taylor, 1983).

Despite the internal crisis and the unfavourable economic developments, the Community did not lose its power of attraction, at least with respect to its immediate neighbours. The best illustration was the new applications for membership submitted only a short while after the first enlargement of 1973. The new candidates came from the south of Europe and had only recently emerged from dictatorial rule. The applications of Greece in 1975 and Spain and Portugal in 1977 need to be understood, first and foremost, as an important act of high politics and a search for a *Pax Europea* on behalf of the new democratic regimes (Tsoukalis, 1981). Membership of the Community was identified with democratic institutions, economic prosperity, and active participation in the building of a united Europe in countries which had been for long, both literally and figuratively, on the periphery of Europe.

The presence of an economic bloc on the European continent (the EC had always looked stronger from the outside than from the inside), with an obvious political dimension and objectives, left the small and less developed countries of the periphery with little choice in the long term. There were, however, serious doubts as to the ability of those countries to face the economic challenge associated with their integration in the competitive economic environment of the EC. Moreover, at a time when redistribution was a highly divisive issue inside the Community of Nine, the three southern European countries were seen as potential *demandeurs* and a further drain on meagre EC resources.

The result was protracted pre-accession negotiations during which the Nine adopted the time-honoured tradition of post-poning the day of reckoning. However, the delay finally seemed to serve some useful purpose. By the time the two Iberian countries were admitted as full members in 1986, Greece having preceded them five years earlier, a whole new package of measures had been agreed for the relaunching of integration. In view of the negative experience of the 1970s, the EC appeared this time determined to combine successfully deepening with enlargement.

3. 1992 and Beyond

The 1980s found Western Europe in the midst of a deep economic malaise. Stagnating output, rapidly rising unemployment, and declining export shares of world markets formed the main elements of a dismal picture in the aftermath of the second oil shock. Community countries were manifestly unable to lift their sights above petty budgetary squabbles and the price of pigs which had absorbed for years an inordinate amount of their time and attention. The trade liberalization model, once the solid basis of regional integration, had apparently reached its limits. True, the EC had survived its first major enlargement and the dramatic deterioration of the international macroeconomic environment; but it was considerably weaker as a result and depressingly unable to adjust to the new economic and political realities.

The efforts to extend integration to new areas of activity, launched at the Hague summit of 1969, had largely failed. The twenty-fifth anniversary of the signing of the Treaty of Rome was greeted on a cover of *The Economist* (20 March 1982) with a tombstone for the EC, carrying a very characteristic epigraph: *capax imperii, nisi imperasset* ('capable of power, if only it had not tried to wield it'). These were the years of 'Euro-pessimism' and 'Euro-sclerosis', terms which became popular in the European and foreign press.

And then things slowly began to change. In the course of the 1980s, the transformation of the economic and political climate was absolutely remarkable. European economies slowly re-discovered their old dynamism: growth rates reached levels that had not been experienced for almost fifteen years, investment picked up rapidly and many new jobs were created. Meanwhile, regional integration gained an ever-accelerating momentum and Commission proposals, which had been gathering dust for years in ministerial drawers, resurfaced on the Council table and were rapidly turned into Community legislation. European politicians competed in federalist rhetoric, and for once their words were not totally divorced from their actions.

The relaunching of the integration process became centred around the magic number of 1992. Although representing the target date for the completion of the internal market (which is in fact 31 December 1992), it soon came to symbolize the new phase in European integration; and in the end, it was about much more than the internal market. The rapidly expanding agenda of the EC has included redistributive policies, external economic relations, the so-called social space, and, last but not least, a renewed attempt to establish an economic and monetary union. This promises to be the main economic issue in the post-1992 phase. Changes in the political and institutional sphere were bound to follow. Coupled with the spectacular developments in Eastern Europe and the collapse of the old communist order, the acceleration of the integration process in the West has helped to switch international attention back to the old continent.

The Rediscovery of Europe and the Market

The 1992 programme was conceived in the depths of the economic crisis. It came to life when economic recovery was still in its early stages, and grew up rapidly at the time of the boom. Although it is plausible to argue that there has been a close link between the two, the exact nature of the relationship between the child and its environment is virtually impossible to establish. This may be partly answered in the next few years as the young adolescent starts to cope with economic adversity.

Following the second large increase in oil prices in 1978–9, the economies of Western Europe entered into the longest and deepest recession since the end of the Second World War; and unemployment doubled in the course of only five years (Tables 2.1 and 2.2; Fig. 3.1). The comparison with other major industrialized economies, notably the United States, Japan, and the countries of EFTA, made the picture even more depressing. While the Europeans had long been used to unfavourable economic comparisons with Japan, for much of the 1980s they also had to reconcile themselves with a performance in terms of growth rates and unemployment which was clearly inferior to that of the United States. The creation of millions of new jobs in the

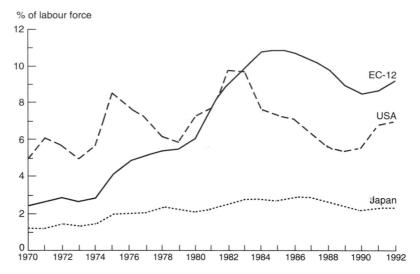

% of labour force

Fig. 3.1. Unemployment Rates, 1970–1992
The figures for 1990–1992 are EC Commission forecasts of
Nov. 1991.

Source: Commission of the EC.

United States was in sharp constrast with the steady growth
of unemployment in Europe; although it could be argued that
'persistent poverty is the American equivalent to persistent
unemployment in Europe' (Dahrendorf, 1988: 149). This was a
novel and highly uncomfortable experience. Furthermore, the
rapid increase in European investment in the United States
during the first half of the 1980s, while investment ratios continued their declining trend in Western Europe, was another
clear manifestation of business perceptions and the lack of confidence in the future of the European economies.

At the same time, the continuing stagnation of intra-EC trade
as a percentage of total trade was coupled with a reduction in
EC world market shares in manufactured goods; and this was
particularly pronounced in the strong-demand sectors such
as electrical and electronic equipment, office machinery, and
information technology (Buigues and Goybet, 1989). These
were times of major technological advances and rapid industrial
restructuring at the global level; and the Europeans increasingly

felt that they were losing out in the international race for the industries of the future. Serious losses of market shares were also sustained in cars and industrial machinery.

These developments were closely interrelated: the gradual loss of international competitiveness of European producers, especially in the rapidly growing sectors, accounted both for the increased penetration of EC markets by third country exporters and the loss of world market shares. The latter was more pronounced than the former, partly due to the protection of European markets.

The next step was to establish a link between the loss of competitiveness and the fragmentation of the European market, due to the existence of NTBs and the policy of national champions pursued by most governments (Albert and Ball, 1983). A consensus view about the existence of such a link gradually developed in the 1980s among key policy-makers and industrialists; and the EC Commission was highly instrumental in this respect. The emphasis was on high technology sectors where economies of scale and gains associated with learning curves were perceived to be particularly important. After all, these were the sectors where the advantages from the creation of a real common market should have been the greatest; and these were precisely the sectors where the common market did not exist.

Bressand argues that 'the early 1980s marked the limits of the national consolidation process in the larger countries in sectors such as telecommunications, electronics, automobiles, etc.' (1990: 50). But the ground had been prepared earlier for the mobilization of the business lobby in this direction. The role of Viscount Davignon, the EC Commissioner for industrial affairs since 1979, was absolutely crucial. He was instrumental in bringing together the heads of the twelve leading European firms in electronics in a 'round table' which provided the basis for the subsequent launching of the ESPRIT programme (European Strategic Programme for Research and Development in Information Technology).

The choice of the electronics sector, and information technology in particular, was no coincidence: there was a European obsession with this rapidly growing sector and the perceived danger of being left behind their American and Japanese competitors. The conclusion drawn was that there should be close intra-European collaboration in research and development

(R. & D.) and an early end to the fragmentation of the large European market; that is, a rejection of the old strategies of national champions (Sharp and Shearman, 1987).

The first full phase of the ESPRIT programme ran between 1984 and 1988, followed by a second five-year phase. It was preceded by a modest pilot scheme. The emphasis was on cross-border collaboration in pre-competitive research. The importance of ESPRIT was much greater than the actual sums of money involved, although those sums have grown substantially over the years. It constituted the first major effort to promote close co-operation among European firms which had been until then prominent national champions, starting with R. & D. and gradually extending into other areas. Co-operation in research meant a sharing of costs and uncertainty; it also facilitated the adoption of common European standards, thus eventually eliminating one of the most important NTBs. Cross-border co-operation among firms, which had been in the past mostly of a transatlantic nature, was consistent with the new corporate search for extensive 'networks' (Bressand, 1990). At the same time, it laid the basis for a powerful lobby militating in favour of a large European market. Yet the logic of global markets and global competition, combined with the continuing weakness of leading European firms in this sector, imposed serious constraints on the strategy of intra-European co-operation; and those constraints were soon to become apparent.[1]

In the search for intra-European 'networks' and co-operation among firms, the EC Commission often played the role of a marriage-broker. This marked a significant departure from an earlier more interventionist approach in industrial policy which had led absolutely nowhere.[2] The early positive response of

---

[1] One of the founding members of Vicount Davignon's 'round table' in the electronics sector was later to be taken over by a Japanese company (majority acquisition of the British ICL by Fujitsu), while other European champions actively sought partners outside Europe's borders. On the other hand, another indication that old national champion habits die hard were French government plans in 1991 to create a large conglomerate consisting of Thompson and the Commissariat à l'Energie Atomique (CEA) which would allow the subsidization of the ailing electronics firm.

[2] In Sept. 1983, the French Government submitted a memorandum for the creation of a 'European industrial space' (Pearce and Sutton, 1986). This was the culmination of many efforts made by France to convince its partners about the merits of close co-operation in the industrial field. Although recent EC policies in the field of high technology bear the marks of French influence, they are, however, both less interventionist and, arguably, also less protectionist in their approach.

European firms to the ESPRIT programme led to the adoption of similar initiatives in other areas such as RACE (Research in Advanced Communications for Europe) and BRITE (Basic Research in Industrial Technologies for Europe) and the proliferation of joint R. & D. programmes. Following another French proposal, which was a defensive reaction to President Reagan's Strategic Defence Initiative (SDI), a new collaborative mechanism was set up in 1985 in the form of EUREKA (European Research Co-ordinating Agency). The membership of EUREKA included all Western European countries. Thus, gradually and in a modest way, European regional co-operation was extending beyond the sphere of trade.

The growing 'Europeanness' of the perceptions and strategies adopted by the large European firms[3] was also manifested through the creation of the Round Table of European Industrialists which has acted both as a powerful lobby behind the scenes and a means of promoting closer ties and co-operation among the heads of some of the largest firms. Such moves were arguably not unrelated to the wave of mergers, acquisitions, and co-operation agreements across national borders, which was to follow a few years later. The composition of this Round Table, first led by Mr Gyllenhammar, chairman of Volvo, clearly suggests that the emerging European corporate reality was not much affected by the EC-EFTA frontier. It was, however, abundantly clear to everybody concerned that only the EC could deliver a large European market free from internal barriers. On the other hand, the elimination of intra-European barriers was not always seen as incompatible with higher protection *vis-à-vis* the rest of the world.

The growing perception among business and governments of the large costs of the fragmentation of the European market was conveniently married with the constantly increasing appeal of supply-side programmes and economic deregulation, particularly evident in telecommunications and financial services. Mainly imported from the United States of President Reagan, with Mrs Thatcher acting as the main and highly energetic European agent, those ideas were gradually adopted by other European leaders, although with a mixture of anticipation and embarrassment

---

[3] Another example is the report published by the president of Philips, calling for a programme to achieve a single European market (Dekker, 1985).

characteristic of young virgins. The adoption of such ideas implied at least a partial rejection of old-established notions about the mixed economy in Western Europe.

The shift towards supply-side measures, also strongly and consistently supported by the EC Commission in its annual economic reports, was in turn a reflection of a more general shift to the right in terms of economic policies, evident in most countries of Western Europe during the first half of the 1980s; and this was true almost irrespective of the colour of the political parties in power. A major example of this emerging consensus was the shift of the French Socialist Government towards more market- and European-oriented policies in 1983, after the failure of 'Keynesianism in one country' (Sachs and Wyplosz, 1986). This shift marked a decisive defeat for those in the Socialist party who had been advocating protectionist measures and an independent monetary policy. It was also a major turning-point for the Community as a whole, in view of the central role subsequently played by France in the relaunching of the integration process. Economic orthodoxy and a strong market orientation similarly characterized the policies pursued by the Spanish Socialists after their arrival in power in 1981; unlike their comrades in Greece.

The new consensus also found an expression in the area of macroeconomic policy, with an increasing convergence towards restrictive monetary policies and budgetary consolidation; and the price was unprecedented levels of unemployment which continued to rise until 1985 (Fig. 3.1). The new consensus meant a collective rush away from the basic Keynesian ideas which had provided the foundation of post-war economic policies in Western Europe.

The first important step had been taken back in 1979 with the setting-up of the European Monetary System (EMS) and the creation of the ECU (a weighted basket of EC currencies), undoubtedly the most important economic event of the previous decade for the Community. It was the crowning act of the close co-operation between Valéry Giscard d'Estaing and Helmut Schmidt, from which Britain had, once again, decided to remain aloof. The main aim was to reduce exchange rate instability among EC currencies and to take another step towards the creation of a regional currency system in a world of generalized floating. The setting-up of the EMS also signified the implicit acceptance by France and Italy (the smaller countries never had much choice) of German macroeconomic policy priorities. The emphasis was now clearly on the fight against inflation.

After an early turbulent period, when the EMS came under serious strain due to diverging economic policies (it was the time when the newly elected Socialist Government in France adopted an expansionist stance against the general trend), subsequent years have been characterized by remarkable exchange rate stability based largely on a convergence of monetary policies. The turning-point was the realignment of March 1983 and France's decision to remain in the exchange rate mechanism. The EMS remained an asymmetrical system, although, arguably, asymmetry was an important element of its success. As with the mini-snake in the 1970s, the Federal Republic remained the leader of the system, with the Deutschmark providing the anchor for the other currencies. The existence of this asymmetry was in itself a further manifestation of German success and growing predominance in the economic field.

Although with some big exceptions, such as Greece and Italy, there was also a clear tendency towards budgetary consolidation in Western Europe. Budget deficits were progressively reduced and there was a reversal of an earlier trend towards constantly rising government expenditure as a percentage of GDP which lasted until 1989 (Fig. 3.2). This in turn led to a reduction of public debt in many countries: another sign of the shift in European government priorities during this period (Commission of the EC, 1989a, 1991a).

Supply-side measures, the emphasis on greater flexibility of labour-markets and restrictive macroeconomic policies went hand-in-hand with a weakening of trade-union power. This was a function of the adverse economic environment, the large increase in unemployment, and the changing composition of the labour force; thus a very different situation from the one prevailing in the 1960s and early 1970s. The result was a considerable reduction in the share of wages and salaries out of total income and a substantial increase in business profitability. This again lasted until 1989 (Commission of the EC, 1991a). Interestingly enough, the path followed by some EFTA countries was different for most of the 1980s.

On the other hand, the transformation of the international political scene in the early 1980s also contributed to the collective

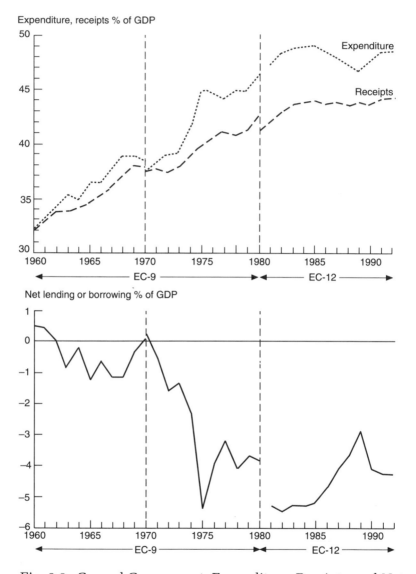

Fig. 3.2. General Government, Expenditure, Receipts, and Net Lending (+) or Borrowing (−) in the EC, 1960–1992
EC-9: EC-12 excl. Greece, Spain and Portugal. The figures for 1990–1992 are EC Commission forecasts of Nov. 1991.

Source: 1960–1970: OECD, 1970–1992: Commission of the EC, 1991*a*.

readiness of Western European countries to proceed further in the field of regional economic integration. It would not be, after all, the first time that essentially economic initiatives were taken at the European level with long-term political objectives in mind. Fears of a revival of the cold war between the two superpowers and a growing divergence of views across the Atlantic increased the uneasiness of Europeans, while exposing yet again the element of schizophrenia which had traditionally characterized their attitude towards the American protector, at least as long as the perception of the need for a protector persisted.

President Reagan's aggressive style of diplomacy and the new arms race strengthened their latent enthusiasm for a stronger European pillar of the alliance; and this would eventually find an expression in attempts to revive the Western European Union (WEU) as well as in new initiatives at the EC level. Paradoxically, some years later the unexpected and rapid dismantlement of the Soviet empire in Eastern Europe also appeared to act as a catalyst for new moves towards further regional integration. But, this time, it was the fear of an eventual withdrawal of US forces from Europe and, even more so, the reappearance of the 'German problem', if only in a different form.

Last but not least, the Community was faced once again with the prospect of further enlargement, this time taking into its fold countries with weaker economies and fragile democratic regimes. The first phase of this new enlargement, with the accession of Greece in 1981, was not particularly smooth; and this was interpreted as a sign of more troubles to come. The accession of new members added considerably to the urgency of strengthening Community institutions and policies, and the objective was to avoid further dilution.

The Long Virtuous Circle

The relaunching of the integration process was preceded by long and difficult negotiations regarding the reform of existing policies and institutions and the identification of new areas for common action. The 'British problem' still remained at the centre of the negotiations; but unlike earlier years, when there had been a search for *ad hoc* arrangements, a general agreement seemed to emerge slowly that a more permanent solution to the 'British problem' should be part of a wider reform.

The successful implementation of new initiatives was predicated on the resolution of two outstanding issues which had plagued the Community for years: budgetary reform and a more effective control of CAP expenditure. The rapid depletion of 'own resources' (customs duties, agricultural levies, and 1 per cent of VAT), coupled with the continuous increase of CAP expenditure and the legal impossibility for deficit financing, had radically changed the framework for negotiation. It was no longer a question of looking for partial compensation measures to placate the British. New financial resources needed to be found, and this required the revision of the relevant treaty articles and ratification by all national parliaments. On the other hand, the search for new 'own resources' could not be divorced from a thorough discussion of the CAP, which accounted for approximately two-thirds of total EC expenditure, and the prospects for policy reform.

In the 1980s, the need for reform of the CAP became more generally accepted. The policy of price restraint, practised for some years, had not succeeded in eliminating surpluses; nor had it operated as an effective constraint on ever growing expenditure. Large increases in labour productivity meant that supply continued to grow much faster than domestic demand. The resulting surpluses, combined with tight international markets, led to an ever-increasing financial cost, while farmers' incomes remained stagnant. Attachment to *status quo* policies thus became increasingly untenable. The rapid approach towards the ceiling of the Community's own resources, used by the UK as a powerful negotiating weapon, combined with growing pressures from third countries and the prospect of further enlargement, which was expected to create new surpluses and thus add considerably to costs, acted as the main catalysts for reform. In fact, it was only in the early 1980s that the very word reform started appearing in official documents.

Redistribution also formed an integral part of the wider debate on budgetary reform. Widening intra-EC income disparities and the economic crisis had brought the issue of redistributive policies to the top of the political agenda. The accession of new, economically weaker, members gave this issue even greater prominence.

On the other hand, the extreme slowness of the EC decision-making mechanism, often a euphemism for a complete standstill caused by the search for unanimity in the Council of Ministers, and the prospect of further enlargement, which could only make the situation worse, led to a wide-ranging debate on institutional reform. This also included the problem of the so-called 'democratic deficit' of the Community (the lack of sufficient democratic control and accountability due to the legal emasculation of the European Parliament and the inability of national legislatures to perform effectively this role), and the artificial division between normal EC business and foreign policy co-ordination in the context of EPC. The Genscher–Colombo proposal of 1981 for a new European Act, which would replace the existing treaties, and the draft treaty on European Union (Bieber, Jacqué, and Weiler, 1985) adopted by the European Parliament in 1984 were the most important initiatives in this area.

On the other hand, the need to reconcile the growing political and economic heterogeneity of the Community, largely related to its geographical expansion, with further integration also led to the consideration of various models incorporating a degree of diversity in terms of the rules and obligations applying to different members. Hence the proposals for a two-tier Community, 'variable geometry', and 'graduated integration' which became popular at the time.[4]

The Fontainebleau summit of June 1984 was an important turning-point in the long intra-EC negotiations. The package agreed then included new measures to control the growth of agricultural surpluses and the creation of new 'own resources'. The decisions taken at Fontainebleau offered a clear indication of the course of policy reform in the agricultural field. The tackling of surpluses would not be based on the pure market solutions advocated by most economists. Instead, the policy of price restraint, as opposed to large price cuts which would have been, of course, wildly unpopular among farmers, was to be combined with an increasing reliance on other measures, such as produc-

---

[4] The idea of a two-tier or two-speed Community could be traced back to Chancellor Brandt and the Tindemans report on European Union (Tindemans, 1975). 'Variable geometry', a French idea, essentially implied variable membership from one issue area to the other (Commissariat du Plan, 1980). Another variation on the same theme was the proposal for 'graduated integration' (*abgestufte Integration*) later put forward by German academics (Grabitz, 1984).

tion quotas and 'co-responsibility levies'. Those measures were intended to limit production and the scale of EC intervention, while also trying to pass on to producers some of the financial cost of storing and disposing of surpluses.

One important aspect of the Fontainebleau agreement was the introduction of a permanent mechanism for a partial compensation to the UK, based on the difference between its VAT contribution and its overall receipts from the Community budget. Special provision was also made for the reduction of the net contribution of the Federal Republic. However, the new agreement, which was heralded at the time as a major breakthrough by all participants, proved to be extremely short lived. When the new VAT rate of 1.4 per cent was introduced in 1986, the upper limit had already been reached because of the intervening rapid growth in expenditure. Thus new negotiations on the budget had to start again almost from scratch.

In 1985, a new instrument was created through the Integrated Mediterranean Programmes (IMPs), intended for the less developed regions of France, Italy, and the whole of Greece. The creation of IMPs was a recognition of the special development problems of these regions and the relative bias of the CAP against southern agricultural products. The Iberian enlargement acted as the catalyst: IMPs were, in fact, seen as a compensation to the existing members of the Community of Ten for the expected negative economic effects of the accession of Spain and Portugal.

Meanwhile, a new Commission, under President Delors, had taken office. Looking for a new driving force for the integration process and the main plank of the Commission's strategy for the next four years, President Delors quickly opted for the completion of the internal market as an objective which appeared to have relatively good chances for gaining the support of member governments. This was announced to the European Parliament in January 1985 when the target date of 1992 was also mentioned for the first time.

The elimination of remaining barriers to the free movement of goods, services, persons, and capital, arising mainly out of different national regulatory frameworks, and the creation of conditions approaching as much as possible those in a domestic market was certainly not a new idea. The ground had been prepared in earlier years through various Commission communications to the Council; and there was already a rich collection of parliamentary resolutions and European Council declarations in this respect. Furthermore, an 'internal market' Council had been created in 1983. Interest in the completion of the internal market was a reflection of the negative developments of the previous decade, the stagnation of intra-EC trade, and the growing popularity of supply-side measures and economic deregulation. Yet the decision to adopt the internal market as the first priority of the new Commission and as the principal instrument of European integration for the next eight years was a major strategic decision. It was the Commission that took both the initiative and the risk; and therefore the one which justifiably received the praise for its subsequent success.

Having been given the green light at the Brussels summit of March 1985, the Delors Commission then presented the next European Council in Milan with the White Paper entitled *Completing the Internal Market* (Commission of the EC, 1985a). The White Paper spelt out with remorseless logic the consequences of the political commitment to eliminate all remaining intra-EC barriers. It included a long list of measures (279 precisely) and a timetable for their adoption, extending to 31 December 1992. The completion of the internal market provided the unifying concept for a set of very disparate measures, ranging from rules on animal sperm and the right of political asylum to the approximation of indirect taxation rates. Many of the proposed measures had been, in fact, lying on the Council table for years in the form of Commission draft directives and regulations. The implementation of the White Paper would, therefore, depend on the existence of political will which until then had apparently been in short supply.

The White Paper also introduced a new approach to the elimination of NTBs arising from different national rules and regulations. Instead of the old, time-consuming, and highly ineffective attempt to harmonize at the EC level, which had earned the Commission the reputation of specializing in mayonnaise labels and the noise of lawn-mowers, the White Paper proposed that the Community should rely as much as possible on the principle of mutual recognition. Whenever harmonization was deemed to be necessary, this should be limited only to essential objectives and requirements, thus leaving the task of defining the technical specifications to private standardization bodies (Schmitt von Sydow, 1988).

The idea was simple but also revolutionary; it built on indications of a more favourable political climate in national capitals and some earlier judgements of the European Court, especially the famous *Cassis de Dijon* which established the general principle that all goods lawfully manufactured and marketed in one member country should be accepted also by the other member countries, while also recognizing the need for some exceptions related to 'the effectiveness of fiscal supervision, the protection of public health, the fairness of commercial transactions and the defence of the consumer' (VerLoren van Themaat, 1988). This new and more flexible approach was considered as the decisive blow against the old Gordian knot of NTBs which had proved the main obstacle in the integration of mixed economies. It was also evident with respect to indirect taxation: the emphasis was now on the approximation and not the harmonization of tax rates.

There are certain characteristics of the White Paper, and the underlying strategy of the Commission, which proved to be crucially important for the eventual success of this new initiative. The first is the supply-side nature of the programme (at least this was the way it was generally perceived) which made it compatible with the prevailing economic and political climate. The second is the existence of a clear timetable for the adoption of the different measures, which helped enormously to provide a focus for national politicians and a clear target to aim for. In this respect, the White Paper followed the successful precedent of the Treaty of Rome.

The third and fourth characteristics were also extremely important, although at least partially misleading. The internal market was presented as a programme for the elimination of barriers, hence with no apparent financial cost. As Pelkmans and Winters put it: 'the emphasis is on rules, not money' (1988: 9). Given the long intra-EC budgetary disputes and the strong resistance of many countries to an increase in Community expenditure, this constituted a major attraction of the internal market programme. The subsequent emphasis on redistributive policies suggests that this was not exactly true.

Another characteristic was the technical and low-key nature of the White Paper, considerably aided by the personal style of

Lord Cockfield, the EC Commissioner for the internal market. Only passing reference was made to what later came to be known as flanking policies. Thus the White Paper, and the concept of the internal market in general, appeared as less threatening to governments, especially in terms of national sovereignty. This was in sharp contrast to EMU as the main economic initiative of the 1970s. A more careful reading of the political implications of some of the measures proposed and, perhaps, a certain prophetic ability to foresee the new momentum in European integration, initiated by the internal market programme, would have made some governments much more reticent about committing themselves to the 1992 target.

The next important step was taken by the European Council in Milan in June 1985. A decision was reached then, against the opposition of Britain, Denmark, and Greece, to convene an intergovernmental conference with the task of preparing a revision of the existing treaties. The division inside the Community was centred on the question of institutional reform which included the more extended use of majority voting in the Council of Ministers, the powers of the European Parliament, and the role of the EPC. The division between the six original members of the EC and the latecomers, with Ireland sitting on the fence, had earlier become apparent inside the Dooge Committee which had been asked to prepare a report on institutional reform. Furthermore, the countries with minimalist views on the EC could hardly welcome a new constitutional debate on Europe.

The issue was of major political importance, and the decision of the Italian Presidency to go ahead on a majority vote was both unexpected and without precedent. It thus became clear that the majority of member countries were now determined to push ahead with institutional reform and the relaunching of the integration process (Taylor, 1989). The White Paper and the completion of the internal market had provided the new element which, perhaps, tipped the balance.

The end result was the Single European Act (SEA) signed by the representatives of all twelve members in February 1986. It came into force in July 1987, after the holding of popular referenda in Denmark and Ireland. The political momentum created by the intergovernmental conference and the fear of being relegated to the second tier forced the three opposing countries to play an active part during the negotiations. In the case of Britain, negative factors were combined with a keen interest in the completion of the internal market which needed to be reconciled with a strong dislike of any institutional changes, especially as regards majority voting in the Council. The final product was in some respects close to the lowest common denominator. In the words of Mrs Thatcher: 'Part of our task the whole time has been to diminish their expectations and draw them down from the clouds to practical matters' (*The Times*, 6 December 1985; see also Colchester and Buchan, 1990; Moravcsik, 1990). The SEA was the result of a compromise among the now twelve members of the Community. It also provided the legal framework for the emerging new package deal in the new phase of regional integration.

A central element of the SEA was the formal adoption of the internal market as an objective to be achieved by 31 December 1992. The internal market was defined as 'an area without internal frontiers in which the free movement of goods, persons, services and capital is ensured' (Article 8*a*), thus adding a political dimension to the 1992 target. On the other hand, in a separate declaration, member governments insisted that the target date 'does not create an automatic legal effect'; this clarification was meant to preclude any such interpretation in the future by the European Court.

The harmonization of national rules and regulations, as part of the implementation of the internal market programme, would be based on qualified majority voting[5] in the Council of Ministers (Article 100*a*). Combined with the new approach towards harmonization and the increasing reliance on mutual recognition, this was intended to break the familiar deadlock in the Council arising from the frustrating search for unanimity. In fact, qualified majority voting was extended to some areas where the unanimity rule used to apply under the old treaties: they included capital liberalization and air and sea transport.

On the other hand, there was a certain *quid pro quo* offered to member countries for abandoning the security of the national veto. A safeguard clause was introduced which would allow member countries, under special circumstances, to continue applying national provisions after the adoption of new EC rules

---

[5] There is weighted voting in the Council of Ministers and the weights for members range from 10 for the biggest countries (France, Germany, Italy, and the UK) to 2 for Luxembourg. In the Community of Twelve, 54 votes out of a total of 76 are needed for qualified majority.

(Article 100*a*, par. 4). It was meant as an insurance against the lowering of standards for the protection of the environment and working conditions. There were also three important exceptions where unanimity would still be needed, namely fiscal policy, the free movement of persons, and the rights and interests of employees. The exceptions were a reflection of the political sensitivity of the issues involved and the existing divergence of views and actual practices among individual member countries.

In recognition of the heterogeneity of the Community and the different levels of economic development among member countries, Article 8*c* opened the possibility for special provisions and derogations of a temporary nature. This constituted a modest step towards some form of differentiated policies; but certainly not an acceptance of the two-tier and variable geometry models of integration which had been aired during the negotiations. There was more to cater for the interests of the weaker economies. Title V, under the heading of 'economic and social cohesion', was an attempt to link the objective of 'harmonious development' and the reduction of regional disparities, objectives which were mentioned in a very general manner in the original treaties, with specific policy instruments. The reference to economic and social cohesion also constituted a formal recognition of the growing political importance of redistribution. The new Article 130*d* called for the effective co-ordination and rationalization of the activities of EC Structural Funds and the Commission was invited to submit proposals in this direction.

The EMS and the ECU became for the first time legitimate children of the Community through a new article included in the SEA. However, this was considered to be a mixed blessing, since Article 102*a* also specified that any institutional change in the field of economic and monetary policy would require a new amendment of the treaties; and this was considered highly unlikely for the foreseeable future. Who could, in fact, anticipate the calling of a new intergovernmental conference in December 1990 in order to prepare the legal ground for a complete EMU?

The SEA also contained provisions extending EC powers in the field of social policy, especially as regards the health and safety of workers, as well as research and technological development, and the environment. Those new provisions were a clear manifestation of the expanding role of the Community in new policy areas.

As a partial answer to the 'democratic deficit' of the Community, a new 'co-operation procedure' was introduced; it was meant to strengthen, albeit in a modest way and only on these issues where qualified majority applied, the input of the European Parliament in the legislative process. The new 'co-operation procedure' also, indirectly, enhanced the central role of the Commission in decision-making, by reinforcing its powers of mediation between the Parliament and the Council of Ministers. Through the SEA, the European Council acquired for the first time a legal status as an institution of the Community. Similarly, EPC became part of the *acquis communautaire*. The relevant provisions were essentially an official recognition of already existing practices, except that they did add some words on security, thus paving the way for a further extension of EC competences in this highly sensitive area.

The SEA was the culmination of many years of difficult negotiations intended to provide the legal framework for the transformation of the European Communities into a 'European Union' (always a very imprecise concept), in accordance with the Solemn Declaration of Stuttgart of June 1983. It constituted the biggest ever revision of the original treaties and an indication of the expanding agenda of the EC. Its most important provisions referred to the completion of the internal market and the extension of qualified majority voting. Yet the importance of the latter could only be tested in practice. After all, provisions for qualified majority voting in several areas of decision-making had already existed in the original treaties; but they had been very infrequently applied.

The Luxembourg agreement of 1966 and the subsequent trend towards increasing intergovernmentalism had forced the Council to search continuously for a consensus among its members, even on relatively minor issues. The question, therefore, was whether the political climate had changed sufficiently in order to allow the new rules to have any real effect on decision-making. The first signs of change became evident even before the entry into force of the SEA as member countries started showing greater readiness to resort to voting. Since 1986, the number of votes in the Council of Ministers has slowly but steadily increased, even in policy areas totally unaffected by the revision of treaty articles.

The SEA was hardly greeted by anybody as the major political event which could change the course of the EC. As *Le Monde*

(5 December 1985) put it: 'Ces améliorations ne sont-elles pas de celles qui enthousiasment les foules.' Compared with the ambitious initiatives of previous years, the draft treaty of the European Parliament, and the time and effort put into its preparation, the SEA was generally considered as a letdown. It was like the mountain giving birth to a mouse. The fact that nationalist leaders, like Mrs Thatcher, did not perceive any serious threat in the revised treaties was seen as yet another proof of the limited political significance of the SEA.

With the benefit of hindsight, it can be argued that the importance of the SEA did not lie so much in the institutional provisions and the new objectives legally enshrined in the new Act, but more in its rather intangible role in the development of a momentum for the deepening and extension of regional integration. It certainly added a great deal to the credibility of the 1992 objective and thus helped to create a virtuous circle, involving both governments and the market-place. The negotiations and the signing of the SEA coincided with a steady improvement of the economic environment. The modest, export-led recovery, started in 1983, gradually turned into a self-sustaining process, with a steady increase in investment. The shift towards a more dynamic macroeconomic environment made governments readier to accept the adjustment costs associated with the 1992 target. The political readiness to go ahead with what was basically perceived as a supply-side programme contributed in turn to a further improvement of the economic climate. There was a close similarity with what had happened back in the 1950s and 1960s.

With the internal market, the Community had once again found a *force motrice* for further integration; and this time the economic objective was accompanied by changes in the decision-making process which were meant to facilitate the implementation of the internal market programme. The Commission was very quick in the production of draft legislation submitted to the Council in accordance with the White Paper timetable. Things also started moving at the Council level. The number of decisions taken by qualified majority steadily increased, although this can tell only part of the story. The mere knowledge of the possible recourse to voting has frequently forced countries to make virtue out of necessity by siding with the majority or simply remaining silent as soon as they realized that they could not command the necessary number of votes in order to block a decision. Schmitt von Sydow had argued that 'the principle of majority voting leads to unanimous decisions [or simply decisions, one might add], while the principle of unanimity leads to no decisions at all' (1988: 98). Paradoxical as it may sound, this statement is an apt description of actual EC experience.

The next crucial step in the credibility game for the internal market programme came with the German Presidency during the first half of 1988, and especially with the adoption of the 'Delors package' at the Brussels summit in February of that year. The Delors package consisted of three main elements, namely the creation of new budgetary resources, CAP reform, and the strengthening of redistributive policies: old issues which, however, remained as pre-conditions for the successful implementation of the internal market programme. The package of measures finally adopted included an agreement on the growth of EC expenditure until the end of 1992; the introduction of a new source of revenue on a GNP key; a compensation mechanism for the UK and the Federal Republic; a limit on the growth of CAP expenditure; new measures to control agricultural surpluses; the doubling of resources for the so-called Structural Funds; and a major reform in the operation and co-ordination of the latter.

At long last the Community had reached an agreement which could provide a solid basis for the relaunching of the integration process. The broad outlines for the direction of agricultural policy and the budget had been settled for the next few years, thus, hopefully, withdrawing from the negotiating table two main issues of contention. Provisions were made for sizeable side-payments for the economically weaker members. And there was already a specific programme for action with precise timetables which would take the Community well into the 1990s; its feasibility being further strengthened by the institutional changes already introduced. Last but not least, the Twelve were agreed that no further enlargement of the Community would take place before 1993 (obviously nobody had ever thought of the German Democratic Republic being a candidate).

The Brussels package played a crucial role in unblocking the process of integration and thus led to a much faster implementation of the internal market programme. The German Presidency of the first half of 1988 acted like a bulldozer reaching agreements on some difficult and sensitive issues which had remained blocked for many years. The decision to proceed with the complete liberalization of capital movements, after virtually no progress in this area for almost twenty-five years, was of paramount importance: it sent a clear message to markets that European governments actually meant business; and this further boosted the credibility of the internal market programme.

The rapid pace continued in subsequent years, and this coincided with the dramatic changes in the political map of Europe. The acceleration in the rate of Council decision-making was sometimes truly impressive: twelve months for the adoption of legislation on the harmonization of technical rules regarding machine safety compared with seventy months in the past for the adoption of a directive on the noise of lawn-mowers; less than three years for the mutual recognition of all university degrees compared with eighteen years in the past for an agreement on the freedom of establishment of architects (Commission of the EC, 1989*b*).

The Commission followed closely the timetable in the submission of proposals provided for in the White Paper. By early 1992, more than three-quarters of the measures had been agreed upon by the Council of Ministers. New EC legislation included the liberalization of financial services, the strengthening of competition policy, the first important measures in the areas of transport, telecommunications and public procurement, and several directives with respect to technical standards. Already then, the internal market programme could be declared a political success. Yet, progress was not uniform in all policy areas; serious difficulties in reaching political agreement had been experienced with respect to company law, taxation, financial services, and animal and plant health controls. Not surprisingly, they included several areas where the unanimity rule still applied.

Major progress has also been registered in more recent years with respect to the transposition of EC legislation into national measures, following a slow start by countries such as Italy which has traditionally been a laggard in the implementation of EC directives (Commission of the EC, 1991*a*). By that time, the continued credibility of the 1992 programme depended essentially on the actual implementation of measures agreed upon in the Council of Ministers and the effective enforcement of EC rules. In fact, the growing number of infringements in earlier years was not entirely encouraging (Pelkmans and Sutherland, 1990; Siedentopf and Hauschild, 1988).

During the second half of the 1980s, Western Europe experienced one of the longest booms of the post-war period, although growth rates remained lower than those experienced during the 'golden age' of the 1950s and 1960s (Table 2.1). Yet, by the standards of the more recent period, the economic performance was truly spectacular. In the last two years of the decade, the average rate of growth of EC economies approached 4 per cent, while most of the weaker economies (Ireland, Portugal, and Spain) grew faster, thus narrowing the intra-EC gap. High growth was also accompanied by some increase in inflation rates which had remained at historically low levels for some years (Table 2.3).

Growth was largely investment led: in 1990 the amount of investment in equipment made by Community firms was about 50 per cent higher in real terms than that of 1984 (Commission of the EC, 1991a: 3). During the same period, EC countries became by far the biggest recipients of foreign portfolio and direct investment which registered a sharp increase (Bank of International Settlements, 1991). There was something reminiscent of the early years of the EEC back in the late 1950s and early 1960s.

The rapid increase in investment ratios was accompanied by a process of major restructuring in the manufacturing and services sector. There was a wave of mergers and acquisitions, including an increasing number across national borders. For the first time, European integration was extended to the production level. The internal market programme and the new approach to EC industrial policy were absolutely consistent with developments at the business level; it was, therefore, only natural that the two processes would become mutually reinforcing.

During the same period, millions of new jobs were created and unemployment fell by two percentage points (Fig. 3.1), while unit labour costs continued their downward slope, also contributing to a further change in income shares in favour of profits. This was a further confirmation of the important change in political and economic conditions in Western Europe, which was in turn closely linked to the relaunching of the process of integration. Meanwhile, the share of intra-EC trade registered a sharp increase (Fig. 2.1).

It is clearly impossible to establish a precise link between political initiatives at the European level and the sharp improvement in macroeconomic conditions during the second half of the previous decade. Business confidence was undoubtedly influenced by exogenous factors which also contributed to the growth of world output during the same period. However, the role played by macroeconomic stability and supply-side measures, including the internal market programme, in shaping business expectations in Western Europe should not be underestimated. An early survey of Community firms showed that 'the "internal market" is having a not inconsiderable effect on the expectations of firms and, in particular, on their investments' (Commission of the EC, 1989a: 183). Although the use of the double negative in the Commission document is, perhaps, a sign of embarrassment of the anonymous economist who cannot apply in practice his familiar *ceteris paribus* assumption, there is now little doubt that 1992 has had a positive effect on investment and growth, thus closing the door to the 'Euro-pessimism' of earlier years. A repetition of the virtuous circle experienced almost thirty years earlier? The deceleration of economic growth which started in 1990, although initially cushioned by the positive economic effect of German reunification, may suggest that this virtuous circle had by then come to an end. This, however, remains to be seen.

The Post-Maastricht Phase

Although the completion of the internal market continued for some years as the main driving force in this new momentum of European integration, the political agenda kept on expanding. There was certainly an element of the spill-over effect on which functionalist theories of integration had so much relied; but there was also a conscious attempt made by the Commission and individual member countries to use the internal market and the new favourable climate in order to expand EC activity into new areas (see also Keohane and Hoffmann, 1991; Moravcsik, 1991; W. Wallace, 1990).

The doubling of budgetary resources and the reform of Structural Funds were presented as a pre-condition for the smoother implementation of the internal market programme and as a means of avoiding a political backlash from potential losers. The new initiatives in the social field had a similar logic; they also constituted an attempt to incorporate the trade union movement in the 1992 process. But by far the most important step was taken with the re-introduction of EMU into the European political agenda. It was first linked to capital liberalization agreed upon in June 1988. At the Hanover summit which took place the same month, a high level committee was set up, under the chairmanship of Mr Delors, to reopen the issue. Its report created a new momentum for the establishment of EMU. Against the strong resistance of the UK government, the European Council which met in Madrid in June 1989 agreed that the first stage of EMU should start in July 1990 and that a new intergovernmental conference should be called in order to prepare the legal and institutional ground for a complete EMU. This was confirmed at the next meeting of the European Council in Strasbourg which set the date for the beginning of the intergovernmental conference.

Meanwhile, the collapse of the political and economic order in Eastern Europe had introduced several new and important factors into the equation. The early celebratory mood caused by the collapse of communist dictatorships was soon followed by anxiety and concern about possible instability in the region and even more so by the re-emergence of the old 'German problem', albeit in a new form. The prospect of a unified Germany, leading perhaps to an important shift in the intra-EC balance of power, unsettled the minds of many European politicians. Following the old established French (and European) logic of '*la fuite en avant*', of which Schuman had been the first teacher, an attempt was made to link German union with the further strengthening of the EC and its federal traits in particular. This lent more support to EMU; but it also helped to bring the subject of political union on the surface.

This general term was intended to cover institutional reform, the extension of EC powers into new areas, and last but not least, foreign and defence policy. The combined efforts of Messrs Delors, Kohl, and Mitterrand, supported by the large majority of other EC leaders, led to the decision to convene another intergovernmental conference which would operate in parallel with the one on EMU. This decision was reached at the Dublin summit in June 1990. Clearly, a long distance had been covered since the 'Euro-pessimism' of the early 1980s. Now, the momentum of economic and political integration appeared to be almost unstoppable and the Europeans were already negotiating the main elements of the new package which would enable the Community to go beyond 1992 and the internal market. This momentum was partly attributable to the success of the internal market programme, but external factors had also played a major role.

The two intergovernmental conferences started in December 1990 and were concluded exactly one year later at the European Council of Maastricht. In the meantime, Mrs Thatcher had been sacrificed by her own party on the altar of European integration; or at least, this was one of the most important reasons for her forced resignation from the post of Prime Minister. The final text of the new treaty was signed in February 1992, and it incorporated a more radical revision and extension of the old treaties than anything tried before, including the SEA.

The provisions made for the creation of EMU were by far the most specific and far-reaching. The date for the entry into the final stage was set as 1 January 1999, or two years earlier if a majority of member countries were deemed by the Council to fulfil the necessary conditions. The final stage would involve the irrevocable fixity of intra-EC exchange rates, the adoption of a common currency, the ECU, and the establishment of a federal system of central banks (the European System of Central Banks — ESCB) with the European Central Bank (ECB) in the middle. The latter would be responsible for the conduct of monetary policy and exchange rate policy for the union as a whole. Article 107 of the new treaty as well as the accompanying protocol on the statute of the ESCB and the ECB referred in no uncertain terms to the independence of the new institution. Measures in the same direction would also have to be introduced for national central banks which had been until then under the tutelage of their political masters.

There would be no corresponding centralization of budgetary policies. Yet, provisions were made for the strengthening of the system of multilateral surveillance ranging from close monitoring of national policies by the Commission to the issuing of policy recommendations by the Council of Ministers and ultimately, as a measure of last resort, to the imposition of fines upon recalcitrant members.

For the intervening period, the European Monetary Institute (EMI) would be set up in order to strengthen the co-operation between national central banks and the co-ordination of monetary policies, thus preparing the ground for the final big step. And it would be a very big step indeed. With the new treaty, the

Community acquired for the first time a macroeconomic dimension which would at least match the provisions for the creation of the common market contained in the original Treaty of Rome. The ultimate stage of virtue (read EMU) was defined in great detail in order to satisfy the German demand for common institutions and policies which would match those at home. This was married with the French insistence on a specific date for the final stage which would legally bind the Germans to EMU. The road to virtue would not be very long, although, perhaps, it was not very clearly marked out.

Strict criteria were set for the participation of member countries in the final stage of EMU. In view of the long distance which some of them would need to cover during the transitional period in order to fulfil the necessary conditions, the further institutionalization of a two- or multi-tier Community thus became a concrete possibility. This was accentuated by the 'opt-out' clause for the UK which retained the right to stay out of EMU; and a milder version of this clause was reserved for Denmark.

The differential treatment of member countries was not restricted to monetary affairs. In an even more unorthodox agreement, member countries, with the exception again of the UK, decided to proceed further in the field of social policy. They would use Community institutions to legislate as eleven countries with the aim of strengthening the social dimension of economic integration. To this purpose, qualified majority voting among the Eleven was extended to new areas. Thus, the inability to reach a compromise on this highly controversial issue increased the isolation of the UK, despite the more conciliatory attitude adopted by the government of Mr Major, while also creating a legal minefield inside the Community.

EC responsibilities were extended to new areas such as culture, public health, consumer protection and the development of trans-European networks in the areas of transport, telecommunications, and energy infrastructures. A new Title XIII was added on industry in a very modest attempt to fill the old vacuum in this area. However, the emphasis on open and competitive markets and the retention of the unanimity principle suggested that an active and interventionist industrial policy at the EC level remained highly unlikely. Community powers in the fields of the environment and research and technology were also strengthened. However, this extension of EC powers was coupled with the first explicit reference to the principle of subsidiarity, according to which 'the Community shall take action . . . only if and in so far as the objectives of the proposed action cannot be sufficiently achieved by the Member States and can therefore, by reason of the scale or effects of the proposed action, be better achieved by the Community (Article 3*b*). Such a general statement can obviously command wide acceptance; it is only when applied to specific cases that disagreement tends to arise.

The new package would have been impossible without some reference to redistributive measures. The section on economic and social cohesion, together with the separate protocol on this subject, provided for the creation by the end of 1993 of a new Cohesion Fund for the poorer countries, which would specialize in the fields of environment and trans-European networks in the area of transport. Furthermore, a commitment was undertaken to review the operation of the already existing Structural Funds, leading to the transfer of more resources to the less developed countries and regions of the Community.

Despite the hopes entertained in the early stages of the negotiations and the link established between EMU and political union, the changes introduced under the latter heading were in the end relatively modest. Supranationality did not appeal much to several European governments, including the French who seemed to have reconciled, at least to their own satisfaction, support for political union with a strong preference for inter-governmental co-operation.

The new treaty established a European Union, and the three European Communities would constitute the most important pillar of the new edifice. The other two pillars represented a common foreign and security policy and co-operation in the fields of justice and home affairs; intergovernmental co-operation would be the guiding principle in both cases. Thus, the new European edifice would consist of three very uneven pillars of which the main one related to the traditional *acquis communautaire* and the changes introduced under the new treaty with respect to EMU and other areas of EC policy. The structure created at Maastricht would certainly not win any prizes for symmetry or even legal refinement and consistency; but this was the price to pay for a difficult political compromise among twelve governments which had decided to take new and important steps towards further integration.

All the changes described above with respect to EMU and the other areas of policy were part of the EC pillar. Institutional reforms were rather modest: the extension of qualified majority voting in a few areas and, potentially more important, new powers in the legislative process granted to the European Parliament through the system of co-decision (Article 189b). As the new treaty does not win any prizes for symmetry, the European legislative process does not distinguish itself either for its simplicity. With each treaty revision, the process has become even more complicated and cumbersome. On the other hand, the changes introduced at Maastricht fell far short of the expectations entertained by federalists for a significant strengthening of the central institutions of the Community, and especially the executive. However, a window of hope was opened for them: a commitment was already undertaken under the new treaty to call another intergovernmental conference in 1996.

The new treaty also introduced the concept of the citizenship of the Union which extended the right of movement and residence across member countries to all persons, and not only working people. This included the right of all citizens of the Union residing in another member state to participate in local and European elections. Citizenship of an idea? Perhaps, but also with some tangible benefits.

The other two pillars were kept separate from traditional EC business, a concession to those who had insisted on intergovernmentalism as the method to be applied to new areas which were dangerously close to the heart of national sovereignty. The twelve countries decided at long last to tackle an old taboo subject, namely defence, which had been until then the preserved domain of NATO. The signatories resorted to some legal acrobatics in order to try and reconcile the expressed wish for a European common defence policy and continued membership of the Atlantic alliance. However, the inclusion of defence in the new treaty was more than symbolic, and it was directly linked to the dramatic developments on the eastern side of the old Cold War divide.

As for the third pillar of the European Union, namely co-operation in the fields of justice and home affairs, this was directly linked to the objective of the free movement of persons inside the Union and the elimination of all intra-EC frontiers. The search for a common asylum and immigration policy *vis-à-vis* the nationals of third countries constituted the main driving force in this area. The constantly increasing number of asylum seekers and potential immigrants had added a great deal to the urgency of this task.

Thus, one year before the target date for the completion of the internal market, the Twelve had agreed on a new treaty which mapped out the road for further integration. In economic terms, the post-1992 phase should be marked by the establishment of EMU. However, at the time of writing the puzzle is still far from complete, and some of the necessary pieces may prove difficult to find. The treaty needs to be ratified by twelve national parliaments, and in two of the member countries, popular referenda are also a necessary part of the ratification process. The first such referendum which took place in Denmark in June 1992 produced a small 'no' majority, despite the support given to the treaty by all major political parties, business and trade union leaders as well most of the leading newspapers. A popular revolt against decisions taken at the top? The popular vote in Denmark bears a close resemblance with another negative vote against the EC produced in a referendum in Norway back in 1972. True, the Danes (together with the other Scandinavians) have never been fervent supporters of the European unification process, even though both the accession treaty of Denmark and the SEA had received comfortable majorities in similar referenda in the past. But the Danish vote cannot be treated as an isolated phenomenon. It risks unleashing nationalist forces in different member countries, which could endanger the whole ratification process. In times of economic recession and serious political difficulties in several European countries, the Maastricht treaty could thus provide the focus for public discontent. However, the most likely scenario is that the treaty will be eventually ratified by all the other members, which then leaves open the question of Denmark's membership of the European Union. The process of ratification was originally expected to be completed by the end of 1992, although this now appears as a rather optimistic target.

Judging from the experience with the SEA and the remarkable transformation of the economic and political climate which had led to the acceleration of the process of integration in previous years, the other missing pieces of the puzzle should include a new agreement on the budget (the old one will come to an end in 1992), and the definition of a policy towards the growing number of applicants for EC membership. As the main decisions with respect to deepening have already been reached, it is now the turn of enlargement to be tackled at the highest level. The Community appears being very close to a radical reform of the CAP, and GATT negotiations have acted as a catalyst in this respect. On the other hand, the success of the new phase will greatly depend on macroeconomic conditions which are only partially under the influence of European governments. We shall return to those issues in subsequent chapters.

4. From Customs Union to the Internal Market

Non-Tariff Barriers and the Mixed Economy

Following a historical overview of the process of European economic integration, this chapter and the next will concentrate on some issues related to the establishment of the internal market. We shall start with a brief examination of the barriers still existing at the time of the publication of the White Paper: a measure of the incomplete nature of the common market, twenty-seven years after the establishment of the EEC. We shall then go on to examine the strategy behind the internal market programme and the expectations associated with it. Some preliminary conclusions will also be drawn regarding the early stages of the implementation of the programme. The next chapter will venture into the open sea by tackling the question of economic regulation and the interaction between the State and the market as they are affected by the latest phase of European integration.

The list of remaining barriers to the free movement of goods, services, persons, and capital, as contained in the White Paper on the completion of the internal market, was indeed very long, and this could be interpreted as a sign of failure to fulfil the main objectives of the Treaty of Rome. Undoubtedly, many of the remaining barriers were the result of only a partial implementation of treaty provisions and the highly incomplete state of common policies, a situation which had become worse during the prolonged period of the economic recession.

Many textbooks still refer to the five stages of economic integration as defined by Balassa (1961). They include the free trade area, the customs union, the common market, economic policy harmonization, and the complete economic union. The different stages are usually presented in the form of a ladder which can be climbed one step at a time and leading ultimately to the state of eternal bliss, namely complete economic integration. However, this categorization, together with the traditional theory of international trade, are very misleading because both basically ignore the reality of mixed economies where state intervention is not limited to border controls or macroeconomic policy (Pelkmans, 1980 and 1984).

In the context of such economies, a complete customs union or a common market can be nothing short of total economic integration; and this has become increasingly apparent in the case of the EC. The long list of remaining barriers contained in the White Paper of 1985 provided perhaps sufficient evidence of earlier failure to achieve a complete common market; but it was also a reminder of the enormity of the task undertaken and a testimony to the omnipresence of the State in European mixed economies and its pervasive role as a regulator of economic activity. This list was long although not exhaustive, and the authors of the White Paper recommended a new determined effort to tackle those barriers, an effort which would rely on a new, more flexible approach and the political will to take the Community to a higher stage of economic integration. However, a number of highly sensitive political issues regarding the choice among different methods of eliminating those barriers and the likely implications for countries, regions, and interest groups were left partially or completely unanswered.

All remaining barriers were divided in the White Paper into three main categories: physical, technical, and fiscal. But with the exception of fiscal barriers, the other two categories have little economic meaning and are only a convenient form of presentation for a mixed bag of disparate issues and measures. This is particularly true of technical barriers which could have been, perhaps more appropriately, referred to as 'other'.

The category of physical barriers contains essentially all frontier controls which had remained despite the establishment of the customs union in July 1968. Some of those controls were the result of gaping holes in the common commercial policy. The most notable example was the survival of national quantitative restrictions on various imports from Japan, which in turn necessitated the application of intra-EC frontier controls in order to establish the origin of goods. Furthermore, EC quantitative restrictions on some imports from third countries, such as textiles and clothing in the context of the various multifibre arrangements (MFAs) and steel products under the crisis measures, were divided into national quotas for individual member countries; and this again brought about the need for intra-EC controls. In all those cases, the EC basically operated like a free trade area and not a customs union; hence the need to apply rules of origin and to preserve frontier controls.

The uncommon nature of the CAP was another important reason for the existence of physical barriers. The system of common prices, expressed in European units of account and later in ECUs, did not survive the breakdown of the Bretton Woods system and the frequent changes in intra-EC exchange rates (the problem having been attenuated but not completely resolved after the establishment of the EMS). The myth of unity was kept only through the creation of fictitious exchange rates for agricultural products and the introduction of border taxes, otherwise known under the more neutral name of monetary compensatory amounts (MCAs) which often varied from one product to the other. The elimination of controls would imply the return to real common prices for agricultural goods, a goal which has eluded the Community for over twenty years.[1]

The Treaty of Rome contained provisions for a common policy for transport. But no progress in this area was registered until the adoption of the 1992 target for the completion of the internal market. This meant the perpetuation of licences and national quotas for intra-EC road transport and the effective prohibition for non-resident carriers to operate transport services in another member country (the so-called *cabotage*); hence, the need for controls at the frontier, but also higher transport costs.

Different rates of indirect taxation and the application of the principle of destination (to be discussed in more detail below)

---

[1] The elimination of MCAs has been facilitated by the decision adopted in 1984 that, in the case of exchange rate realignments, common prices which are denominated in ECUs, are adjusted in relation to the strongest currency. By creating a *de facto* DM standard for common agricultural prices, the EC has eliminated the problem of price reductions (or negative MCAs) for the strong currency countries while introducing an upward tilt for prices in the other countries. Intra-EC exchange rate stability and further progress towards monetary union would obviously do away with the problem of artificial exchange rates for agricultural products.

raised the need for border tax adjustments and, hence, fiscal controls at the frontier. And so also did different health regulations, especially with respect to food products.

The task of creating a Europe without frontiers extends much beyond the field of economics; and this was the explicit objective of the White Paper and the Commission's strategy. The elimination of controls for individuals would have to be based on a far-reaching agreement among member governments on issues such as the control of drugs, immigration, and national security more generally; major issues which touch sensitive nerves of national sovereignty. Differences in terms of visa requirements for nationals of non-member countries and the need for intra-EC frontier controls are the political counterpart of the problem created by remaining gaps in the common commercial policy for the intra-EC movement of goods: gaps in the common external policy leading to internal controls.[2]

The proposals for the elimination of technical barriers contained a motley assortment of measures. Different technical regulations, which are legally binding in each country, and standards, which are voluntarily agreed codifications written by national standardization bodies and considered as an indicator of quality, had long been regarded as a major factor behind the fragmentation of the EC market. They also directly added to costs by forcing producers to adjust their products to the requirements of each national market; that is, whenever those barriers could be effectively overcome, which was not always the case.

Different technical regulations and standards were particularly important in specific sectors such as mechanical and electrical equipment as well as transport goods; virtually all high technology sectors were especially affected. That was also true of other sectors such as food and drink, and pharmaceuticals. Different regulations and standards may be the result of legitimate differences between countries regarding the trade-off between economic efficiency on the one hand, and health and environmental considerations on the other. Other factors such as technological development and administrative efficiency also play an important role. But regulations and standards can also be a covert, albeit very effective, means of protection against foreign producers. The dividing line between legitimate and illegitimate

---

[2] Further steps on this area will be taken as a result of the new treaty signed at Maastricht.

use of different technical regulations and standards is usually blurred.

The experience of the first twenty-seven years of the EEC had not been very encouraging. Despite the efforts made by the Commission and the Court of Justice, the role of the Community had been effectively limited to slow, rearguard action against the spreading of different technical regulations and standards in individual member countries. Their proliferation reflected growing public concern for the protection of consumers and the environment; but they were also an instrument of external protection in times of recession and growing unemployment. On the other hand, the process of harmonization (or approximation) at the EC level had proved to be both inefficient and ineffective. Its snail's pace compared unfavourably with the speed with which national authorities introduced new laws and regulations (Pelkmans, 1987).

The new approach referred to in the White Paper was, indeed, radical: instead of trying to harmonize everything, the emphasis from now on should be on mutual recognition of technical regulations and standards which should open the way for the unrestricted circulation of goods inside the EC. Recourse to legislative harmonization should be limited only to essential cases. Furthermore, this harmonization, whenever deemed necessary, should be restricted to basic health and safety requirements, while the task of setting detailed technical specifications should be left to specialized European organizations such as CEN (Centre Européen de Normalisation) and CENELEC (Centre Européen de Normalisation Electrotechnique). The task of harmonization was to be made much easier through the subsequent replacement of unanimity by qualified majority voting with the new Article 100*a* of the SEA.

The strong reliance on mutual recognition, which extends much beyond technical regulations and standards in the general strategy for the establishment of the internal market, is supposed to be compatible with the federal nature of the EC and the so-called principle of subsidiarity which has been increasingly referred to by the EC Commission in recent years as a means of pacifying those worried about the progressive erosion of national sovereignty. The principle of subsidiarity basically means that only those powers that cannot be efficiently executed at lower levels will be transferred to a higher level of decision-making (Wilke and Wallace, 1990). Such a general statement can obviously command wide acceptance; it is only when applied to specific cases that disagreement tends to arise. Mutual recognition was also presented as something compatible with the political, economic, and social diversity inside the EC. But it was certainly not devoid of problems: for example, how does one ensure that mutual recognition and the ensuing 'competition among rules' do not lead to a general erosion of standards? The definition of efficiency and the optimal distribution of power between different levels of political authority are not the kind of issues on which most people easily agree. Can competition and the market determine the optimal level of regulation? This subject will be discussed further in Chapter 5.

Discriminatory public purchasing has been traditionally one of the most powerful and frequently used instruments of industrial policy and an effective means of promoting national champions. Even in the case of the United States, where industrial policy has been almost a dirty word, public purchasing, and defence contracts in particular, has always played such a role. Various EC directives adopted in the past seem to have had virtually no effect on government practices. The comparison of the import content of government purchases (less than 4 per cent for the large member countries according to EC Commission data) with that of the private and public sectors combined is very revealing of the extent of the discrimination involved.

Public purchasing in general accounts for approximately 16 per cent of Community GDP, and the contractual part of it, usually called public procurement, is estimated at 7–10 per cent of GDP. Especially in some sectors with a strong high technology content and sizeable economies of scale, such as telecommunications, the public sector has been by far the biggest client. Hence, the inevitable fragmentation of the EC market and the high costs of production. The White Paper aimed at greater transparency and the extension of competition to four sectors which had been excluded from previous EC directives, namely water, energy, transport, and telecommunications.

One important aspect of EC competition policy deals with state aids. Article 92 of the Treaty of Rome contains a general prohibition of such aids, because of the distortions created in intra-EC competition; and this prohibition is then followed by a long list of exceptions. In the past, the Commission had adopted a very cautious (tolerant would be, perhaps, a more appropriate term) approach on this subject in view of the strong resistance of national governments to outside interference. Political realism had thus overridden any desire to attempt a more strict application of treaty provisions. The period of the economic recession saw an increasing resort to state subsidies as a means of covert protection of domestic producers; the example of steel being one of the most prominent. According to the first reports published by the Commission on this subject, the amount of state aids remained quite significant in the 1980s; as a percentage of gross value added in the manufacturing sector, they were highest in the less developed countries and regions of the Community such as Greece, Portugal, Ireland, and South Italy (Commission of the EC, 1990*b*). The White Paper, consequently, pointed to the need for a more rigorous application of Community discipline in this area.

Distortions in intra-EC competition can also arise from different company laws, while also causing major difficulties for cross-border co-operation among firms, including mergers and joint-ventures. The creation of a common legal framework, included in an old Commission proposal for a European company statute, would help to deal with this problem. Similarly, the development of a Community trademark would help to tackle the difficulties arising from different intellectual property laws.

The Treaty of Rome talks about the freedom of establishment and the free movement of services in the context of the common market. Yet reality had remained for many years far short of the full achievement of those objectives. The gap became more noticeable as many services gradually entered the category of 'tradeables', while the size of the whole sector was steadily expanding as a percentage of GDP (already approaching 60 per cent for the EC as a whole).

The main barriers resulted from the existence of different regulatory frameworks in the member countries. Thus the cross-border provision of insurance was virtually excluded for this reason. Various restrictions had also survived with respect to the exercise of the freedom of establishment in both banking and insurance. With regard to financial services in general, exchange controls were another major impediment to free intra-EC movement. Road and air transport remained under heavy governmental control: with respect to the former, by imposing restrictions on capacity and access and operating a system of licences for non-national hauliers, and for the latter, through a system of bilateral agreements in which designated carriers provide services whose cost, capacity, and conditions were directly or indirectly regulated. For new services, such as audiovisual, information, and data processing, government regulations and different standards were the most effective barriers. And the list was almost endless.

Despite the relevant provisions in the Treaty of Rome, the progress made towards the liberalization of capital movements had been extremely slow until the second half of the 1980s. The effect of two directives adopted in the early stages of the EEC was effectively nullified by the extensive use of safeguard clauses, especially during the years of recession and monetary instability, which followed the breakdown of Bretton Woods and the advent of the oil crisis. As for labour, the main restrictions were lifted by July 1968, and considerable progress was also registered with respect to the harmonization of social security legislation. Yet this had little effect on the cross-border movement of many skilled workers and professionals. One, but certainly not the only, reason for this was the lack of harmonization of professional qualifications on which the White Paper laid great emphasis. The Commission proposed to deal with this particular Gordian knot in exactly the same way as with different technical regulations and industrial standards, namely through mutual recognition of university degrees and vocational training.

Distortions arising from different fiscal systems have been an old concern of economic integration theory; and this was duly acknowledged in the relevant provisions made in the Treaty of Rome. The main emphasis has always been on indirect taxation and its effects on the allocation of resources within a customs union or a common market. Taxes on income and capital had received until more recently relatively little attention, because of implicit assumptions about the low mobility of factors of production compared with the movement of goods; and this was also true of the White Paper. However, with the rapidly growing mobility of capital, the validity of such assumptions has become more and more questionable.

The EC succeeded in replacing different turnover taxes with a single system of indirect taxation, namely the VAT. But the harmonization of the system did not extend to the taxable base and the rates used in the different member countries, despite repeated efforts made by the Commission in the past. The differences in excise taxes were even more dramatic; those taxes are usually levied on tobacco, alcohol, and fuel. In order to avoid discrimination between different national producers, member countries applied the so-called destination principle both to their intra-EC and international trade. This meant that at any consumption point the same tax was levied irrespective of the origin of goods. And this in turn meant that exports were zero rated. The application of the destination principle did, however, raise the need for frontier fiscal controls and border tax adjustments. This was precisely what the White Paper would like to put an end to: the final aim being to tax sales across borders in exactly the same way as sales within a country.

The abandonment of the destination principle would raise two fundamental problems, namely trade distortions arising from different national rates and a redistribution of revenue among national tax authorities. The way to deal with the former would be, according to the Commission, through the harmonization of the taxable base and the alignment of different rates. Based on the US experience, the Commission argued that a complete harmonization of VAT rates and excise taxes would not be necessary; instead, an approximation within a 5 per cent margin around a target rate (or rates) would be sufficient to avoid large distortions. As for tax revenue, the Commission would aim at the continuation of the *status quo* through the creation of a 'clearing house'. The differences in rates of indirect taxation reflect different traditions, economic realities as well as political and social preferences. Therefore, resistance to change should be expected to remain strong; hence also the insistence on unanimous voting in the Council of Ministers, which survived the treaty revisions introduced through the SEA. We shall return to this subject in the next chapter.

The White Paper referred basically to what are generally known as NTBs: barriers and distortions created by different forms of domestic government intervention. It was largely about services and factors of production, areas in which very little progress in terms of integration had been achieved until then. And the reason was simple: the most effective barriers were not to be found at the border; they were, instead, the result of different regulatory frameworks which created the notorious NTBs. Thus the White Paper signalled a qualitatively new phase in the process of integration.

Domestic government intervention is part and parcel of the established political and economic order in each country. The White Paper was presented as a set of technical measures, presumably intended to assuage nationalist fears. However, a more careful reading of the text and the various measures put forward in the accompanying annex would lead to a different conclusion. Many of those measures touched at the very heart of national economic sovereignty; and modern politics is very much about welfare issues. Fiscal harmonization, monetary policy and capital movements, state subsidies, and even industrial standards are the basic material of which the economic role of the State consists; and also the instruments through which governments can influence the direction of votes. Such issues have become increasingly prominent as the old trade model of European integration reached its limits. As it did so, the contradiction between trade liberalization and the mixed economy, which had been lurking for years in the background, slowly came to the surface.

The White Paper and the internal market programme were an attempt to tackle this problem, and the method adopted reflected the new political consensus in Western Europe and the shift towards more competition and the market. It could, perhaps, be argued that the main emphasis was on negative integration, following the old distinction drawn by Tinbergen (1954) between negative and positive integration. According to the Dutch economist, negative integration refers to the elimination of obstacles to the free movement of goods and factors of production, while positive integration refers to the harmonization of rules and the adoption of common policies; the latter, therefore, being more consistent with the mixed economy. Yet this distinction has become increasingly blurred as the Community began to deal with NTBs. The elimination of the latter requires a combination of negative and positive integration measures: deregulation combined with a certain harmonization of national rules and an agreement on new regulatory frameworks for the economic union as a whole. Although the White Paper tried to set some guidelines (would the principle of mutual recognition prove to be a euphemism for negative integration?), the final mixture was impossible to determine in advance. Much would depend on subsequent negotiations in individual areas and the implementation of the measures adopted. This subject will be taken up in Chapter 5.

Supply-Side Economics and the Marketing Campaign

The publication of the White Paper and the adoption of the 1992 target led to the birth of a new, booming industry which has been enthusiastically welcomed by the economics profession in all member countries. Its main function has been to explain and analyse the consequences of the internal market and predict the likely economic effects. Numerous quantitative estimates have been produced for individual sectors, national economies, and the EC as a whole, and the range of those estimates is very wide, thus catering for different tastes and expectations.

Among the various *ex ante* studies of the likely effects of the internal market, the Commission work on 'The economics of 1992', otherwise known as the 'Costs of non-Europe' or as the Cecchini report, (Commission of the EC, 1988a),[3] has received wide publicity. It still remains the most comprehensive work on this subject, and the quantitative assessments contained in it have been regularly used by the Commission and member governments as the scientific support for the internal market programme and an important instrument to influence market expectations. Of course, several micro- and macroeconomic variables have changed since the publication of the 'Costs of non-Europe', and therefore the quantitative assessments in the latter need to be seen in a new light. This does not, however, mean that this work should be now only of historical interest. Although most of the measures have already been adopted by the Council of Ministers, there is in most cases a considerable time-lag between the adoption of legislation and the implementation of measures (some of them, in fact, not taking effect before 1993). Thus the Commission's predictions still remain to a large extent predictions. Furthermore, interest in this work also lies in the implicit economic logic, and the strategy revealed through its pages. They are essentially the economic logic and the strategy of the Com-

[3] Its popular version can be found in Cecchini, 1988.

mission associated with the implementation of the internal market programme. The 'Costs of non-Europe' has been followed by several other studies undertaken by the Commission services, including a more substantial piece on the likely effects by industrial sector and individual national economies (Commission of the EC, 1990c).

The completion of the internal market constitutes yet another phase in the process of integration and the elimination of intra-EC barriers. However, integration is already much beyond the stage of tariff liberalization, and the orthodox theory of customs union, still the most developed part of economic integration theory, has not much to say about it. Surely, there will be further trade creation and trade diversion, leading to even more openness of the EC economies and a further increase of intra-EC trade. More trade should be associated with welfare effects, although when dealing with the multiplicity of NTBs, the analysis inevitably becomes much messier. In fact, most studies of the internal market, and the 'Costs of non-Europe' in particular, point very clearly to an ever widening gap between traditional theory and the real issues at stake. It is not only the non-tariff nature of the new phase of integration and the central role of services which account for this gap; the emphasis is now clearly on the so-called dynamic effects on which, yet again, our knowledge remains rather limited.

According to the Commission study (Commission of the EC, 1988a), the elimination of the remaining barriers, such as frontier controls and different technical standards, should lead to a reduction of costs, which can then be translated into either a widening of profit margins or a lowering of prices or a combination of both. In fact, the strengthening of competitive forces, through the further opening of frontiers and the pursuit of a vigorous competition policy, should help to turn cost reductions into lower prices. Stronger competition at the EC level should also lead to a further lowering of costs through the reduction of the so-called X-inefficiency (poor allocation of resources inside a firm due to weak external competition). Uncompetitive producers, previously protected through various NTBs, would be pushed out of business, while the more efficient ones would be able to benefit from economies of scale through increased production; hence, even lower costs.

The Commission study laid great emphasis on the scope for

further economies of scale as well as learning economies. It referred to the 'minimum efficient technical scale' (METS), and argued that in many sectors the actual firm size is significantly smaller than the estimated METS. Earlier studies on the effects of integration had pointed to sizeable economies of scale (Owen, 1983). A positive relationship was also assumed to exist between stronger competition on the one hand, and technical progress and innovation on the other.

The Commission's views on competition were certainly not those of the old textbook model. The following passage is very illustrative of the approach adopted:

European integration would thus assist the emergence of a virtuous circle of innovation and competition — competition stimulating innovation which in turn would increase competition. This is not to say that the desired form of competition corresponds to the theoretical and simplified model of perfect competition. The relationship between competition and innovation is not linear and indeed there exists an optimal level of competition beyond which competition has an adverse effect on innovation because of the difficulty of allocating gains and the greater risks which obtain in highly competitive markets. *The optimum market structure from the standpoint of innovation ought rather to promote strategic rivalry between a limited number of firms'*. [italics added] (Commission of the EC, 1988a: 129)

The word 'competition' may crop up all too often in the above quotation. Yet the Commission's approach to it was more akin to the oligopolistic reality of many contemporary markets. 'Strategic rivalry between a limited number of firms' constituted the essence of the Commission's approach; and this did not exclude various forms of inter-firm collaboration as, for example, those promoted by Brussels in various high technology sectors in recent years. They formed part of the new, more flexible approach to industrial policy.

According to the Commission study, a clear indicator of the importance of the remaining barriers, causing the fragmentation of the EC market, were the observed price differentials between member countries. A good example was the financial sector: according to the figures provided by Price Waterhouse, prices in different EC countries often differed by more than 100 per cent and this was attributed largely to the lack of intra-EC competition (Commission, 1988a: Table 5.1.4, 91). The next step was to argue that the elimination of barriers and the new competition created would lead to an alignment of prices downwards, which was precisely what the Commission expected or hoped for.

There should also be important macroeconomic effects. Investment was expected to rise because of stronger competition, industrial restructuring, and the lower cost of borrowing resulting from the liberalization of financial services. In this respect, the credibility of the internal market programme should be absolutely crucial in creating favourable expectations among economic agents, which would then be translated into new investment decisions. The Commission seems to have been perfectly aware of the need to keep those expectations well nourished. On the other hand, lower prices should lead to gains in competitiveness and to an improvement of the external trade balance. Gains in competitiveness and in domestic purchasing power, combined with higher investment, should lead to an increase in aggregate demand and higher growth; and this, in turn, to lower public sector deficits and the creation of new jobs. Each step of the argument, however, depended on some crucial assumptions relating to the behaviour of exchange rates and the functioning of labour-markets among others.

The 'Costs of non-Europe' contained seventeen horizontal and vertical studies which provided the basic material for an assessment of the overall effects of the internal market. They were followed by two overall quantitative assessments based on a microeconomic and macroeconomic approach respectively. The results obtained from both were remarkably similar and are summarized in Table 4.1: additional growth of 4.5 per cent of GDP, reduction in prices of 6 per cent, creation of 1,750,000 new jobs, reduction of public sector deficits of 2.25 per cent, and improvement of the Community's external balance of 1 per cent of GDP. According to the Commission, the positive effects in terms of GDP and employment could be significantly increased if member governments were to take advantage of the new margin of manœuvre created by lower inflation, smaller public sector deficits, and the improvement in the external balance by adopting more expansionary macroeconomic policies. This could then lead to additional growth of the order of 7 per cent and 5 million extra jobs. The Commission expressed a strong preference for a more active macroeconomic policy as a means of creating a more favourable environment which would in turn facilitate the implementation of the internal market programme.

Table 4.1. Macroeconomic Consequences of the Completion of the Internal Market

Microeconomic approach	Welfare gains as % of GDP 4.25–6.5				
Macroeconomic approach	GDP as %	Prices as %	Employment in millions	Public balance as % point(s) of GDP	External balance as % point(s) of GDP
Without accompanying economic measures	4.5	−6	1.75	2.25	1
With accompanying economic measures	7	−4.5	5	0.5	−0.25

Note: Margin of error ± 30%.
Source: Commission of the EC, 1988*a*: 167.

Several questions have been raised regarding the main assumptions and the methodology employed in the Commission's study; and often, indirectly, about the economic strategy behind the internal market programme. We shall concentrate here on some of them. The direct benefits expected from the elimination of frontier controls were, according to the study, relatively small, while on the contrary there were big expectations associated with the secondary, so-called dynamic effects resulting from economies of scale, restructuring, and greater competition, always assuming a full implementation of the internal market programme. However, the more important the expected effects, the more difficult it is to produce anything more than educated guesses.

The importance of economies of scale, which remained to be exploited in the context of the internal market, has been challenged by several economists (Dicke, 1989; Grimm, Schatz and Trapp, 1989). One would expect that at least for the big countries of the EC, the large size of the domestic market should have already allowed firms in most sectors to take advantage of any potential economies of scale. Is the Commission dreaming of European champions to replace their largely inefficient national counterparts? Geroski (1989) argues that 1992 will lead, if anything, to more diversity of goods instead of the small range of mass produced items which seems to be implicitly assumed in the Commission's approach.

The liberalization of financial services occupied an absolutely crucial position in the overall macroeconomic effects of the study, accounting for one-third of the expected additional GDP growth. Thus, whatever happens in the financial sector should have a major impact for the internal market programme as a whole. High price differentiation was interpreted by the Commission as evidence of the fragmentation of the EC market, expected to disappear through the completion of the 1992 programme. But this price differentiation may also be the result of oligopolistic markets divided along national lines, which are unlikely to disappear overnight, even when government-created barriers are eliminated.

More generally, there were two fundamental and politically relevant questions about the completion of the internal market to which the Commission's study gave only partial answers at best. The first refers to the expected overall economic impact of the proposed programme. The authors of the Commission's study admitted to the existence of a wide margin of error for their quantitative estimates, although this warning did not always find its way into the more popular presentations of the results intended for the wider public. The majority of similar studies undertaken by both private and official organizations in the different member countries have produced significantly lower estimates than those found in the Cecchini report, and this has usually been attributed to lower expectations regarding the dynamic effects. A notable exception in this respect has been the attempt by Baldwin (1989) to incorporate the dynamic effects of the internal market on savings and investment, leading him to conclude that the final effect could be as much as double the estimate produced by the Commission.

With respect to the numerous quantitative estimates of economic integration made in the past, the following sober conclusion has been drawn by Sellekaerts (1973: 548):

All estimates of trade creation and diversion by the EEC which have been presented in the empirical literature are so much affected by *ceteris paribus* assumptions, by the choice of the length of pre- and post-integration periods, by the choice of the benchmark year (or years), by the methods to compute income elasticities, changes in trade matrices and relative shares and by structural changes not attributable to the EEC but which occurred during the pre- and post-integration periods . . . that the magnitude of no single estimate should be taken too seriously.

If econometric studies can provide relatively poor explanations of past economic performance, their predictive capacity leaves even more to be desired. In view of the very large degree of uncertainty and the multiplicity of factors involved, the various *ex ante* estimates of the internal market effects should be treated at best as very rough indicators of the direction of those effects and the broad orders of magnitude. To attach any greater importance to them would be completely unjustifiable on the basis of previous experience; although perhaps justifiable for political reasons.

The other question, left almost completely unanswered by the Commission's study, was about the likely distribution of costs and benefits between countries, regions, and social classes, the answer to which would be absolutely crucial for the continued political acceptability of the internal market programme. Here, the Commission's study had little to offer in terms of prediction, apart from acknowledging the problem and expressing the hope that EC redistributive policies would provide adequate means for compensating potential losers or, even better, for helping weaker economies and regions to face the strong winds of competition unleashed by the elimination of barriers.

Modern theories of international trade, with their emphasis on imperfect competition, economies of scale, and the dynamic effects of innovation suggest that a very unequal distribution of gains from integration is a very concrete possibility (Krugman, 1987; Helpman and Krugman, 1985).[4] Comparative advantage is no longer seen as something determined by the particular factor endowments of a country. Instead, comparative advantage is created through deliberate policies directed at investment, education, and research and development. Countries pursue strategic trade policies in order to capture an ever-increasing share of dynamic sectors where demand is growing and where the benefits of scale economies, advantages of experience, and innovation can be reaped; in economic terms, sectors where there are good prospects for 'rent'. In such a world, which is thousands of miles

---

[4] There is a close parallel between the new 'paradigm' in international trade theory and the literature in regional economics. Both concentrate on various forms of market failure and try to explain the development of inter-regional and inter-country disparities in the context of free trade. See also Ch. 8.

apart from the assumptions made by traditional theories of international trade, a highly unequal distribution of gains and losses from integration can by no means be excluded. Krugman (1987: 130), for example, argues that trade based on economies of scale, which seems to be largely the case of intra-EC trade, 'probably involves less conflict of interest *within* countries and more conflict of interest *between* countries.'

After the publication of the 'Costs of non-Europe', the Commission has tried to go further by identifying those industrial sectors which should be more affected by the elimination of remaining barriers and more detailed country studies (Commission of the EC, 1990c, 1991b). Since the completion of the internal market is largely about NTBs (discriminatory public procurement being one of the most important), it is not entirely surprising that many of the sectors found to be most affected were high technology sectors (telecommunications, computers) and more traditional, capital-intensive sectors (railway equipment, shipbuilding, electrical engineering). But the list of the most sensitive sectors to the 1992 programme also included certain mass-consumer products, such as textiles and clothing, where technical and other barriers continued to have a significant effect on intra-EC trade. A more careful study of those sectors led the Commission to the conclusion that their share of industrial employment was highest in the two least developed countries of the EC, namely Portugal and Greece (Commission of the EC, 1991b).

These studies also pointed to the inter-industry dimension of intra-EC trade which has grown significantly due to the greater economic heterogeneity of the EC, in turn a function of successive enlargements. This is very much the case of trade between Germany and Greece or Portugal (see also Neven, 1990). The adjustment costs from liberalization are expected to be greater in the case of inter-industry trade, and thus adjustment in relation to the internal market could be more painful than that experienced in the years following the establishment of the EEC back in 1958 when the Six constituted a fairly homogeneous group. One crucial question is whether the less developed countries of the EC will be able to climb up the ladder of the division of labour or whether they will continue specializing in more labour-intensive products where competition from Third World countries has been steadily growing.

The distribution of gains and losses from trade and integration is, certainly, not pre-ordained by any economic law. A great deal depends on the policies and the internal flexibility of participants; and this has been clearly demonstrated by historical and more recent experience. However, there are structural factors which may also have a significant influence in this respect. There is a clear tendency for heavy geographical concentration of economic activity in high technology sectors which could be further accentuated in a European market without internal barriers. This concentration is usually explained in terms of significant location economies associated with the availability of skilled manpower, proximity to administrative and financial centres, and other linkage effects (Padoa-Schioppa *et al.*, 1987; see also Chapter 8). The example of electronics and information technology immediately comes to mind. Do countries like Portugal or Greece have many firms to compete in those areas and are they likely to attract in a future European market without internal barriers anything more than the odd assembly plant, with low value added, from a large multinational? Even if they were to follow successfully the example of Ireland, R. & D. and the more knowledge-intensive activities in those sectors would most likely remain in the more advanced countries.

A similar problem can arise with respect to the financial sector which occupies a central position in the internal market programme. The Commission argued that the less developed countries of the periphery, which include the more recent entrants to the EC, could benefit more than the others from the liberalization and deregulation of financial markets. More competition should lead to considerably lower financial costs in the heavily protected and relatively inefficient markets of southern European countries. This seems to imply that protection and heavy regulation in those countries have been part of an economically irrational policy, and thus rationality imported from Brussels should have beneficial effects. This is certainly not an implausible argument; but the alternative hypothesis cannot be altogether excluded either. The old policies of protection may have been, at least partly, based on a well-founded fear that liberalization would be accompanied by strong centripetal forces and the small national champions being swallowed up by the large firms of the more advanced countries of the centre. Historical evidence seems to suggest that financial activities tend to be highly concentrated, although there is always room for the local banks competing for the money of the small depositor (Grilli, 1989).

On the other hand, the internal market is expected to lead to lower transport costs, and this should equally apply to telecommunications equipment and services. Both will be crucial in reducing the economic distance for geographically peripheral countries and regions, and naturally the Commission has placed great emphasis in this respect.

Economic studies and business surveys reflect different attitudes and preoccupations among member countries, regions, sectors of economic activity, and social classes. Thus the concern expressed in economically less developed and geographically peripheral countries of the Community about the possible centripetal forces of integration, often combined with fears about further concentration of production and ownership which would work against the small and medium sized enterprises (SMEs) of those countries, has been partly matched by the concern of German and Dutch trade unions about a possible shift of production in search of lower wages and less costly working standards (the so-called *Standortdebatte*; see also Chapter 6). On the other hand, the general optimism of German industry has been strongly qualified in the case of specific sectors such as energy, telecommunications, transport, and insurance. After all, each country has its own sacred (and vulnerable) cows.

Business expectations have also varied from one country to another and from one year to the next, as shown in the surveys regularly undertaken by the Commission. But the mood of the market is like Verdi's *Rigoletto* (la donna è mobile), and it would make little sense to draw any medium- or long-term conclusions on this basis. Until 1990, while confidence indicators in the Community followed an upward slope, together with profit ratios, the adoption of the internal market programme coincided and was also strongly influenced by a favourable macroeconomic environment. During this period of economic boom in Western Europe, more new jobs were created than those predicted in the Commission's study as a result of the completion of the internal market. On the other hand, the rapid acceleration of investment was accompanied by a wave of mergers and acquisitions, including an increasing number of intra-EC ones. Rapid economic growth and the reduction of unemployment attenuated the adjustment problem associated with the internal market, and hence the resistance to the elimination of protectionist barriers.

At the same time, the majority of the Community's weaker economies experienced faster economic growth than the EC average; and this was mainly due to a large increase in investment, with substantial capital inflows being added to domestic savings. Did this mean that earlier fears about the unequal distribution of benefits from market liberalization, at the expense of the weaker countries and regions, prove unjustified? Or was it still too early to draw any definitive conclusions? Perhaps, at this level of generalization, there is not much that can be said with confidence.

The EC Commission has tried to establish a link between the internal market and the observed increase in investment, distinguishing between the short-term effects associated with positive expectations and the medium-term effects which would result from the actual implementation of the internal market programme. Thus, since there were hardly any measures that had already been implemented at the time, the Commission argued that the much higher than forecast investment of 1988–9 could be partly explained in terms of the '1992 effect'; and business surveys also seemed to point in this direction (Commission of the EC, 1989a).

This is a very plausible argument, although one which is virtually impossible to prove decisively. Interestingly enough, such an anticipation effect had not been predicted earlier. One may also talk about the 'Schumpeterian' effect of 1992,[5] the latter being generally identified with a boost to entrepreneurship producing in turn a shift in the behaviour of private economic agents. Investment behaviour may, indeed, have been influenced, if economic agents were convinced that the internal market would lead to a significant change in their environment, irrespective of any economist's views as to whether this was 'objectively' true. Thus, the internal market should not be seen only as a supply-side programme; it has also been an attempt to influence expectations and strengthen the European orientation of firms. Hence, the emphasis on marketing. In this respect, the Commission, and to a lesser extent national governments, seem to have done an

---

[5] This is the term used by E. Malinvaud in a series of lectures given at the European University Institute in Florence in spring 1990.

excellent job. The internal market and 1992 have been given extraordinary publicity. There has been, undoubtedly, some hyperbole in the official discussion about the significance of the internal market. But equally, this hyperbole seems to have had a positive effect on market expectations, contributing to this new confidence in the future of the European economy.

Kay (1989: 28) described 1992 as 'the most successful marketing campaign of the decade'. On the other hand, a survey for the Bank of England (1989: 18) identified:

a tendency among some corporate consumers, as well as suppliers, of financial services to feel they ought to be 'in on 1992'... This could lead to corporate reorganisations... based less on a realistic assessment of opportunities and threats than on concern 'not to miss the boat', or for reasons of 'public profile', or simply out of fear. Some non-EEC... institutions also appeared to feel pressure to 'do something' about '1992'.

So much for 'rational' economic behaviour. The 1992 effect will be important, if markets believe it to be so; and arguably much of this effect has already happened. However, the durability of the effect and the policies behind it still remain to be tested, especially under less favourable macroeconomic conditions.

5. The Political Economy of Liberalization and Regulation

The elimination of the remaining barriers and hence the gradual transformation of an incomplete customs union into a real internal market, which promises to be a long-lasting process extending beyond 1993, raises some fundamental questions regarding the relationship between the State and the market; in other words, the very nature of economic order and the mixed economy in individual member countries and the Community as a whole.

The gradual opening of public procurement, the liberalization of financial and transport services, the new approach to technical standards, and the renewed attempts at tax harmonization will have a major impact on the role traditionally played by national governments. By rendering policy instruments partially or totally ineffective, the creation of the internal market will have a negative effect on the ability of national governments to influence the allocation of resources within and between countries. The 1992 programme has often been presented as an essentially deregulatory exercise in which the elimination of the remaining physical, technical, and fiscal barriers will be tantamount to a reduction of the role played by public authorities and, consequently, the erosion of the mixed economy. This aspect of the internal market has received different emphasis in individual member countries (compare, for example, the stress on deregulation in Britain with the more subdued approach adopted in this respect by the German government in the domestic debate), depending very much on different traditions, prevailing ideologies, and the political balance of forces.

At least in some cases, the creation of the internal market will, inevitably, involve the adoption of a new regulatory framework at the EC level and, therefore, the transfer of powers to Community institutions. There will be more market in the new economic order, but also more European state; and the trade-off between the two will be determined in the future by a combination of economic and political factors, both internal and external to the EC. This is what we may call the political economy of liberalization and regulation. It is about the distribution of economic power between different levels of political authority as well as the distribution of economic power between private and public agents, the two being closely interrelated. But it is also, more indirectly, about the allocation of resources; choices between efficiency and stability or between production and the protection of the environment; it is about relations between producers and consumers; and, last but not least, it concerns the ownership and control of the means of production. These are all highly political issues which come under the more general theme of economic order. Thus the internal market programme is part and parcel of *Ordnungspolitik*, as the Germans would call it. This is a vast and still insufficiently explored territory which straddles different disciplines and specializations. This chapter will attempt to give tentative answers to some of the questions raised above by concentrating on four case-studies.

The first one will examine the recent wave of mergers, acquisitions, and co-operation agreements which are in the process of changing radically the corporate map of Europe. In view of the constantly increasing transnational nature of takeovers and co-operation agreements, national frontiers will no longer be marked by heavy lines on this map. Does this restructuring stem from the search for rationalization and the further exploitation of economies of scale, the attempt to secure a strong presence in what were considered until recently 'foreign' markets, or are we witnessing, instead, the process of cartelization of European industry? And then, what are the policies developed at the European and national level to deal with this phenomenon? We shall be interested not only in the emerging allocation of functions between EC and national institutions but also in the whole approach adopted by public authorities regarding industrial structures and market competition.

The second case-study will concentrate on the financial sector, and especially banking. This sector has been traditionally characterized by strong government intervention and different national regulatory frameworks which, precisely because of their difference, have acted in the past as an important barrier to cross-border activity. Intra-EC liberalization goes together with the adoption of some common rules regarding the supervision of financial markets. Will the establishment of a new regulatory framework and the competition among different national systems, implied in the principle of mutual recognition, turn liberalization simply into a deregulatory exercise, and what could be the more general effects on the functioning of financial markets?

Tax harmonization is an old issue dating back to the early days of the EEC. Fiscal barriers were among the three main categories of remaining barriers singled out in the White Paper of 1985. On the other hand, fiscal issues are at the very heart of national sovereignty. How important are the distortions created by different tax regimes and how far are national governments prepared to go in relinquishing their fiscal autonomy? Changes in this area can have important effects in terms of the provision of public goods and the internal redistributive function of government budgets. This will be the third case-study on the political economy of liberalization and regulation.

Different technical standards and regulations have an important effect on the allocation of resources within and between countries, even when this does not happen to be the primary objective of governments and private institutions in the use of those policy instruments. The new approach to standards and the emphasis on mutual recognition are intended to deal a decisive blow against an important technical barrier. The last section of this chapter will examine the progress made in the first six years after the adoption of the White Paper, and then draw some tentative conclusions about general developments in this area and the wider implications of the choices made.

The relationship between the State and the market is not static. The post-war economic history of Western Europe bears witness to this changing relationship. We have argued earlier that both the internal market initiative and the relaunching of regional integration are, at least partly, a reflection of new economic ideas and, consequently, a conscious attempt to strengthen the market forces in Europe. This chapter is intended as a provisional balance sheet of the movement in this direction, and an early snapshot of

the changing economic order in Western Europe. It will be complemented by subsequent chapters on the labour-market, macroeconomic policies and redistribution, and EC relations with the
rest of the world.

The Emergence of Euro-Champions: Competition or Industrial Policy?

The second half of the 1980s was marked by a rapidly growing
number of mergers, acquisitions, joint ventures, and other forms
of co-operation agreements between firms. Although there are
serious difficulties in terms of the accuracy and the comparability
of data, stemming from the different ways in which monitoring
organizations deal with problems of coverage, definition, and
transparency in this area, all evidence available points in the
same direction, namely a steadily rising trend which coincided
with the early stages of the implementation of the 1992 programme and the economic boom which accompanied them. This
activity was not confined to national boundaries, unlike earlier
periods of restructuring and concentration (e.g. the 1960s) which
had seen the emergence of national champions.

Cross-border alliances, ranging from co-operation agreements
in R. & D. to outright mergers and acquisitions of majority
holdings, have been prominent in recent years. The joint Anglo-
German bid of GEC and Siemens for Plessey in the electronics
sector, the Franco-American agreement between CGE and ITT
in the same sector, the marriage of Asea and Brown Boveri, the
Swedish and Swiss national champions in the electricity power
plant industry, the takeover of Rowntree in the UK by the Swiss
multinational Nestlé which was subsequently engaged in a big
battle with the Agnelli family over the Perrier water, the link
between Renault and Volvo in the car sector, between BMW and
Rolls Royce in aeroengines, and the joint venture between
Thompson and Philips, the French and Dutch electronics groups,
for the production of the first high-definition television (HDTV)
in Europe have all caught the public eye. Meanwhile, the takeover bid of Societé Générale by the Italian predator, Mr de
Benedetti, provoked nationalist reaction in Belgium, which
finally sent the Belgian national champion to the arms of the

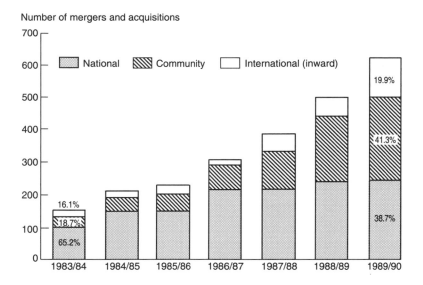

Fig. 5.1. Mergers and Acquisitions of Majority Holdings in Industry, 1983/84 to 1989/90
(Operations involving at least one of the 1000 largest firms in the Community)
Source: Commission of the EC, Annual Reports on Competition Policy.

French group Indosuez, while the Continental and Pirelli saga in the tyre sector exposed the peculiarities of industrial cultures and institutions in both Germany and Italy. Such activity, which is only the tip of the iceberg, is in the process of transforming production structures and thus extending European interdependence beyond the area of trade.

According to Commission data, based on the operations of the 1000 largest firms in the Community, the total annual number of mergers in industry rose from 115 in 1982/83, to 208 in 1984/85, 492 in 1988/89 and 622 in 1989/90 (Fig. 5.1), with chemicals and food leading among sectors. The share of intra-EC mergers has been rising very rapidly, and for the first time in 1989–90 their total number was higher than that of mergers taking place within national boundaries. A similar trend has been experienced with respect to joint ventures and the acquisition of minority holdings.

Number of mergers and acquisitions

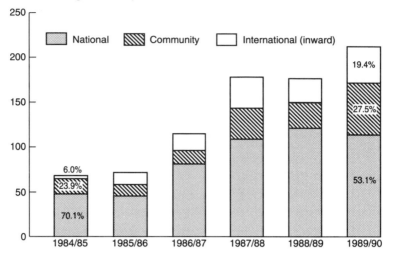

Fig. 5.2. Mergers and Acquisitions in Services, 1984/85 to 1989/90
(Distribution, banking and insurance; involving at least one of the 1000 largest firms in the EC)

Source: Commission of the EC, Annual Reports on Competition Policy.

As far as the services sector is concerned, the total number of mergers rose from 67 in 1984–5 to 211 in 1989–90 (Fig. 5.2); thus, the same trend again, albeit less pronounced. The large majority of mergers was registered in banking and insurance and, compared with industry, the share of intra-national mergers remained considerably higher (for further discussion, see below the section on financial services).

Large firms in the Big Three of the Community have been particularly prone to the merger fever. Corporate behaviour has varied considerably among countries, and there are some interesting observations that can be made in this respect (Woolcock, 1989, Commission of the EC, 1991c).[1] British firms provided a favourite shopping area for predators from other European

---

[1] Observations in this area are also based on reports in the financial press and specialized publications such as *Acquisitions Monthly*.

countries and also across the Atlantic. At the same time, they accounted for a large share of the total of transfrontier acquisitions, including a relatively high percentage of non-European deals. This is another illustration of the outward-looking nature of the British corporate sector and its financial institutions. Cross-border mergers of French and German firms were predominantly intra-European. The aggressive and outward-looking approach of many French firms, especially those under state control, which characterized the late 1980s, contrasted sharply with past behaviour. Therefore, corporate behaviour became very consistent with the strong 'Europeanization' of French foreign policy in the Mitterrand era, and the two trends may not have been purely coincidental in view of the close links between political and industrial élites in France. On the other hand, the strong presence of German firms in intra-European trade was not matched by a similar activity in terms of cross-border mergers and alliances. Thus, the corporate managers of the most powerful European economy remained more inward-looking, or less keen on the mergers and acquisitions game, than some of their counterparts in other European countries.

Mergers and acquisitions are not the only form of inter-firm collaboration. In fact, another important manifestation of the latter has been the steadily growing number in recent years of co-operation agreements which often extend beyond national boundaries. In those sectors characterized by new technologies, intra-European co-operation in R. & D. has been actively promoted by the various programmes launched by the EC Commission. The success of the first ESPRIT programme in information technology brought about a 'collaboration bandwagon' (Sharp and Shearman, 1987: 60), with the development of new programmes which concentrated usually on the pre-competitive stage of research (BRITE for the development of advanced technologies, RACE in telecommunications, etc.). Although non-EC firms have not been completely excluded from Community programmes, the launching of EUREKA has helped further to extend collaboration to the wider area of Western Europe.

All those programmes have created a favourable environment and the material incentives for the establishment of intra-European strategic alliances in the field of high technology (Mytelka, 1989). Faced with high research costs and uncertainty in increasingly oligopolistic markets, industrial managers saw important advantages in close inter-firm collaboration. The Commission acted as a marriage broker, also providing part of the dowry to the newly-weds. Undoubtedly, the big firms have continued to call the shots, with a dominant presence in the various research programmes. However, through greater inter-firm co-operation, some of the smaller ones may have improved their chances of staying in the game, even when reduced to a satellite role *vis-à-vis* the giants, in markets which have been characterized by increased concentration. On the other hand, the role and participation of some of the less developed countries of the EC has remained marginal and usually limited to research institutes.

The matching game has not been restricted to the EC. The active participation of firms such as Volvo, Asea, Brown Boveri, and Nestlé suggests that the national champions from some of the EFTA countries have been playing an active part; and this has also been manifested in terms of increased investment undertaken by those countries, and especially Sweden, in the EC area. Similarly, the second half of the 1980s was characterized by the rapid growth of US investment in Western Europe and a very active participation of US firms in the mergers and acquisitions game, especially in the later years. This was also true, although to a much lesser extent, of Japanese investment. In fact, mergers and acquisitions were not only a European but an international phenomenon of the second half of the 1980s. The surge of foreign direct investment (FDI), which rose worldwide by almost 30 per cent annually between 1984 and 1989 (United Nations Centre on Transnational Corporations, 1991), has been seen as part of the 'globalization' of business and the changing structure of the world economy. DeAnne Julius (1990: 36, 40) argues that:

As a means of international economic integration, FDI is in its take-off phase, perhaps in a position comparable to world trade at the end of the 1940s . . . [We] are in the middle of a decade (1985–95) when FDI gains its maturity as a major force in international economic integration. It is in this sense that quantitative increases in FDI flows have reached the threshold where they create a qualitatively different set of linkages among advanced economies.

This process of restructuring of the world economy can be linked to the sustained economic recovery which started around 1984 and, even more so, to major technological changes which have characterized the recent years. According to Sharp (1990), this was the phase of widespread diffusion of new technologies accompanied by major institutional innovations. Restructuring at the global level also needs to be considered in conjunction with the establishment of wide 'networks' and co-operation agreements among firms, covering the whole range from R. & D. to marketing and distribution. Production rationalization, specialization, 'customized' services, and improved access to markets and technological skills have been among the primary motives in this respect. Thus economic internationalization rapidly extended beyond trade (Bressand, 1990).

How does one then disentangle international from purely European factors? Is 1992 and the relaunching of European integration an important explanatory factor for the wave of mergers, acquisitions, and co-operation agreements or are they simply part of an international phenomenon? Although the isolation of the '1992 effect' is clearly impossible, there are a number of factors which suggest that this effect has not been insignificant, especially in the geographical orientation of the strategic alliances pursued by European firms. Several business surveys indicate that both the investment behaviour and the external strategies of firms have been influenced by the prospect of a truly unified market, and this has been manifested in the increased merger and 'networking' activity of firms, both within and beyond national boundaries. The strengthening of the European and global presence has been, in fact, often preceded by the consolidation of a firm's position in the national market (Commission of the EC, 1989a, Buigues and Jacquemin, 1989).

The internal market and the various programmes launched in the field of high technology had a dual effect: on the one hand, they created expectations about the creation of a single European market and, on the other, they strengthened the European orientation of firms and their links with potential partners in other countries. Hence, the growing emphasis on intra-EC links as opposed to a search for partners across the Atlantic or, alternatively, in Japan, although with notable exceptions such as the takeover of British ICL by Fujitsu in 1990, which served as a reminder of the limitations of a purely European strategy in the electronics sector and of the discrepancy between rhetoric and action. Production and business alliances may have become increasingly global, but in recent years there has also been a significant strengthening of its European dimension, which is, indeed, a very new phenomenon. Cross-border alliances between former national champions in sectors which should be particularly affected by the internal market programme, such as telecommunications and power plants, cannot be totally unrelated to EC developments. Furthermore, the sudden surge of US (and also Japanese) foreign investment in Western Europe and the purchase of many European firms cannot be independent of expectations about the elimination of intra-EC barriers and also fears about 'Fortress Europe' (a subject which will be discussed in more detail in Chapter 9). In this respect, the behaviour of US firms would be very similar to actions undertaken in the early years of the EEC.

It is undoubtedly true that there is a strong element of 'herd instinct' (which could be called, more politely, fashion) in corporate behaviour and which may partly explain the link between 1992 and the recent wave of mergers and acquisitions. Previous experience suggests that many of them are unlikely to survive the test of time. Yet the change witnessed in European industrial structures has been quite significant, and at least some of it should be irreversible.

These developments have been generally welcomed, when not actively encouraged, by the Commission and some of the member governments. They have been regarded as entirely consistent with the logic of the internal market. After all, economies of scale and the efficiency gains from restructuring had received much emphasis in the Cecchini report. On the other hand, it is extremely difficult to decide where efficiency ends and monopoly power begins or even when strategic alliances turn into collusive behaviour. This is the domain of competition policy. The problem was aptly summarized by McGowan and Seabright (1989: 308), with reference to the boom of inter-firm co-operation agreements and alliances in the airline sector, following the first steps towards liberalization: 'At times it has even seemed as if European airlines would pass directly from the regulated state to a situation of concentration, not bothering to pause for a period of liberalisation in between.'

Modern microeconomic theory has gone beyond simple measurements of market concentration (the definition of the relevant product and geographical market being a difficult task in itself) and, instead, lays the emphasis on the 'contestability' of markets. Factors such as import penetration, entry barriers and the elasticity of demand are taken into consideration in deciding the trade-off between efficiency and monopoly power. Obviously, mergers in rapidly growing sectors with strong international competition raise fewer problems in terms of potential monopoly power than those taking place in mature industries with low import penetration. There have been examples of both cases in Europe (Jacquemin, Buigues, and Ilzkovitz, 1989).

The Treaty of Rome laid the basis for a common competition policy to be conducted by the Commission. Article 85 deals with collusive behaviour, Article 86 with the abuse of dominant position, and Article 92 with the problem of state aids, where the general prohibition is followed by a long list of exceptions. No direct powers for an *ex ante* control of mergers were given to the Commission, although provision for such powers had been made in the Treaty of Paris. In view of the long history of cartelization in the steel sector, this was understandable. No such powers were deemed necessary for the EEC and, in fact, the establishment of the latter in 1958 was followed by a process of economic concentration in individual European countries, which saw the birth of many national champions. Member governments and the Commission adopted at the time a relaxed approach on the subject, since large size was generally considered as the way of facing the competition of US-based multinationals. This was, after all, the time of the alleged 'défi américain' (Servan-Schreiber, 1967).

Britain introduced anti-merger legislation in 1965, and was followed by the Federal Republic of Germany in 1973. This was the year when the Commission also decided to initiate proposals for similar legislation at the Community level. It would have to wait for sixteen years before the Council of Ministers decided to move in this direction. National governments, at least those with legal powers in this area, were not at all enthusiastic about transferring them to the EC. In fact, it is worth noting that by 1989, when Community legislation was adopted in this area, only three countries (Britain, France, and Germany) had effective instruments for the control of mergers; while Ireland and the Netherlands also had some legislation in this respect. If size and the high degree of dependence on international trade might be considered as effective substitutes for domestic rules on market concentration as far as the small and open economies were concerned, this argument would be less valid in the case of Italy and Spain. As a matter of fact, the adoption of EC legislation has triggered, in the meantime, similar responses at the national level.

Three different developments gradually brought about a change of attitudes on behalf of both governments and private firms on the subject of the desirability of EC legislation for an *ex ante* control of mergers. One was the jurisprudence from the European Court of Justice: the decisions issued on the cases of Continental Can in 1973 and Philip Morris-Rothmans in 1987 opened the possibility for the use of Articles 85 and 86 for an *ex ante* control in some narrowly defined cases. And this immediately brought the Commission into the picture, thus creating a grey area in terms of legal powers between the national and the EC level. On the other hand, the rapidly increasing cross-border activity of European firms in more recent years meant that this activity could be subject to the jurisdiction of many different countries, thereby creating a nightmare for the firms concerned and a very lucrative business for competition lawyers. For instance, the joint bid by Siemens and GEC for Plessey had to be notified to several national bodies, both inside and outside the EC. Hence the yearning for what came to be called the 'one-stop shop' which meant that one single authority, at least within the Community, should be competent to decide on the desirability of a cross-border merger. Thus EC legislation in this area became relevant only when the merger activity started to spill over national frontiers. The third factor which made such legislation politically feasible was the gradual convergence of national attitudes in terms of competition policy.

The long and difficult negotiations revolved around three main issues: the threshold, expressed in terms of the combined turnover of the firms concerned, beyond which Commission powers would be activated; the line of demarcation between EC and national competences; and the criteria to be used when judging whether a proposed merger would be acceptable. On the question of threshold, the division was between big and small countries and, more precisely, between countries which had effective legislation in this area and the others which did not. The former, naturally, were in favour of high thresholds which would maximize the catchment area for national bodies. A similar division appeared on whether the Commission should have the final say on a proposed merger. Thus Britain and Germany fought to restrict as much as possible the powers of the Commission. As for the criteria to be used, the division between countries was a reflection of different traditions and attitudes towards industrial policy. While Germany, with the most developed anti-trust policy among member countries, insisted on the exclusive use of the competition criterion in deciding on the desirability of a proposed merger, countries such as France and Italy wanted to add other considerations related to industrial and regional policy objectives.

The regulation finally adopted in December 1989, and which has been applicable since September 1990, set a threshold of 5 billion ECUs for the combined worldwide turnover of the firms concerned. This threshold can be reduced to 2 billion ECUs, when a review of the policy takes place (planned for 1994). Two additional conditions were set for a proposed bid to come under Community jurisdiction, namely that at least 250 million ECUs of the combined turnover should be realized within the Community and that less than two-thirds of the combined turnover should be in a single member country. German influence on the new legislation was very strong indeed: Commission decisions should be based primarily, on competition grounds, although reference to 'efficiency' criteria in future Commission decisions could not be altogether excluded (George and Jacquemin, 1992). The Germans also succeeded in diluting the objective of 'one-stop shop' by opening the way for joint EC and national decisions in at least some narrowly defined cases. Double control was intended to minimize further the risk of the Commission adopting other than competition criteria in allowing big mergers to go through.

Thus the increasing number of cross-border mergers and acquisitions finally led to the adoption of a new regulatory framework and the transfer of more powers to the Commission. Between the entry into force of the new regulation and the end of 1991, the Commission had received a total of 75 notifications of mergers. In six cases, the Commission decided to launch a detailed investigation because of fears that those mergers could undermine competition in the internal market. It was the de Havilland case which brought the Commission's powers under the new regulation to the front pages of the financial press. The proposed takeover by ATR, a joint venture operated by Alenia (Italy) and Aerospatiale (France) of the Canadian turboprop aircraft manufacturer was blocked by the Commission, because of the dominant position which the merged firm would acquire in the world market.

The decision was adopted against the strong, and highly publicized, opposition of several Commissioners, not to mention the companies concerned and the respective governments behind them. As a result, the old debate about competition versus industrial policy was revived: French and Italian politicians challenged the powers of the Commission in this field, while the German Commissioner for industrial affairs attacked the 'competition ayatollahs' for not taking into account economic realities, which meant, according to him, the need for European firms to achieve global competitiveness (*Financial Times*, 12 February 1992). On the one hand, the de Havilland decision has provided yet another sign of the strengthening of the Commission's competition muscles, which has undoubtedly brought the new Community policy closer to the German model. The record fines imposed on several firms (Solvay, ICI, Tetra Pak) in application of Articles 85 and 86 in recent years is consistent with this trend. On the other hand, the decision also served as a reminder of the old ideological divisions lurking in the background about the appropriate mix of competition and industrial policy, despite the apparent hegemony of economic liberalism.

The Commission has been trying to encourage cross-border inter-firm co-operation through its various projects in the high technology field, its internal market programme and its persistent attempts to harmonize company legislation. The new regulation on the control of mergers can also be seen as a means of facilitating European mergers, by reducing as much as possible the area in which national and Community competences overlap. Cross-border co-operation among firms has been considered as a means of realizing the objective of the single European market and improving international competitiveness. It would also help to strengthen the economic and political base for European integration, by gradually shifting political allegiances to the European level and thus cutting the umbilical cord linking large firms to their national governments.

The creation of Euro-champions as the first stage towards European political union? It would be a godsend for underemployed Marxists and an excellent case-study of the link between changes in the economic base and the political superstructure. The answer, however, needs to be strongly qualified. On the one hand, measures to strengthen the European dimension of an international process towards global production and 'networking' have been accompanied by a growing emphasis on antitrust policies both at the national and the European level. This reflects a certain shift in attitudes, although not always entirely consistent, towards more competition and the realization that large size by itself is no guarantee of international competitiveness. On the other hand, unlike national governments in the earlier stages of integration, the EC does not have many instruments at its disposal in order to actively promote the development of Euro-champions, nor is it likely to obtain them in the foreseeable future. This has been further confirmed by the new chapter on industrial policy included in the Maastricht treaty, which shows that there is still little consensus for an active, interventionist role for the Community in this area. A weak state also implies a weak industrial policy, even if the intention were there, which is still not the case for many. This could prove a serious disadvantage in a world of strategic trade interaction, although views differ widely in this respect.

The adoption of new legislation for the control of big cross-border mergers still leaves a very large area for decisions by national authorities. Commission proposals for harmonization of national legislations and the process of osmosis may gradually bring about a closer convergence of national competition policies. There is already a similar experience in relation to the application of Articles 85 and 86 of the Treaty. Nevertheless, this process is likely to be extremely slow, and this in turn raises the question of possible distortions in intra-EC competition among firms. The same observation applies with respect to different company laws and their effect on corporate activity and the allocation of resources between countries. But then, how far do the Community and member countries want to go in terms of harmonization and the transfer of powers to EC institutions in order to eliminate economic distortions which can be legislated away only through negative integration measures? Sometimes, the countries complaining about those distortions and the unequal terms of competition are unwilling to pay the political price of reduced autonomy resulting from any serious attempt to tackle them. This is likely to be a perennial problem, as more power is being slowly transferred to the centre.

A major problem has arisen as regards the legal conditions applying to takeovers of firms in individual member countries, which have then led to pressures for the introduction of Community legislation in order to deal with the distortions created. Most cross–border takeovers have been taking place in the UK, and this has been especially true of hostile bids. This concentration of takeover activity in one country has been attributed to different legal and institutional conditions which largely determine the functioning of the 'market for corporate control'; and such differences are more than apparent in the EC of Twelve.[2]

To start with, the size and role of stock exchanges as providers of finance capital and as a means of correcting managerial failure vary enormously between the different countries of the Community. Furthermore, different regulations with respect to the functioning of stock exchanges as well as differences in competition policy, company law, and labour law can influence decisively the freedom to effect changes in the ownership and control of private firms. The large capitalization of the London Stock Exchange and the active 'market in corporate control' which operates in the UK through the buying and selling of shares in the stock exchange contrasts sharply with the situation prevailing in most other European countries. It is not only the significantly smaller size of continental bourses, including those of Paris, Milan, and Frankfurt, which acts as a constraining factor. The legal restrictions on hostile bids in France, the importance of family shareholdings and non-voting shares in Italy, the predominant role of German banks, not to mention German legislation on workers' participation in the supervisory boards, impose major restrictions on changes of company ownership and control.

Mainly due to British demands for the creation of a 'level playing field' (similar conditions applying in all countries) in terms of the takeover of firms, the Commission proposed the adoption of legislation aiming at the elimination of various technical obstacles that exist in different member countries. This legislation follows closely the British model which offers the most open 'market for corporate control'; perhaps, on the basis of the same principle that German cartel rules have served as the model

---

[2] The various technical barriers to takeovers were examined in two consultancy reports prepared by Coopers & Lybrand and Booz Allen in 1989.

for EC legislation for the control of mergers. But here, national practices are still very different, and a considerable distance will need to be covered before anything resembling a 'level playing field' can be achieved. Differences between countries in terms of the relationship between capital markets and corporate control are the result of both legal and institutional factors; they are also a reflection of very different industrial cultures. While the Anglo-American model puts the stress on the 'market for corporate control' as a means of correcting managerial failure, the German model (most other European countries are closer to it and so is Japan) puts the emphasis on long-term relations between investors, managers, and employees. According to Franks and Mayer (1990: 215), 'there is a tradeoff between correcting managerial failure and promoting investment'.

In fact, there is little evidence to suggest that the Anglo-American model is more efficient in the long term (the evidence seems to point more the other way) and that the whole Community should therefore move in this direction; except, perhaps, for a greater transparency of rules which should be more generally welcome. Even if a new EC takeover code were to be adopted, this would be unlikely to bring about a significant change in the market, since legal rules are only a small part of the overall environment determining changes of ownership and control in the corporate sector. And this cannot be transformed overnight. Here again, the creation of an EC regulatory framework is likely to coexist for a long time with strong national idiosyncrasies which will in turn continue to create economic distortions. A case of competition between different economic systems? This has been, in fact, advocated by several authors (Siebert, 1989).

Another aspect of EC competition policy concerns state aids. Trying to avoid a direct confrontation with national governments, the Commission adopted a very cautious approach in this respect for many years after the establishment of the EEC. The problem grew rapidly in size, however, in the years of the prolonged recession and growing unemployment, when state aids became an important instrument of protectionism and a means of keeping 'lame ducks' afloat. The steel industry is the best example in this respect.

Following the publication of the White Paper in 1985 and the adoption of the internal market objective, the EC Commission has been flexing its muscles in this area too. The growing number of notifications given by the Commission to individual member countries and the handling of difficult cases, such as those of Renault and Rover, to mention only two, are good examples of the much tougher stance adopted by the Commission. A sign of this new attitude was also given with the publication of the first comprehensive reports on state aids. Nationalized firms present the most awkward problems, especially because of the need to distinguish between investment (which is legitimate) and aid (which may not be), a distinction which is not at all obvious in the public sector. After all, if strictly private criteria were to apply in trying to draw the demarcation line between the two (and this has been increasingly the Commission's position on this issue), what would then be the economic logic of a publicly owned firm? This is obviously not a technical question. A tough stance by the Commission with respect to nationalized firms could contribute further to the trend of privatization, although not all national governments should be expected to give in too easily.

There are some general conclusions that can be drawn regarding the development of the common competition policy and the recent wave of merger activity. A close link seems to exist between policy changes and the market place, with changes in one feeding into the other. Policy initiatives have contributed significantly to changes in corporate structures, while the wave of cross-border mergers and acquisitions has precipitated the adoption of a new EC regulatory framework and hence the transfer of some powers to Community institutions. On the other hand, the distribution between EC and national powers in this area is very unlikely to remain static, in view of the political tensions and market distortions arising from different national rules. There is evidence of functional spill-over, with economic integration demonstrating a cumulative logic of its own, as had been predicted by earlier functionalist theories (Haas, 1958; Lindberg, 1963).

While national barriers are eliminated, we are witnessing an acceleration of the process of economic concentration at the European and the international level. There are also signs of a geographical concentration in terms of ownership and control, with large firms in the economically more developed countries of the Community being active buyers in the corporate market; if continued unabated, this trend might produce eventually a political backlash. To the extent that there is a fledgling European policy on industrial structures, the emphasis seems to be on its anti-trust dimension, thus bringing Western Europe closer to US practice. This happens partly by choice and partly by default. The stress on competitive markets is consistent with the prevailing economic ideology and the internal market programme itself. On the other hand, an active industrial policy would be impossible because of the combination of intra-EC economic interdependence and the highly decentralized nature of political power in the Community. National governments become increasingly unable to perform such a role, while EC institutions lack the legitimacy and political power to fill the vacuum. Thus, even if the economic fashion were to change in the foreseeable future, an interventionist policy would be extremely difficult to pursue for political and institutional reasons.

Deregulation in Financial Services

The 1992 programme is more about services than goods, and the reason is simple: barriers to the cross-border provision of services are mainly the result of government intervention inside the border, intervention which takes many different forms creating barriers which are usually lumped together under the general category of NTBs. Until the publication of the White Paper and the subsequent adoption of the SEA, very limited progress had been registered in this area, thus allowing for little intra-EC competition in most service sectors. Since then, the picture has changed drastically, with the flurry of legislative activity in Brussels already having concrete effects on the behaviour of economic agents.

Financial services occupied a prominent place both in the White Paper and the 'Costs of non-Europe' study. In fact, according to the latter, the positive effects expected from the liberalization of financial services accounted for one-third of the overall macroeconomic gains associated with the completion of the internal market. The large and rapidly growing size of the sector, the closed nature of national markets, manifested through persisting large price differentials, and the indirect effect on investment resulting from the expected lowering of the cost of financial services, and hence on growth through the multiplier effect, are the main explanatory factors for the emphasis laid on the liberalization of this sector in the context of the internal market programme.

On the other hand, the elimination of intra-EC barriers in a heavily regulated sector, such as financial services, can provide an interesting case-study of the link between liberalization and regulation as part of the 1992 programme. The effects of liberalization go, in fact, much beyond the functioning of financial markets. They extend to macroeconomic policies and wider issues of economic sovereignty. We shall return to them below. In this section, we shall concentrate on the role of public sector intervention in the new EC regulatory framework and the likely effects of the change brought about by the 1992 programme on the functioning of financial markets. The main emphasis will be on banking where EC legislation was most advanced at the time of writing, although references will also be made to other financial services, especially since the demarcation line between banking, insurance, and investment services has become increasingly blurred.

When the White Paper was published, the main barriers in the financial sector were the product of exchange controls and different regulatory frameworks in individual member countries. The former were by far the most important, while the latter acted mainly as a barrier to the free cross-border provision of services, thus reinforcing the negative effect of capital controls. There was little overt discrimination against foreign banks in national markets, although some obstacles still remained for the full exercise of the freedom of establishment. They included authorization procedures, capital endowment requirements, and restrictions on foreign acquisitions.

Considerable caution had been expressed in the Treaty of Rome with respect to the objective of capital liberalization, despite the fact that this was supposed to constitute one of the so-called four fundamental freedoms of the EC. Fears about potential instability in the capital markets and balance of payments considerations were behind the cautious attitude expressed by the authors of the treaty. Preserving the autonomy of national monetary policy was the other unspoken reason. Collective caution remained the catchword for many years. Two directives were adopted in 1960 and 1962, marking the first step towards the liberalization of capital movements. They laid the emphasis on transactions related to the movement of goods, the right of establishment, direct investment, and the purchase of listed shares in stock exchanges. They therefore went only slightly further than 'necessary to ensure the proper functioning of the common market' (Article 67 of the Treaty of Rome). No further step was taken in this area for a long time. If anything, the 1970s saw a regression, when an increasing number of countries tightened their system of capital controls, essentially abusing the safeguard clauses included in the above mentioned directives.

Capital controls were used as a means of preserving some stability in intra-EC exchange rates and widening the margin of independent manœuvre in terms of monetary policy. National responses in this area were far from uniform: thus while Britain, Germany, and the Netherlands had abolished virtually all restrictions to international capital movements by the end of the 1970s, other countries such as France and the whole of southern Europe retained for years a heavy armoury of controls directed mainly at short-term capital movements; and these in turn restricted the free flow of financial services (Commission of the EC, 1988*b*).

The financial sector, and banks in particular, have a long history of heavy regulation, with the 1929 Great Crash being a major turning-point in it. Government intervention can be first of all justified with reference to what economists call market failure (Baltensberger and Dermine, 1987). Banks usually borrow short and lend long, and this asymmetry between assets and liabilities depends on a highly volatile element called confidence, that is, the confidence of the depositor. In view of the limited information that the average customer is expected to have of the quality of a bank's asset portfolio, not to mention his or her ability to judge such quality, government regulation is intended to protect the customer and more generally the savings of the population. On the other hand, because of the potentially unstable nature of public confidence and the ease with which doubts about the liquidity of one particular bank can turn into panic and a general run on banks, governments have felt this need to intervene in order to protect the stability of the whole financial system and at the same time to avoid major negative repercussions for the real economy. The two sides of this intervention are the various forms of deposit insurance schemes and the function of the lender of last resort undertaken by the central bank in times of crisis and illiquidity of the banking sector.

However, protective measures in the form of government

guarantees of bank liabilities and the lender of last resort function create the problem of moral hazard. This means that banks may take excessive risks in their lending or over-expose themselves in the search for higher profits, in the knowledge that if things turn out badly, they can always rely on the safety net provided by the central bank. Hence, therefore, the need for more direct supervision and control of the activities of the banking sector, which has been translated into numerous prudential and preventive measures.

An additional reason for government regulation stems from the role played by banks in the money creation process and the financing of public sector deficits, which in turn raises wider issues regarding macroeconomic stability. The link between monetary policy and the latter has been the subject of a never ending controversy among economists. Reserve requirements, deposit rate ceilings, and quantitative credit controls have been among the main instruments used in this area.

Last but not least, fears of excessive concentration, the close links between banks and industry, and the frequent identification of financial with political power explain many of the restrictions imposed by governments on the activities of banks. Resistance to large mergers and acquisitions, monitoring of the identity of shareholders, and limitations imposed on the degree of control by banks over the non-financial sector have figured prominently among the instruments used by governments.

It is clear that regulation and the various forms of restrictions applied have a negative effect on the efficiency of the banking sector by limiting competition and adding to the costs of operation. Governments have thus always been faced with a trade-off between efficiency and stability, and the choice has varied enormously between countries and time periods, reflecting among other things changes in economic fashion and different economic structures. On the other hand, the resulting differences in regulatory frameworks have helped to preserve the high walls separating national markets.

One characteristic which distinguishes most European countries from the United States is the predominance of banks in capital markets; in other words, the relative underdevelopment of non-intermediated finance. But beyond that, the differences between banking structures in individual European countries are more than substantial (Bisignano, 1990; Dermine, 1990). Those differences manifest themselves in terms of the role of publicly owned banks (mainly in France, Italy, Greece, and Portugal), the openness to foreign competition (particularly strong in the UK, Belgium, and Luxembourg), and the degree of market concentration (not surprisingly, pronounced in some of the smaller countries). Important differences can also be detected in terms of the range of activities undertaken by banks, from the universal bank model in Germany, France, Italy, the Netherlands, and, more recently, also the UK, to the more traditional role of banks which has excluded any direct involvement in the stock exchange. Many of those differences can be directly attributed to regulatory policies.

The decade of the 1980s was marked by a process of deregulation of banking as well as the financial sector more generally, coupled with a further strengthening of the internationalization process, the two being mutually reinforcing (Aglietta, Brender, and Coudert, 1990). They were a response to the continuing internationalization of trade and production, the revolution in communications, financial innovation, and the intensified competition among financial centres. The liberalization measures introduced in both London and Paris in the second half of the 1980s need to be understood in that context: a gradual conversion to the market as an efficient allocator of resources was further strengthened by the need to survive in international competition. Liberalization measures in individual countries paved the way for the legislation subsequently adopted at the EC level.

During this period, which was also marked by the rapid growth of money, bond and equity markets despite the temporary setback caused by the crash of 1987, the presence of foreign banks in national markets grew steadily and so did the size of Eurocurrency markets. Deregulation and internationalization brought about some convergence in national banking structures. But this convergence remained limited: international banking activity concentrated mainly in inter-bank deposits and the wholesale end of the market occupied by large corporate customers. The big bulk of retail banking is still confined within national boundaries and controlled by national banks operating in conditions of very imperfect competition. This explains the term dual economy used to describe the banking sector in several European countries: one segment being part of a competitive international market, while another one is operating under heavy government regulation and protection from foreign competition (Bisignano, 1990).

Deregulation and internationalization were not, however, devoid of problems. Considerable financial instability was one of the main characteristics. Problems of moral hazard also became prominent. The international debt crisis of the 1980s, which brought to public attention the irresponsible exposure of some of the world's largest banks, and the collapse of the US thrift industry raised doubts about the stability of the financial system, while also imposing high costs to the taxpayer. There was the fear that the financial system was becoming deregulated in its assets and guaranteed in its liabilities, a combination which would not necessarily be acceptable to everybody.

The EC strategy for the liberalization of financial services was first expounded in the White Paper of 1985 (see also Commission of the EC, 1988b; Zavvos, 1988). It started with the complete liberalization of capital movements, as a necessary condition for the creation of a European financial area. A necessary condition but not sufficient: the second plank of the strategy referred to the harmonization of regulatory frameworks. Following the radical innovation introduced in the White Paper, this would be based on the harmonization of only essential rules combined with the application of the principle of mutual recognition (thus using the famous *Cassis de Dijon* decision as the model). As a belated recognition of the dangers of widespread tax evasion and avoidance in a world of unrestricted capital movements, a third plank was later added, aiming at the harmonization of tax rates on savings and investment income. There had been no reference to this subject in the White Paper.

With respect to capital movements, the developments of recent years have been, indeed, spectacular. A new directive was adopted in 1986, which signified the first modest step towards liberalization since the little remembered directives of 1960 and 1962. It was followed by another, much more ambitious directive in 1988 which marked the decision of EC countries to proceed to completely free capital movements, including the abolition of two-tier exchange markets in Belgium and Luxembourg. The decision entailed the elimination of all remaining capital controls in eight member countries by 1 July 1990. This deadline was met by all the countries concerned. The timetable for the elimination of capital controls extends to the end of 1992 for Spain and Ireland, while Greece and Portugal may be given an extension of three more years in view of the fragility of their balance of payments. The different timetables envisaged in the 1988 directive testify to the new flexibility towards the integration process, which is itself a recognition of the economic heterogeneity of the Community of Twelve.

The 1988 directive includes provisions for emergency measures in case of threat to the monetary or exchange rate policy of a member country. The safeguard clause can only apply to capital movements liberalized under the new directive and for a maximum period of six months, subject to the Commission's approval. There can be, however, some serious doubts as to whether this safeguard clause could become operational after the whole armoury of controls has been dismantled.

This is, undoubtedly, one of the most remarkable decisions of recent years, after a long period of virtual stagnation in this area. The decision to proceed with complete liberalization by countries such as France and Italy, which had a long history of extensive capital controls, represented a search for greater efficiency for national financial sectors in an increasingly more competitive international environment. National authorities have also tried to make virtue out of necessity, by recognizing the increasing ineffectiveness of capital controls.

The internationalization of financial markets and increased capital mobility, facilitated through the abolition of controls, feed into each other. The 1988 EC directive is part and parcel of this logic. Although liberalization represents an international trend and the new directive makes no attempt to discriminate between intra- and extra-EC capital movements (a sign of economic realism?), it is clear that the Community as such has, for the first time in many years, influenced decisively national policies and attitudes in this area.

In the case of the banking sector, the harmonization of essential rules and mutual recognition have been combined with the principle of home country control as the three main elements on which the process of EC liberalization has been based. Home country control means that the responsibility for prudential control and supervision of all domestic and foreign branches of a bank rests with the regulatory authorities of the country of origin. This principle had been already adopted in the First Banking

Directive of 1977, although with no concrete effects because of the lack of progress in the area of harmonization. The real break-through came only with the adoption of the principle of mutual recognition which reduced the need for harmonization down to essential requirements.

In 1989, the Community adopted three directives which form the basis for the liberalization of the banking sector and the model for the liberalization of financial services as a whole. The Directive on Own Funds was adopted in April 1989; it dealt with some general principles and definitions regarding bank capital. It was followed by the Directive on Solvency Ratios and the Second Banking Directive which were adopted in December of the same year. The former contained an agreement on a weighted risk assessment of bank assets and off-balance sheet items and a harmonized minimum solvency ratio of 8 per cent for all credit institutions.

The Second Banking Directive is the most important in this triad of directives which form the basis for the new EC regulatory framework for the banking sector. The first main element is the creation of a single banking licence. This will enable any bank which has received authorization by the competent authorities of a member country of the EC to provide services across the border and to open branches in any other member country with-out the need for further authorization. Banks will be allowed to engage in any of the activities listed in the annex of the Second Banking Directive; and this list is based on the model of the universal bank, including investment services, but not insurance. The new regulatory framework also includes provisions in terms of minimum capital for authorization and the continuation of business; control of major shareholders and of banks' participa-tion in the non-banking sector; and sound accounting and control mechanisms. The responsibility for prudential control, including the foreign branches of a bank, is given to the authorities of the country where the parent company is based (home country), while the authorities of the host country will share responsibility for the supervision of the liquidity of branches in their own territory as well as for measures related to the implementation of national monetary policy. Provisions are also made with respect to reciprocity in relations with third countries. The new EC legislation takes effect on 1 January 1993.

The significance of the above cannot be easily overestimated.

With one stroke, EC countries went further than a federal country such as the United States in creating the legal conditions for an integrated banking sector. Furthermore, in adopting the model of the universal bank, the EC has opted for a more liberal approach as regards the permitted activities of banks than either the United States or Japan. The adoption of the single banking licence is intended to eliminate all remaining restrictions and obstacles associated with authorization requirements for the opening of branches in any EC country. The emphasis on home country control is meant to facilitate prudential control on a consolidated basis, thus including the branches of a bank in other EC countries. Through home country control, any discrimination by national regulatory authorities against the branches of a bank from another EC country established in their own territory will become virtually impossible. Furthermore, by creating the conditions for regulatory pluralism within each national market (since banks will be subject to prudential control and supervision by different regulatory authorities), the EC has also set the ground for competition among regulatory systems.

As for the essential requirements for prudential control and supervision, a pre-condition for the application of the principle of mutual recognition, they were the subject of an interesting and delicate exercise in international financial diplomacy. The main emphasis was laid on minimum capital requirements, and this raised the need for agreement on definitions of bank capital, weighted risk assessments, and solvency ratios. Here, the ground had been prepared by the Basle Committee on Banking Regulations and Supervisory Practices of the Group of Ten. Yet, what were only recommendations in the case of the latter, applying to 'internationally active banks', became mandatory rules in the case of all EC-registered banks. A minimum solvency ratio of 8 per cent has been adopted for all EC banks. There were also some differences between the Basle agreement and the directives adopted by the EC in 1989, as, for example, with respect to the definitions of capital.

Reflecting a wider international trend, the EC approach relies heavily on high capital–asset ratios as a means of preventing an over-exposure of banks, thus replacing many specific restrictions adopted by regulatory authorities in the past. But there is also an EC-specific explanation for this strong reliance of prudential control on one element. A policy of liberalization based on the harmonization of essential requirements and mutual recognition needs to rely heavily on a few transparent and generally accepted rules and less on the discretionary powers of different national authorities. Hence the emphasis on high capital–asset ratios, but also the risk of a certain rigidity in the regulatory approach.

It may be one thing to agree on general principles and something totally different to try to apply those principles in concrete situations. In view of the continuous expansion of banking activities, it has become increasingly difficult to apply the same regulatory rules to very different activities which may also produce different risks. Should the same capital–asset ratio apply, for example, to traditional banking loans and investment brokerage which may involve very little capital risk for the bank concerned? This became a major issue of contention precisely because of the different banking structures and regulatory traditions of individual member countries. What has been at stake is the search for a formula which would primarily reconcile the interests of investment houses in the City of London with those of universal banks in the Federal Republic. If the same capital–asset ratio were applied to investment activities, it might force many specialized investment firms in London out of business. If, however, a lower ratio were to be adopted, then this would discriminate against large German banks where the whole range of activities comes under one roof. In this respect, the negotiations were not about an optimal rate of regulation, based on some fancy economic theory; they were, instead, a search for a political compromise between different economic interests which are in turn a product of history, economic structures, and regulatory traditions.

Another important issue of contention has been about the reciprocity provisions with respect to third countries. Here, the liberal approach of Britain and Luxembourg, with important international financial centres, clashed with French 'economic nationalism' which was now turned into 'European nationalism'. The size of the financial sector as a percentage of GDP in the UK and Luxembourg is more than double that of the average EC country, and this, combined with the strong international orientation of their financial sector, clearly explains the concern of those two countries about any decision which might lead to a loss of business in this area. This may also explain, at least partly, their reluctance to accept any transfer of powers to Brussels, especially as regards policy *vis-à-vis* the rest of the world. The final compromise, included in the Second Banking Directive, was much closer to *laisser-faire* principles. This subject will be discussed in more detail in Chapter 9.

With the new legislation adopted, the Community aims at liberalization both in terms of branches and the range of activities of banks. This will increase inter-bank competition both across and within national frontiers. It will also accelerate the move towards universal banking by eventually forcing regulatory authorities to lift existing restrictions in order to prevent unfair competition against their national banks in their own market. It is clear that the effects of liberalization will be felt more strongly in those countries where the banking sector has been subjected until now to the heavy regulatory hand of national authorities and limited competition, the two usually going together. This will be particularly true of southern European countries where EC liberalization is expected to act as a catalyst for the introduction of their banking sectors to the modern era. It is not, therefore, surprising that in the Price Waterhouse study which covered eight EC countries, the biggest price reductions, associated with liberalization, were expected to take place in Spain and Italy (the theoretical, potential price reductions were 34 per cent and 29 per cent respectively; Commission of the EC, 1988a).

The magnitude and time-scale of the effects of liberalization on the banking sector in the Community are very much open to speculation, given also the relatively long time-lag between the adoption of legislation at the EC level and its implementation by national authorities. In the 'Costs of non-Europe', the authors expect significant changes. On the other hand, there are various observers who stress the oligopolistic nature of national markets and the existence of important economic barriers to entry, associated mainly with imperfect competition, extensive branch networks, and the long-established reputations of banks at the local and national level. They therefore expect a very slow process towards a genuinely European market (Grilli, 1989). This will apply particularly to the retail end of banking where the average household and the small local business are the potential customers. On the other hand, the effects of liberalization are expected to be more pronounced in terms of big deposits and specialized services provided to large corporations which are likely to be more mobile in their search for the least costly provision of services required. Thus the largest players would be expected to gain the most. The wholesale end of the market is also the one where liberalization may lead to further geographical concentration of banking activity due to thick-market externalities.

The growth of merger and co-operation activity among banks, experienced in the late 1980s (see Fig. 5.2), was closely associated with the anticipation effect of 1992. This seems to be confirmed by various business surveys, much though the expectations of firms about 1992 appear to be generally exaggerated (Bank of England, 1989; Commission, 1989c). Here again, perception is more important than reality. Compared, however, with the experience with non-financial institutions, the merger and co-operation itch has been mainly confined within national boundaries. The main aim seems to be the consolidation of a firm's position in the national market in order to face more successfully the expected intensification of international competition resulting from EC liberalization. In Italy, the restructuring of the banking system is underway, and this is even more true of Spain. In the latter country, where competition from outside is expected to produce a mild storm in the tranquil waters of an oligopolistic, protected and also lucrative market, mergers between *'las grandes siete'* have been actively promoted by the central bank in the role of the undisputed chief of a quarrelsome orchestra.[3]

Majority acquisitions of foreign banks are generally expected to remain limited in number, because there are not that many potential acquisitions in the market. The large number of state-owned banks, the difficulty of making an effective entry into long-established oligopolistic markets, and the sensitivity of national authorities to transfers of ownership of large banks to foreign hands are likely to continue acting as serious impediments to a surge of mega-merger activity across borders. The purchase

---

[3] In January 1990, the Governor of the Bank of Spain effectively appointed the chairman, the two deputy chairmen and several members of the board of the biggest private bank, Banco Bilbao Vizcaya, itself the product of a merger which had not worked out very smoothly. It was a clear reminder of the enormous influence of the national regulatory authority and its wide discretionary powers on the eve(!) of the creation of a single European financial area. Two mega-mergers followed in 1991: the first one grouped together Spain's state banks into a large holding company, the Corporación Bancaria de España; this also seems to have acted as a catalyst for the subsequent merger between Banco Central and Banco Hispano-Americano.

of the whole branch network of the Bank of America in Italy by Deutsche Bank has been the exception rather than the rule. This may, however, change in the future as the national character of markets is slowly diluted. In the meantime, cross-border co-operation agreements will help to facilitate access to each other's national networks, and there has been a great deal of them in recent years. On the other hand, various authors have pointed out that economies of scale in the banking sector are usually exhausted at relatively low levels of turnover, although economies of scope, diversification of risks, and, last but not least, international recognition and prestige may still act as important factors conducive to merger activity (Gilibert and Steinherr, 1989).

Does 1992 imply a policy of deregulation for the banking sector? The answer is a qualified yes. The single banking licence, coupled with mutual recognition, will lead to some competition among regulatory authorities and, consequently, to lower levels of regulation. The EC is bending with the international wind, although the new 'regime'[4] is unlikely to lead to a deregulatory race. The basic common rules already agreed upon will provide the bottom line. In fact, the minimum capital standards adopted were higher than those applied by most banks when the new legislation was introduced, which therefore raises the need for further capitalization before the deadline of 1993. In addition, the quality of regulatory sources is likely to remain an important aspect of the international competitiveness of banks in a business where confidence is a crucial element.

Unlike competition policy, the new EC regulatory framework for the banking sector does not involve any direct transfer of powers to the Community level. Mutual recognition is the name of the game. But the experience so far suggests that there is no general agreement as to where harmonization stops and mutual recognition begins. The real substance of some of the common rules adopted can only be tested in practice. Furthermore, many important questions have been left unanswered. The EC liberalization of financial services moves in the same direction as the more general trend towards deregulation and internationalization.

---

[4] Regimes became a fashionable concept of international political economy in the late 1970s and 1980s. They were supposed to provide a substitute for hegemonic stability after the alleged fall of the American hegemon. International regimes were defined as 'principles, norms, rules and decision-making procedures around which actor expectations converge in a given issue-area' (Krasner, 1982: 185).

It is impossible to predict how far this trend is likely to go. The experience of the 1980s and the early 1990s does not offer much comfort about the efficiency and inherent stability of financial markets. The problems experienced in several European countries, and most notably the Scandinavian countries, as the economic recession and the decline of asset prices began to hit over-exposed banks in more recent years, bear witness to it.

Financial deregulation has not been accompanied by a serious attempt to deal effectively with problems of moral hazard, as many financial institutions showed a tendency towards excessive risk-taking. The Bank for International Settlements, and its General Manager, Mr Lamfalussy, have repeatedly expressed serious concern about the fragility of international financial markets (see for example, Bank for International Settlements, 1991). Will a future repetition of the 1987 crash, this time at a bigger scale, lead to a reversal of the trend towards what Strange (1987) calls 'Casino Capitalism' and to renewed attempts for closer regulation and supervision of financial markets? Even if the political will were there, it is doubtful that, once opened, the Pandora's box could be closed again.

On the other hand, the move, albeit slow, towards a truly European market in financial services is bound to raise, sooner rather than later, the issue of a European regulator. Reference to such a role for the European Central Bank has been made in the new treaty adopted at Maastricht. Mutual recognition and the uncertain dividing line between the responsibilities of home and host authorities are unlikely to provide a satisfactory answer for very long. This may be another example of a dynamic disequilibrium as a driving force of European economic integration.

Tax Harmonization: This Side of Eternity?

Tax harmonization issues have figured prominently on the European integration agenda, although there has been little correlation between inputs (man-hours spent in negotiations) and outputs (EC legislation adopted). The subject has been highly controversial, and proposals emanating from Brussels have often provoked strong reactions in national capitals.

Tax matters were dealt with in a general manner in the Treaty of Rome, although the harmonization of indirect tax rates was envisaged in Article 99 of the treaty. Harmonization was seen as necessary only to the extent that different national taxes could create distortions in the free movement of goods, services, persons, and capital. Hence, therefore, the emphasis on indirect and, to a lesser extent, corporate taxation, based on the assumption that goods and capital are more mobile across frontiers and thus more sensitive to different tax conditions applying in the member countries.

When the EEC was set up in 1958, the six member countries had very different systems of taxation. The main achievement of the Community in this area has been the adoption of a uniform system of turnover taxes, based on the French system of VAT. This is a non-cumulative tax levied at each stage of production and/or marketing, and it has two main advantages *vis-à-vis* the various forms of 'cascade' taxes previously operating in the other member countries. The VAT system avoids the bias towards vertical integration of firms inside a country inherent in the cumulative nature of different sales taxes. The other advantage is its neutral effect on internationally traded goods. According to the destination principle, generally used in international trade, the tax is levied at the point of consumption and not of production. This means that exports are zero-rated and that tax adjustments (export refunds and payments for imported goods) have to take place at the border. The transparency of the VAT system makes border tax adjustments a straight-forward procedure; unlike the experience with other sales taxes which have frequently been used by governments as a means of indirect protection of domestic producers.

The introduction of a common system of turnover taxes by all members of the EC was not accompanied by a harmonization of the base of taxation nor by a harmonization of rates. The Second and the Sixth VAT Directives, adopted in 1967 and 1977 respectively, failed to bring about a completely harmonized base, due to numerous exceptions and derogations included in those two directives. On the other hand, the wide diversity of VAT rates, in terms of both product categories and countries, has continued all along. Any form of differentiated indirect tax creates distortions in the allocation of resources. However, through the application of the destination principle in both international and intra-EC trade, distortions in trade between countries have been minimized (Robson, 1987).

The continued diversity of rates has been even more true of excise taxes applying mainly to alcohol, cigarettes, and petroleum products. Here diversity reflects different social values, consumption habits, and health considerations, not to mention the financing needs of national governments. On the other hand, it has not been devoid of protectionist impulses which have brought both the Commission and the Court of Justice on the heels of member governments. Discrimination against foreign products has been usually achieved through a narrow definition of product categories and the imposition of widely different levies on each one of them. Thus Scotch whisky can be effectively separated from the Italian grappa, ending up with a very different tax burden incorporated in the final consumer price. The different approach to the taxation of beer and wine between northern and southern countries is another good illustration of innocent(!) discrimination in the use of excise taxes by national authorities (for somebody from a wine-growing country, the idea of taxing wine borders on sacrilege).

A long stalemate also characterized the Commission's efforts to harmonize corporate taxes in an attempt to eliminate distortions in the allocation of direct investment and also as a means of promoting cross-border mergers and alliances. Concern with taxes on investment income is much more recent; it is itself a product of the growing mobility of short-term capital. As for personal taxation and social security payments, they have remained all along beyond the harmonizing ambitions of the EC Commission.

Despite the frustrations of earlier years, the Commission plucked up enough courage and called for the elimination of fiscal barriers in its White Paper of 1985. The main emphasis was, as before, on indirect taxation. The aim was to do away with frontier fiscal controls which involved a direct cost for companies engaging in cross-border trade. An attempt was made at estimating those costs, which produced a figure of 8–9 billion ECUs, representing 0.2–0.4 per cent of Community GDP (Commission of the EC, 1988a). In macroeconomic terms, this figure was not very impressive, although surely not insignificant for firms heavily engaged in cross-border trade. But the motives of the Commission were not strictly pecuniary: the elimination of all frontier controls, of which border tax adjustments constituted a substantial part, was expected to have an important psychological effect. Open frontiers would be, after all, the most tangible effect of the 1992 programme for the ordinary citizen.

The Commission came up with detailed proposals in 1987. It called for a shift from the destination to the so-called common market principle for VAT payments (Cnossen and Shoup, 1987; Biehl, 1988a).[5] This would basically extend the regime applying to transactions within a member country to cross-frontier transactions, which meant that exports would be taxed like any other transaction and that the importer would be able to deduct the VAT payment against the VAT charged on the sale of output in his own or any other EC country. Thus sales and purchases across borders would be treated in exactly the same way as similar sales and purchases within a member country. The abandonment of the destination principle would lead to the elimination of border tax adjustments and the abolition of the Single Administrative Document (SAD) used for cross-border trade. On the other hand, it could also lead to trade distortions as long as the disparity of rates continued. Thus the next logical step, according to the Commission, should be the harmonization of rates. Complete harmonization was not deemed necessary. Instead, the Commission proposed the adoption of two bands of VAT rates for 'basic' and 'luxury' items, within which national VAT rates would be allowed to move. The band for the standard rate would be 14–20 per cent, and 5–9 per cent for the reduced rate. The width of the bands was considered small enough to avoid any serious trade deflection.

But the Commission was not prepared to go all the way along the road of fiscal federalism. Harmonization of rates was one thing, and reallocation of tax revenue another. The replacement of the destination by the common market principle of taxation would lead to a redistribution of tax revenue between member countries. Those with higher tax rates and trade surpluses would gain at the expense of countries with lower tax rates and trade deficits. To avoid such a redistribution, considered to be politically impossible, the Commission proposed the creation of a clearing house system, based on the VAT returns of trading firms (or on

---

[5] A distinction is drawn in the literature between the destination and the origin principle, the latter meaning that the goods are taxed where they are produced. However, it has been shown by the above authors that the shift from the destination to the origin principle would not be sufficient to allow the elimination of border tax adjustments in the case of a multi-stage tax like the VAT.

aggregate trade statistics according to another proposal), which would enable the preservation of the *status quo*. The aim was to minimize as much as possible the changes required, without at the same time compromising on the fundamental principle of the unhindered movement of goods and services. As for excise taxes, the single-stage nature of those taxes and the wide differences between rates applying in the various member countries forced the Commission to come up with a proposal for a uniform rate.

The Commission proposals were not very popular with most member governments. Opposition was centred on two main issues: the harmonization of rates and the clearing house arrangement for the redistribution of tax revenue among national authorities. The UK led the opposition, finding itself once again in the familiar role of the flag-bearer of those defending national sovereignty against attacks from Brussels.

In view of the diversity of rates existing until then, national resistance to change should have been hardly surprising. National sensitivities were at stake; and so were delicate internal political balances which could be upset by tax changes having important sectoral and distributional effects. Asking the British Government to impose even a small VAT rate on food and children's clothing or expecting the Danes to reduce substantially their excise taxes on alcohol and tobacco did not elicit favourable responses from politicians jealous of their power and acutely aware of the sensitivities of their constituents. The harmonization of rates would also restrict considerably the flexibility of governments in the use of fiscal policy for stabilization purposes. On the other hand, changes in indirect tax rates would have important effects in terms of prices, which would then feed into other macroeconomic variables.

Revenue considerations were also very important for some member countries. Although the Commission's proposals were designed to minimize the overall effect, two countries, namely Denmark and Ireland, would suffer a significant reduction in revenue through the lowering of their VAT rates and excise taxes (Centre of European Policy Studies, 1989). The case of Luxembourg was more complicated: the required large increase in rates could be overcompensated in terms of tax revenue by the loss of cross-border purchases from neighbouring countries, since the tax incentive would no longer be there. Tax policies in the Grand Duchy have had an element of the free rider principle analysed in the literature on public goods: an option open to a small country in the geographical position of Luxembourg. The Commission proposals on tax harmonization were intended to put an end to that situation, thus provoking a negative reaction from an otherwise 'European-minded' country.

Last but not least, the proposal for a clearing house crucially depended on the setting-up of an effective system for combating fiscal fraud and tax evasion without adding much to the existing administrative burden. This would in turn depend on the close co-operation and mutual trust between national administrations. But neither could be taken for granted, and this soon became obvious through the reactions of several governments.

The Commission's proposals suffered from a strong imbalance between ends and means; the changes required were too big and costly (at least, in terms of loss of national autonomy), for economic gains that were not expected to be very great. It has often been argued (Centre for European Policy Studies, 1989) that many of those gains could be obtained through politically less demanding means. Finally, the Commission was forced to recognize defeat, and to put forward new proposals which were close to the lowest common denominator. This time, the response of the Council of Ministers was positive. The crucial element of the political compromise was the idea that the elimination of fiscal frontiers, a crucial part of the 1992 process, could happen without the abolition of the destination principle and the harmonization of rates. This could be achieved by shifting fiscal controls to internal tax offices, partly by adopting the system already operated by the Benelux countries. This will also mean passing some of the administrative burden on to traders.

The new system should apply from 1 January 1993, when all fiscal frontiers are expected to be abolished. It is supposed to be a transitional system of four years leading eventually to the abandonment of the destination principle. However, transitional arrangements can last longer than originally planned, if national differences persist. A political agreement has also been reached on a minimum standard rate of VAT of 15 per cent, with a list of exceptions for reduced rates. However, until the time of writing, the UK had not accepted the necessity of a legally binding decision along those lines, arguing that the desired harmonization could be achieved through market forces. Some convergence of VAT rates has, indeed, taken place. But this is likely to be a slow process; and as long as there is no legally binding agreement on minimum rates, this convergence can only be downwards. In the meantime, the elimination of fiscal frontiers will require several special provisions in order to prevent serious trade distortions. They will have to apply to non-VAT registered agents (such as hospitals, local authorities, and small businesses), to mail orders across borders and car sales. Cross-border shopping will also present problems, once all frontier controls have been eliminated. For obvious geographical reasons, these will be felt much more by France, Germany, and the Benelux countries than by Greece, Britain, and Ireland (with the exception of the frontier regions inside Ireland). This, perhaps, explains, together with ideological preferences, the detached attitude of the British as regards tax harmonization from above.

The viability of the new system will ultimately depend on the ease with which fiscal controls are shifted from the border to internal tax offices, the burden of the new administrative procedures, and the vulnerability of the new system to fraud and tax evasion. Serious concern has been widely expressed, especially with respect to the latter.

For goods subject to excise taxes, a new system of interconnected bonded warehouses in the different member countries is intended to reconcile the objective of no frontier controls with the survival of the destination principle and different rates. As with VAT rates, limited progress has been made so far on the harmonization front.

Back in 1975, the Commission had aimed to harmonize corporate taxes, albeit with no success. Trying to make virtue out of necessity, this aim was later abandoned and the Commission emphasized the principle of subsidiarity, pointing to the need to leave as much flexibility as possible to member countries in determining their corporate tax systems (Montagnier, 1991). However, the growing mobility of capital imposes greater constraints on national taxation policies, leading to some convergence, again downwards. The signs of such a downward trend have been there for some time (supply-side policies have also added to this trend). Corporation taxation has been regularly used as a means of influencing the location of investment; and competition among national authorities will become stronger as Europe moves closer to a single market. This was recognised by a group of experts, under Mr Ruding, a former Minister of Finance of the Netherlands, which called again in March 1992 for the adoption of a band for corporate taxes (30–40 per cent) in order to avoid distortions and self-defeating competition among national tax authorities. It remains to be seen whether any serious progress can be made in this area, contrary to past experience. A first sign of some readiness by national governments to co-operate in the field of corporate taxation had been given in June 1990 with the adoption of three directives designed to smooth out tax problems faced by companies operating in more than one country. Those directives had been gathering dust in the Council's drawers for more than twenty years.

The decision to liberalize capital movements has brought to the fore the link between liberalization and the harmonization of taxes on income from interest and dividends. But here again, the belated recognition of the link has proved very difficult to translate into effective action. Under strong pressure from France, a country with high tax rates on savings and which, understandably, feared that capital liberalization would lead to a massive relocation of savings and financial activity in search of fiscal paradises, also supported on this issue by Belgium, Italy, and Spain, the Commission proposed a minimum withholding tax of 15 per cent on all investment income (Lebègue, 1988; Gros, 1990). This proposal did, however, meet with strong opposition from Britain and Luxembourg which feared from their side that the imposition of such a tax would lead to the loss of international business to other financial centres outside the EC. After all, the decision to liberalize capital movements would apply *erga omnes* and, therefore, there would be nothing to prevent EC residents from shifting to more attractive locations from the point of view of tax. The spectre of Switzerland and the various small fiscal paradises in Europe (Liechtenstein, Andorra, and the Channel Islands) loomed large in the negotiations.

The experience of West Germany in 1989 left few doubts about the sensitivity of capital flows to differential tax treatment (Bisignano, 1990). Given the porous nature of EC frontiers towards the rest of the world in terms of capital movements, the UK and Luxembourg have been basically asking their partners in the EC to accept the constraints imposed on fiscal sovereignty by free capital movements. They were certainly not prepared to offer any concessions with respect to the latter.

The precipitate abandonment of the withholding tax in West

Germany in 1989, forced by large capital outflows, gave a deadly blow to EC attempts at harmonization of taxes on interest. As a second best, some countries then tried to secure close co-operation among national fiscal administrations in order to deal with the problem of tax evasion and tax avoidance after the dismantling of capital controls. To make this co-operation effective, there would be a need for the lifting of bank secrecy and blocking laws in various countries (Giovannini, 1989).[6] But this, again, was resisted, especially by Luxembourg which rushed to strengthen its own bank secrecy laws. Thus capital liberalization and the first steps towards the creation of a single European financial area are likely to take place without any serious progress on the tax front. The deadlock reached in the EC eventually led Belgium to appeal to the IMF for an international agreement on withholding taxes. It was an act of despair by a country with a large public debt overhang, but also a recognition of the enormous difficulties encountered in setting up an EC 'regime' against the international trend.

It has been argued that the 'liberalization of capital movements stands to turn tax avoidance into a cottage industry' (Giovannini, 1989: 345). Capital mobility will increase further as the EC moves closer towards monetary union and the elimination of the exchange risk. This may prove to be yet another example of functional spillover. Continued failure to agree on tax harmonization and the means to achieve an effective co-operation among national fiscal authorities is bound to lead to a market-induced harmonization of taxes on interest (and capital more generally?) towards zero. The conclusion is exactly the same as with other areas of taxation.

The unanimity requirement in the Council, which did not change with Maastricht, persisting differences in national taxation systems, the importance of the interests at stake, and international pressures have all acted as important constraints on the Commission's tax harmonization efforts until now. Referring

---

[6] Giovannini has argued in favour of the adoption of the so-called worldwide principle of taxation which means that domestic residents are taxed on all their investment income, irrespective of the country where the investments are located. This would be intended to deal with the problem of tax evasion and avoidance in a world of free capital movements. But Giovannini's proposal assumes an effective co-operation among national fiscal authorities, which in turn depends on mutual confidence and the lifting of bank secrecy laws. These assumptions do not appear as yet realistic in the European, not to mention the international, context.

to indirect tax rates, Prest (1983: 82) had argued that to expect to achieve an alignment of tax rates 'this side of eternity' would be like 'crying for the moon'. Similar comments have been made in the past about other areas of economic integration, which were subsequently proved wrong. Although, undeniably, considerable progress has been registered in the context of the internal market programme, it seems that national resistance to harmonization and integration on fiscal matters will survive longer than in many other areas. Strict harmonization may, in fact, be unnecessary; what will be needed in many cases is an agreement on minimum rates of taxation as long as they are not always based on the lowest common denominator.

Failure to co-ordinate at the policy level, implying some pooling of fiscal sovereignty, will, inevitably, lead to forced harmonization through market forces. The higher the mobility of goods and factors of production, the higher the pressure for tax rates to converge downwards, and in some cases towards zero. Short-term capital movements are the most prominent example. But this will also mean the financial weakening of the State, with negative implications both in terms of the provision of public goods and income redistribution. Can this be politically acceptable in the long run? On the other hand, the almost exclusive preoccupation so far of the Community with resource allocation considerations with respect to tax issues may be seen as an indication of a certain bias of the integration process, especially in view of the still marginal role of EC taxation (see also Chapter 8). After all, stabilization and equity concerns should be at least equally important.

The Race between National and European Standards

Different technical regulations and standards were identified in the White Paper as one of the most important technical barriers which had to be eliminated in the context of the internal market programme. According to Mattera (1988: 270), they represented approximately 80 per cent of all remaining barriers in intra-EC trade. Their importance was also recognized in the business survey included in the Cecchini report where different technical regulations and standards ranked high among different barriers in the answers given by European firms (Cecchini, 1988). In fact, according to those answers, a clear distinction could be drawn between northern and southern countries in the Community, with representatives of the south (Greece, Italy, Portugal, and Spain) expressing less concern about the negative effects of this technical barrier on trade. The nature of production in those countries and the relatively underdeveloped state of market regulation can largely explain this difference in business attitudes. Technical regulations and standards are after all a feature *par excellence* of the most advanced mixed economies.

While technical regulations are legally binding rules dealing with the health and safety of workers and consumers as well as the protection of the environment, standards are of a voluntary nature. They are written by national standardization bodies and are usually an indicator of quality or a means of providing information to consumers and ensuring compatibility between different products or systems. In practice, the distinction between mandatory and voluntary is not always clear. Conformity with 'voluntary' standards is often a necessary pre-condition for market entry due to the demands of insurance companies, the need for compatibility with other products or standards specified in public procurement contracts. Conformity assessment, which constitutes the third stage in terms of market regulation, involves tests, inspections, and certification to assess compliance to regulations and standards.

Following the elimination of virtually all tariffs and quantitative restrictions in intra-EC trade by 1968, the Commission and the Council turned their attention to technical regulations and standards as a powerful NTB. Articles 30, 36, and 100 of the Treaty of Rome provided the legal basis for dealing with this problem. Article 30 prohibited quantitative restrictions on imports and 'all measures having equivalent effect', while Article 36 followed with a list of exceptions and the implicit recognition that health and safety provisions, together with the protection of the environment, are ultimately the responsibility of national governments. In an attempt to bridge the gap, Article 100 of the treaty provided for the approximation of national laws and regulations aiming at the elimination of obstacles to the proper functioning of the common market.

The European Court of Justice later gave a broad interpretation to Article 30 in an attempt to erect an effective barrier against the abuse by national authorities of health and environmental considerations for the protection of domestic producers against foreign competition. In the 1974 *Dassonville* ruling, the Court of Justice defined 'measures having equivalent effect' as 'all trading rules enacted by Member States which are capable of hindering, directly or indirectly, actually or potentially, intra-Community trade'. Another landmark ruling was the famous *Cassis de Dijon* decision of 1979 (who would have thought that a court ruling on a relatively unknown liqueur would have revolutionized intra-European trade?). It was followed by similar decisions against the German purity law for beer and its Italian equivalent for pasta (Mattera, 1988, Pelkmans, 1990).

But the problem created by technical regulations and standards could not be completely eliminated by a wide interpretation of Article 30. Derogations still allowed member countries a relatively wide margin of manœuvre. The next line of defence was, therefore, the harmonization efforts undertaken under Article 100. For many years, the results of this collective effort remained mediocre, at the very best (Dashwood, 1983; Pelkmans, 1987). Bureaucratic resistance, the unanimity requirement in the Council, the excessive emphasis on uniformity, and the technical and complex nature of legislation were the main factors behind the failure of the 'old approach'. There was a big discrepancy between time and effort invested on the one hand, and the amount of directives adopted at the EC level. It took, for example, eleven years before the Council could agree on a directive for mineral water. Furthermore, given the slowness of the decision-making process and the pace of technological change, many of those directives were out of date already by the time they were published. There was also no effective link established between the harmonization of technical regulations at the Community level and the production of new standards by private bodies. Meanwhile, national governments continued on their production spree of new regulations, thus constantly adding to the large number of NTBs in this area.

Frustratingly slow progress in terms of harmonization finally led to the development of a new policy, usually referred to as the 'new approach'. The first step was taken with the Mutual Information Directive of 1983 which called for the notification of draft regulations and standards by governments and private standardization bodies before their enactment. The aim of this directive was to strengthen preventive action against the erection of new barriers, also allowing for a standstill of one year at the request of the Commission, and the promotion of European regulations and standards in priority areas.

The essence of the 'new approach' can be found in a Council resolution adopted one month before the publication of the White Paper in June 1985. There were four main elements:

1. harmonization should be limited to essential safety requirements;
2. the task of drawing up technical specifications in relation to the essential safety requirements established by the Council should be left to European standardization organizations ('reference to standards');
3. the new European standards should be voluntary;
4. but governments would be obliged to presume that products manufactured according to those standards are in conformity with the essential requirements set out in the relevant directives; if producers chose not to manufacture according to European standards, they would need a certificate of conformity from designated bodies.

The 'new approach' was confirmed in the White Paper. The aim was clearly to unblock the legislative process in the Community by transferring the responsibility for technical details to specialized private organizations, while at the same time establishing a direct link (and a division of labour) between legislators and standardizers. Presumably, a more cynical interpretation could be that politicians were passing the buck to the private sector.

The 'new approach' relies heavily on the principle of mutual recognition which means that any good legally manufactured and marketed in one member country should be allowed to circulate freely in the rest of the Community (the *Cassis de Dijon* decision). Any attempt to restrict free movement should be properly justified with reference to health and safety provisions and also be proportionate to the objective. Only when obstacles created as a result of different national regulations and standards proved insuperable would the process of harmonization be set in motion. Thus the emphasis is on mutual recognition and not harmonization, although the line of demarcation between the two will always remain blurred and subject, as a last resort, to adjudication by the European Court.

The next important step was taken with the new Article 100*a* of the Single European Act, which replaced unanimity by qualified majority voting. As a counterbalance and in response to the fears expressed about a possible erosion of standards, the new article referred to the need for 'high levels of protection' while also offering a window of escape in the form of a safeguard clause. To complete the picture, new articles in the Single European Act covered health and safety in the working environment and created the legal basis for an EC environmental policy.

Unlike banking and financial services in general, there are no master directives setting out the new regulatory framework in this area. Regulatory provisions are inevitably more specific, varying from one sector to the other, thus rendering more difficult any attempt at generalization. The technical nature of the subject is an additional barrier to entry for the non-initiated. To quote an expert in this field, 'the empirical economics of (the removal of) technical barriers are largely still uncharted territory' (Pelkmans, 1990: 108). It is too early to pass final judgement, and the following lines are intended as a provisional and rather fragmented balance sheet of the 'new approach' as a means of eliminating technical barriers in the single market.

The Mutual Information Directive has acted as an important restraining factor on the production of mutually incompatible technical regulations and standards. Notification and consultation of other national authorities and the Commission and the provision for standstill have led to frequent amendments as well as the withdrawal of draft regulations and standards. On the other hand, the 'new approach' and the more favourable environment created by the 1992 process have led to a real breakthrough in terms of European legislation, although serious delays have been recorded in the transposition of EC directives into national legislation. The breakthrough has been true of several proposals requiring 'old approach' harmonization included in the White Paper.

Major examples of EC directives in the field of harmonization are those dealing with pressure vessels, toys, building products, and machines. Compared with previous efforts, the progress achieved has been quite remarkable. Serious progress has also been registered with respect to 'horizontal' directives in the food sector, dealing with labelling and additives. At the same time, a more rigorous application has been made of the principle of mutual recognition, thus opening more widely the national gates in intra-EC trade.

Criticism of the 'new approach' has concentrated on three main issues: the danger of excessive deregulation and erosion of standards, the slowness of European standardization, and the even greater slowness of tackling the problem of certification and testing. Since EC directives deal only with 'essential requirements', there is the risk that European legal provisions will be limited to generalities and some rather anodyne statements based on the lowest common denominator. For example, the machine safety directive, adopted in 1989, packed into seventeen pages almost half of the engineering sector. In order to have a real effect, this directive will need to be complemented by several hundred European standards. This is the task entrusted to private organizations like CEN and CENELEC, the membership of which consists of the national standardization bodies of EC and EFTA countries. The rules of decision-making of those organizations have been changed as a result of the new 'reference to standards' approach adopted by the EC and the need to accelerate the production of new standards at the European level. Qualified majority voting has thus been adopted along the lines of the EC Council of Ministers. Standards organizations of EC countries carry the same weights of votes as in the Council of Ministers, with corresponding weights given to EFTA members (EC countries have seventy-six out of a total of ninety-six votes). Provisions have also been made for separate voting to take place among the representatives of EC countries, when qualified majority cannot be obtained for the membership as a whole (Woolcock, Hodges and Schreiber, 1991, Ch. 4).

The Community dominates those organizations both in terms of finance and in terms of the standardization process which largely follows the mandates given by the EC Commission in relation to directives adopted in Brussels. Given the small size of the secretariats of CEN and CENELEC, much of the work has been in fact in the hands of the organizations representing the large EC countries, such as the Deutsches Institut für Normung (DIN), the Association Française de Normalisation (AFNOR), and the British Standards Institution (BSI). Concern has been expressed with regard to the overpowering presence of producers in the European organizations, which could lead to inadequate provisions for the protection of consumers and the environment.

On the other hand, the relative weakness of the Commission, in terms of administrative capacity and expertise, does not enable this body to act as an effective guardian of the 'public interest' in this respect.

The production of European standards has expanded enormously in recent years. Yet, given the nature of decision-making and the severe financial limitations of the relevant organizations, the production of national standards in the large EC countries is still a multiple number of the corresponding production of CEN and CENELEC. The latter will certainly not meet the deadline of 1993 by completing the work of standardization in relation to directives adopted in Brussels. This has been already recognized by the Commission which has called for a radical overhaul of the structures and practices of European standardization through the creation of specialized sectoral bodies to replace the existing umbrella organization. In the meantime, the EC will still have to rely much on the principle of mutual recognition. National organizations have, not unexpectedly, resisted a rapid transfer of powers to the European level. Bureaucratic interests, coupled sometimes with the belief in the superiority of national standards, have worked in favour of the *status quo*. There is also a problem of implementation, since European standards still need to be translated to national standards in order to become effective.

European organizations are, in fact, squeezed between national and international pressures. In some sectors, such as telecommunications and information technology, the rapid globalization of production could make more and more nonsense of attempts to agree on specifically European standards, despite the renewed efforts made in this area. The European Telecommunications Standards Institute (ETSI) was set up in 1988 to draw up several thousand standards in this sector. Barry (1990: 115) argues that: 'European standards are often simply international standards redeployed at a European level.' But standards can also serve as an instrument of industrial policy. The race between the Japanese and the Europeans to have their own standard adopted for high-definition television (HDTV) is a good example.

On the other hand, progress in terms of certification and testing has been even slower. The 'new approach' initially overlooked the need for common conformity assessment procedures. In the absence of common rules for conformity assessment, product safety inspectors in the importing country may have reservations about the reliability of tests performed in the exporting country. They may, therefore, impose a new set of conformity assessment controls, even though the imported good has been manufactured in accordance with the 'essential requirements' of EC legislation and it has already satisfied one series of tests. At the heart of the problem is mutual suspicion about the professionalism and objectivity of testing and certification bodies in partner countries. In July 1989, the Council adopted the 'Global Approach to Testing and Certification' which aimed at defining a single set of conformity assessment procedures to be applied throughout the EC, while also laying down the technical criteria to be met by the relevant bodies. This was followed by the creation of the European Organization for Testing and Certification. Building confidence in partner country standards and testing criteria as well as the establishment of a European accreditation system for testing and certification bodies is likely to prove a long and difficult process. The other side of the coin is the effective protection of the consumers (Goyens, 1992).

The 'new approach', combined with increased cross-border collaboration between firms at the level of production, qualified majority-voting in the Council, and the 'Euro-euphoria' of recent years, have created the foundations for the development of a European doctrine in terms of product safety and the adoption of common European standards. We are still in the early stages of this process which, although rarely reaching the political lime-light, is a key factor in the completion of the internal market. Technical details concerning the dimension of tubes relate to wider political questions regarding the optimum degree of regulation, the trade-off between *de jure* and *de facto* regulation, or between diversity and standardization. Different political traditions and the economic heterogeneity of the EC of Twelve, not to mention important interests at stake, explain why uniform answers to those questions are not easily forthcoming. In fact, German or Danish fears about the possible erosion of consumer safety and environmental protection standards coexist with the understandable apprehension of some of the less developed countries of the Community about the economic costs associated with too high standards, not to mention the implications in terms of certification and testing laboratories. This heterogeneity also points to the limitations of the mutual recognition principle, unless it is believed that the market will find the optimum level of regulation and standardization. Not surprisingly, attitudes differ in this respect.

Joerges (1988: 199) argues that the 'regulatory dilemma of a Community which has to resolve federal questions but is not a Federal State has repeatedly led . . . to attempts to get round the difficulties of substantive decision-making at European level'. In the case of technical regulations and standards, this attempt has been part of the more general strategy, relying on mutual recognition and the harmonization of essential requirements, coupled with the transfer of major powers to European standardization bodies. In terms of eliminating the large number of remaining technical barriers, this strategy has met so far with only partial success. As for the possible erosion of standards, the jury is still out, although the judgement is unlikely to be the same for Denmark and Greece.

6. Social Policies and Labour-Markets

One market where state intervention has always been extensive and where conditions in general bear little resemblance to the standard perfect competition model of traditional economic theory is the labour market. Regulations regarding employment and working conditions, remuneration and social security payments have been standard features of the post-war mixed economy and the welfare state in Western Europe. On the other hand, wages and salaries have been largely determined by power relations between employers and organized labour, power relations which do not necessarily reflect supply and demand conditions in the labour market.

The term 'social policy' covers a much wider area, the boundaries of which are usually, and perhaps deliberately, left undefined. In addition to different forms of regulation of labour-markets, intended to guarantee a certain minimum of conditions for the protection of employees, social policy also includes education and training, housing and health. There is, of course, hardly any agreement as to where government intervention should begin and where it should end. This subject is at the very centre of the ideological divide between right and left. Concern with equity and industrial democracy is meshed with the pursuit of the objective of economic efficiency. For some, all three are complementary, while others talk in terms of a trade-off which needs to be decided at the political level. Political views differ substantially, and so do the historical experiences of individual European countries. However, despite those intra-European differences, which are still significant, reference is frequently made to a 'European model' which is contrasted with US and Japanese practices (Emerson, 1988; Aubry 1989). What divides most

Western European countries in terms of social policies and the regulation of labour-markets appears to be infinitely less important than the wide gap separating them from other advanced industrialized democracies, and especially the United States.

The early stages of European integration coincided with the rapid expansion of the role of the State in the social policy field, aiming mainly at some internal redistribution and the reduction of what Boltho (1982: 2) calls 'microeconomic risk' (individuals' loss of income arising from unemployment, accident, illness, or old age). Social policies were aimed at the incorporation of the working class in the political and economic system and the achievement of a wider consensus for the growth policies of the 1950s and 1960s. The progressive strengthening of trade-union power and the more general acceptance of social democratic values were the logical consequence of this implicit package deal and the long period of prosperity and high levels of employment. With the advent of economic recession, however, the political and social scene changed dramatically. Attention rapidly turned to labour-market rigidities and the excesses of the welfare state. 'Euro–sclerosis' then became the fashionable term and the launching board for supply-side economics and the deregulatory wave of the 1980s; and this was coupled with the steady weakening of trade-union power and the partial reversal of earlier trends which had led to a considerably higher share of wages and salaries.

The relaunching of the integration process in the mid-1980s was closely related to the new political trend prevailing in Western Europe. The desire of big business for the elimination of the remaining barriers and the creation of a large European market was coupled with a gradual shift in the political balance of power, from which organized labour was the net loser. The policies which had laid the foundations of the post-war economic miracle in Western Europe were now perceived as a heavy liability undermining the prospects of economic growth. The emphasis was on more competition and the market, not on regulation and redistribution. And this was particularly true of the labour-market. The internal market programme was born in this political context. Social policies hardly figured at all in the early stages of the 1992 process. But this did not last for long. Pressure soon started to build up for the creation of a European social space as a necessary complement to the establishment of the internal market, and social policy became later one of the most controversial issues of the intergovernmental conferences which led to the treaty revision agreed at Maastricht in December 1991. This was further evidence of the continuous expansion of the European agenda. But it could also be interpreted as a sign of the resurgence of those political and social forces which had earlier laid the foundations for the development of the mixed economy and the welfare state in individual European countries.

Until now, the role of European institutions in the field of social policy has been marginal, with the nation-state acting as a jealous and effective guardian of its powers and competences in this area. It remains to be seen whether this division of power will change much in the foreseeable future, although the odds are against it. On the other hand, the opening of economic frontiers has an impact on the functioning of national labour-markets and the effectiveness of social policies. Relations between capital and labour, their respective negotiating power, and the distribution of the pie will all be affected by the division of powers between central and national institutions. This chapter will concentrate on the European dimension of social policies and the regulation of labour-markets as main themes of the debate on the new European economic order.

Free Movement and Little Mobility

The Treaty of Rome was sprinkled with references to several aspects of social action, although a distinction can be drawn between provisions of more or less binding nature and statements of general intent. Articles 48–51 referred to the free movement of workers, and Articles 52–8 to the freedom of establishment. These two freedoms are an integral part of the customs union, which means that the provisions made in them were of a constraining nature. Articles 117–128 referred specifically to social policy. They included references to the improvement of working conditions, 'equal pay for equal work' between men and women, and paid holidays. There were more specific provisions for the role of the European Social Fund, the main task of which was defined as 'rendering the employment of workers easier and of increasing their geographical and occupational mobility' (Article 123). The ESF was expected to meet 50 per cent of the expenditure of approved programmes for vocational training, the resettlement of workers as well as for aid to workers who have been laid off. The origins of such action were to be found in the ECSC Treaty which had made provisions for aid to workers in the coal and steel industries.

The list was long, but it was clear that, with the exception of provisions for the free movement of workers and the freedom of establishment, the authors of the Treaty of Rome did not envisage a major role for European institutions in the field of social policy. The latter remained the preserved domain of national governments. France had tried hard to secure a commitment to the harmonization of social security payments, because of fears that high social charges would prejudice the competitiveness of French products in the customs union; but the other signatories to the treaty had been adamant. Even the Social Fund was basically seen as an adjunct to the free movement of workers, both offered to Italy and the Mezzogiorno in particular as part of the overall package deal.

During the early years of the EEC, social action at the European level remained low key. The main emphasis was on the achievement of the free movement of labour which proved to be a formidable task in itself. It was not only a question of abolishing frontier controls, restrictive residence permits, and numerous other administrative obstacles for workers and their families. Free movement of labour in the context of mixed economies depends on a much wider set of conditions which need to be fulfilled, including the transferability of social security payments, the mutual recognition of degrees and professional qualifications, and the dissemination of information about jobs through a properly functioning labour exchange. Not surprisingly, progress in many of those areas has proved slow, especially with respect to barriers arising from different national regulations as opposed to physical controls. Several remaining NTBs in this area now form part of the internal market programme. Pursuing the trade analogy further, it should be added that the Community has always been close to a free trade area in the labour field, since there has not been until now a common policy on labour migration from third countries; although this may change after Maastricht.

Free movement of labour was achieved in 1968. It was based on the principle of national treatment for workers from other EEC countries in terms of occupational and social benefits.

Given the nature of remaining obstacles, this freedom was more relevant to young, unskilled workers rather than older professionals. Labour migration continued to play a significant role in economic growth in Western Europe during the 1960s. It relieved labour shortages in the northern countries and provided additional flexibility in tight labour-markets, while at the same time it acted as a safety valve for countries with abundant labour and as a source of much needed foreign exchange through migrant remittances back to the home country. True, it was not an unqualified blessing for either side, and this became gradually more obvious as social tension grew in the host countries with the steady increase of foreign 'guest-workers' who came to stay, while the exporting countries became more aware of the social and economic costs arising from the loss of many of the young and most dynamic members of their labour force. What is very interesting, however, is that the progressive liberalization of intra-EC labour movements coincided with the steadily diminishing role of these movements as part of total labour migration in Western Europe. The following statistic is quite revealing: intra-EC labour flows represented for the original six members of the Community approximately 60 per cent of total flows in the beginning of the 1960s (Italian workers accounting for a large part of them); the corresponding figure was only 20 per cent one decade later (Straubhaar, 1988).

The transitional period envisaged by the Treaty of Rome coincided with the gradual drying up of the labour pool in Italy. Thus the labour shortages experienced in some of the rapidly growing economies of north-western Europe were filled with labour immigrants from other southern European countries. Greece, Portugal, and Spain became the main sources of labour export, and later Yugoslavia, Turkey, and North Africa were added to the list. In more recent years, significant numbers of (political and economic) refugees from Asia and increasingly from Eastern Europe have provided new additions to the number of foreign workers in the EC; and they could multiply further in the near future, as the transition to market economies in the Eastern European countries proves much more painful than originally expected, while demographic factors and economic stagnation produce an explosive mixture in North Africa. Stronger pressure from the East and the South has coincided with a steady return flow of early migrants from southern Europe. By the time

Greece, Portugal, and Spain joined the Community in the 1980s, net emigration from those countries had already come to a halt. Thus, intra-EC labour movements have always remained small.

The accuracy of statistics on foreign workers and foreign residents leaves much to be desired. Work permits for EC nationals have been abolished, and there has also been a relaxation of rules regarding foreign residence permits. This, therefore, makes it difficult to trace labour movements inside the Community. On the other hand, one important feature of labour migration since the early 1980s has been, precisely, its clandestine nature, and this is particularly true of migration into previously labour-exporting countries in southern Europe from the other side of the Mediterranean. It is estimated that the total number of foreign workers in the EC of Twelve is now approximately 4 million, with EC nationals accounting for less than half of foreign workers (Commission of the EC data). Turkish migrants constitute by far the largest group. The number of foreign residents is estimated at 12.5 million, out of whom 4.5 million are from other EC countries. The bulk are to be found in the Federal Republic. As a percentage of total population (342 million in EC-12, including the five new German Länder), the range is very wide: from 26 per cent in Luxembourg, 9 per cent in Belgium and 7 per cent in the Federal Republic (the data are for West Germany), to less than 1 per cent in Portugal (Romero, 1990; OECD, 1987).

With the exception of Ireland, where net emigration has again reached sizeable proportions in recent years, most of it directed towards Britain, all other EC countries have become net importers of labour, with southern Europe acting as the gate for mostly illegal flows from the other side of Mediterranean and also Asia. It remains to be seen whether the lifting of all restrictions on labour movements from the two Iberian countries in 1993, at the end of the transitional period following accession, will have any significant effect on labour movements into the rest of the EC. Previous experience seems to suggest that economic conditions will be more important than any formal restrictions or lack thereof.

Economic 'pull' and 'push' factors, depending on macro-economic conditions and labour-market imbalances, have been much more important in determining labour movements than any changes in formal or other controls. Rapid economic growth and the raising of the standards of living in Italy during the 1950s and 1960s had a dampening effect on the 'push' factors in a traditional labour exporter; the same became later increasingly true for some of the other southern European countries. On the other hand, the deterioration of the macroeconomic scene in the mid-1970s and the ensuing prolonged recession, combined with growing social tensions in the host countries, especially as regards the non-European and non-Christian immigrant populations, had a strong negative effect on the 'pull' factors. Special schemes were introduced in some countries for the repatriation of foreign workers. In the Federal Republic, such schemes led to a substantial reduction of the foreign workforce during the first half of the 1980s. The coexistence of high rates of unemployment and large numbers of immigrant workers in several EC countries nowadays can be explained largely in terms of earlier flows and the nature of jobs performed by most foreign workers. These are mostly low-level, unskilled jobs, and also, usually, unregistered; which means low wages, no social contributions, and no union rights. Illegal immigration into Europe has strengthened the dichotomy between official and unofficial labour-markets.

In fact, it could be argued that the *acquis communautaire* may have had a negative effect on intra-EC movements of labour to the extent that one important advantage of foreign labour from the point of view of employers has been the flexibility and the lower costs associated with the non-application of labour-market regulations. Workers from other member countries enjoy the same protection (social security, trade-union rights, etc.) as nationals, thus making them less attractive to many employers. At the same time, free circulation inside the EC has influenced the nature of intra-EC flows, therefore encouraging short-term employment and also, increasingly, the movement of professionals. This will become even more true after the elimination of the remaining NTBs in this area.

In the meantime, European social policy had been given a new push in the early 1970s, and especially with the social action programme adopted in 1974. The latter referred to the goal of full and better employment, the improvement of living and working conditions, the closer participation of social partners in EC decision-making, and the involvement of workers in management decisions. But, in the end, it proved difficult to translate good intentions into concrete action. The rapid growth of unemployment, combined with the shortage of EC funds and the opposition of most national governments to the transfer of real powers to Brussels, acted as major constraints on the development of EC social policy. This situation continued for many years, with small fragments of EC action here and there as testimony of the Community's subdued role in the social policy field. Collins (1985: 281) argued that European social policy comprised 'a miscellany rather than a social policy informed by a few compelling themes'; and Vogel-Polsky (1989: 179) went even further, talking about a negative consensus in this area: 'le consensus de ne pas faire de politique sociale européenne.'

The Social Fund provided the main policy instrument in this field. Expenditure through the ESF grew considerably over the years (Table 6.1). With the rapid rise in the number of jobless in the 1970s, the emphasis was bound to shift from problems associated with sectoral adjustment in the context of integration and the encouragement of geographical mobility towards measures aimed at the more general problem of unemployment. The ESF concentrated efforts on long-term and youth unemployment, relying mainly on vocational training, while expenditure took increasingly a strong regional dimension. This was confirmed by successive reforms of the ESF in 1971, 1977 and 1983. Absolute priority was accorded to less developed regions, although the regional bias of ESF expenditure remained, until the reform of the structural funds in 1988, less pronounced than that of the ERDF (see also Chapter 8). One reason was, presumably, that the correlation between levels of income and rates of unemployment was far from perfect. On the other hand, labour-training schemes and the ability of national administrations to submit credible proposals to the ESF varied enormously from country to country. The UK, Italy, Ireland (in relative terms), and now also Spain and Greece stand out as major beneficiaries of expenditure through the ESF (Table 6.1).

Despite the substantial increase in ESF expenditure, the amounts of money spent remained small, when compared with corresponding expenditures at the national level. Until the reform of the Structural Funds of the EC in 1988, which will be discussed in more detail in Chapter 8, the Commission had enjoyed little flexibility in the selection of projects, often being restricted to a rubber-stamping role of decisions already taken in national capitals. It is true that its power to shape policy

Table 6.1. ESF Payments to Member Countries, 1980–1990

(1980: Mio EUA; 1981–1990: Mio ECU)

	1980	1981	1982	1983	1984	1985	1986	1987	1988	1989	1990
Belgium	12.1	15.3	16.8	20.6	52.1	49.4	72.9	56.5	32.0	32.8	51.8
Denmark	14.7	18.5	17.6	14.7	68.7	33.5	80.7	31.7	34.3	19.1	38.3
Germany	80.5	72.3	89.9	81.5	63.8	109.8	134.6	131.6	147.1	151.7	186.4
Greece	–	6.6	23.5	20.4	71.3	79.0	107.0	151.9	147.9	217.5	303.2
Spain	–	–	–	–	–	–	174.9	311.5	407.1	469.8	633.9
France	195.8	155.3	119.3	140.5	225.7	255.6	328.4	406.1	292.1	327.7	442.9
Ireland	72.5	60.4	115.0	134.2	131.4	171.6	203.1	247.4	179.6	189.5	204.4
Italy *	194.4	207.1	235.1	221.2	368.5	383.5	462.2	539.2	329.7	457.0	419.5
Luxembourg	0.4	0.6	1.1	0.3	0.5	0.6	1.4	1.7	1.2	1.6	3.4
Netherlands	5.1	14.3	9.0	12.6	14.1	46.0	50.6	52.1	46.5	56.8	68.8
Portugal	–	–	–	–	–	–	109.2	190.5	202.4	215.7	69.5
United Kingdom	159.7	195.4	278.3	244.9	610.2	284.0	596.2	595.1	478.9	536.9	608.3
Total EC	735.2	745.8	905.6	890.9	1606.3	1413.0	2321.3	2715.3	2298.8	2676.1	3030.4

Note: Annual Payments includes payments against appropriations for the current year plus payments against carry-overs from the previous year.
Source: Court of Auditors Annual Reports.

increased over the years, but the starting-point had been very low indeed. On the other hand, there was little evidence to suggest that the money spent by the ESF was in addition to social expenditure which would have been undertaken by national governments in its absence. This refers to the so-called problem of additionality. Like the Regional Fund, the ESF was generally seen by governments as a means of redistribution across national frontiers and not as the instrument of a common social policy (Collins, 1983).

There were also some bits and pieces of European social legislation, which did not, however, change the overall picture. For example, the Community had played a significant role in the legal enforcement of equal pay between men and women referred to in Article 119 of the treaty. It had been mainly the role of a catalyst, especially after the adoption of the three equality directives in the 1970s which gave the European Court of Justice the legal powers to push national administrations into taking effective action against discrimination between sexes in the labour-market. Certainly, discrimination had not completely disappeared as a result, but this was another example of the limits of legislation in changing long-established economic and social practices. There had also been some legislation under Article 117 regarding conditions of collective dismissals, although this had brought about only minor changes to corresponding national laws. It had been essentially an exercise of minimum standard setting. Another area of EC action was related to issues of industrial health and safety which had appeared, from the mid-1970s onwards, with increasing frequency on the agenda of Council of Ministers meetings; it was later given a further boost with the relevant provisions of the SEA.

Social Space: Heavy on Symbolism . . .

Social policy did not form part of the White Paper of 1985. Reference was only made to the remaining obstacles to the free circulation of workers in a programme which concentrated almost exclusively on negative integration measures. The sensitivity of national governments in the social policy field was confirmed once again with the SEA where the 'free movement of persons' and 'the rights and interests of employed persons' were two out

of the three exceptions to the qualified majority principle introduced by Article 100*a* for the achievement of the internal market.
It was clear that many national governments were not ready to
take the risk of being overruled in this area. With respect to
social policy, only measures related to the health and safety of
workers (Article 118*a*) were made subject to qualified majority.
Reference was also made to 'the dialogue between management
and labour at European level which could, if the two sides consider it desirable, lead to relations based on agreement' (Article
118*b*). Last but not least, economic and social cohesion became
an agreed objective of further integration, and this was, in fact,
to serve later as the launching board for further action in this
area.

Since then, interest has grown rapidly, with the Commission,
several member governments, and labour organizations calling
for a social dimension to the internal market and a European
social space. This had not been part of the original package deal,
and such calls have, not unexpectedly, led to strong opposition
from the Conservative government in the UK and the employees
organizations. The subject lends itself to ideological confrontation which has, indeed, characterized the European debate in
recent years. As a corollary to that, the political rhetoric used
by participants in this debate, generally addressing themselves
to constituents back home, is difficult to disentangle from the
actual policies pursued on specific issues. In fact, the discrepancy
between rhetoric and proposed action, or simply between the
words and deeds of the various actors in this game, has been
sometimes truly astounding.

We shall attempt to identify here the main themes of the
debate as well as the motives of the principal actors. If the free
market and competition have been the underlying principles of
the White Paper of 1985 and the internal market programme,
calls for the creation of a European social space have been based
on different ideological foundations. True, there is one aspect, at
least, in which the social space is directly linked to the establishment of the internal market, namely through the creation of
a truly European labour-market and the encouragement of
mobility as a means of economic adjustment. The elimination
of various obstacles to the free circulation of workers and the
adoption of positive measures to encourage it are fully consistent
with the logic of the 1992 process. But the main emphasis with

respect to the European social space lies elsewhere: it is related to concepts of equity, participation, and consensus which are not necessarily compatible with the 'magic of the market-place', to use a famous expression of President Reagan. They form part of a broader vision in which the social and the economic are inextricably linked; though a vision which is not shared by everybody.

Arguments for the creation of a European social space are, first of all, based on the belief that the market mechanism on its own will not ensure an equitable distribution of gains from the establishment of the internal market. It is both a question of solidarity for those who may suffer from the unleashing of the forces of competition and a form of insurance which may be aimed at the prevention of a political backlash from losers who might be tempted to rally behind the nationalist flag.[1] The elimination of the remaining barriers started taking place at a time when unemployment levels were uncomfortably high; despite the creation of many new jobs during the latter part of the 1980s, the average rate of unemployment for the EC-12 was still 9 per cent in 1990. Political concern was further strengthened by the prediction of a J-curve effect on unemployment in the Cecchini report. Unemployment was expected to rise in the first two years after the establishment of the internal market, before it began to fall. Would promises for a better life after death prove convincing to trade-union leaders and the large numbers of unemployed in the attempt to secure from them a political endorsement of the internal market programme? Furthermore, economic growth in the late 1980s coincided with the widening of income disparities in most countries, and an unequal distribution of the effects of the internal market among different social groups or classes would make matters worse.

Labour-markets being highly regulated in all European countries, although the degree and kind of regulation varies enormously from one to the other, there is also the fear that the elimination of the remaining barriers to the free movement of goods, services, persons, and capital will lead to stronger competition among national economic systems and to an excessive

---

[1] The compensatory and redistributive role of social policy was strongly emphasized in one of the early Commission papers on this subject, the so-called Marin report named after the Commissioner in charge of social policy at the time (Commission of the EC. 1988c).

deregulation of labour-markets at the expense of workers, in terms of both remuneration and working conditions. Repeated references have been made by politicians and trade union leaders to the danger of 'social dumping', a term which implies some form of unfair competition based on low wages and poor working conditions.

Labour costs, including direct wage and non-wage costs, differ widely between the various EC economies. According to EC Commission data for 1988, hourly industrial wages in Portugal were approximately one-sixth of those in the Netherlands (Commission of the EC, 1991*b*). The breakdown of total labour costs also differed significantly from one country to the other. However, labour cost differences reflect to a large extent different productivity levels. The relationship is, of course, far from perfect, allowing for the idiosyncrasies of national labour-markets based on historical and institutional factors as well as the domestic balance of power in each country. Thus, differences in terms of unit labour costs are much smaller than may be implied by comparisons of nominal wage costs.

The use of terms such as social dumping and unfair competition implies a deliberate underpricing by certain countries through a downward pressure on their labour costs in the search for external competitiveness and the attraction of foreign investment in their economy. Labour-intensive industries would be the obvious candidates for such a policy, since, for many technologically advanced industries, labour costs are only one, and certainly not the most important, factor determining the location of investment. For labour-intensive industries such as textiles and clothing, southern European countries would, however, have to compete with Tunisia, Morocco, or Turkey where wage costs are much lower, although the choice of location for foreign investors would depend not only on the combination of wages and productivity but also on the height of protective barriers around the EC. The counterpart to arguments about social dumping are the fears frequently expressed in the less developed economies of the EC about the tendency of geographical concentration of high technology industries and financial services in the more advanced countries, because of positive externalities (this subject will be discussed in more detail under regional policy). Each side, naturally, has its own fears and preoccupations.

Unfairness is virtually impossible to substantiate or even define in this case. Furthermore, it is hardly plausible that labour costs can be easily manipulated by governments, in countries with pluralist systems and organized trade unions, in order to obtain this alleged unfair advantage over their economic partners. This has been demonstrated time and time again through the unsuccessful attempts of several European governments to contain wage pressures. Of course, as long as productivity levels remain so different, any convergence in terms of labour costs would make absolutely no economic sense (Donges, 1989; Mayer, 1989). A movement in this direction would, in fact, ensure the worsening of the existing regional problems in the Community, especially if it were to coincide with an acceleration of the process towards EMU which would deprive countries of yet another instrument of economic adjustment, namely the exchange rate. In an EMU, the emphasis will be precisely on greater flexibility of the labour markets. The less developed economies of the EC could hardly afford an upward harmonization of their wages and social security rates which is what is often implied by arguments against social dumping.

This problem may appear to be a red herring, although there is something more substantial hidden in the above arguments. In the context of mixed economies, the distribution of the national pie between capital and labour is the joint product of market processes and power relations. With the progressive opening of economic frontiers, this distribution may be expected to shift against labour for the very simple reason that capital is more internationally mobile. In other words, with the elimination of remaining barriers inside the EC, certain kinds of investment could shift more easily towards locations with lower labour costs and more flexible working conditions, because of the prospect of unrestricted access to the whole European market. Hence, the appeal for some form of joint regulation to complement free trade. According to Rhodes (1992), it is mainly about creating a floor of basic rights. This argument is not very different from arguments for the harmonization of essential rules with respect to technical standards and the regulation of financial markets. It is, therefore, part of the political economy of liberalization and regulation. Can competition between national systems deliver the optimum degree of regulation, and who, after all, decides what is the optimum?

Calls for a European social space also relate to concepts of industrial democracy, participation, and consensus. They refer to the participation of workers in the running of enterprises, to social dialogue and the search for consensual forms of politics; that is, forms of collectivism and corporatism in the words of ideological opponents. But whatever the words used to describe them, those practices have marked the post-war experience of several Western European countries. Organized labour has so far played a limited role in the process of European integration, and this has also been true of the 1992 programme. In fact, a direct link has been sometimes established between the weakening of European trade unions during the previous decade, a function of the recession, high unemployment and structural changes in the economy, and the relaunching of European integration. The addition of a social dimension to the internal market programme would help to integrate more securely the working class into the process of European integration and also avoid the identification of the 1992 programme with the 'Europe of businessmen'.

Ideology is closely related to symbols; and symbols have been very important in the debate on the social space. The latter is meant to symbolize a certain European economic and social model; the modern version of Europe's mixed economy which had proved so successful in the 1950s and 1960s. This may in turn suggest that, despite the apparent ideological shift of the 1980s, the battle continues, or alternatively that symbols are now offered as a substitute for policies which are in the process of being diluted. On the other hand, the social space could help to bring about another reshuffling of cards between national and European institutions. Thus, once again, federalist ambitions are difficult to separate from the intrinsic merits of the proposed policy.

The above discussion of the main motives behind proposals for further European action in the social field suggests that political consensus would not be easily forthcoming. The main pressure in this direction has come from the more prosperous countries of the Community, with strong trade unions, high wage costs, and advanced labour legislation, among which the Federal Republic has played a prominent role. The Government has taken its cues from an intense domestic debate (the so-called *Standortdebatte*) about the risks of industrial investment shifting to the south of Europe in search of lower labour costs and more flexible labour legislations, thus risking an increase in unemployment in

Germany and the progressive erosion of the many social and other benefits enjoyed by German workers (Adams and Rekittke, 1989; Fels, 1988). These arguments have been forcefully put forward by Germany's powerful trade unions, while, from the other side, employers have frequently appealed to the likely effects of 1992 in support of their own calls for wage moderation and lower taxes.

The sensitivity shown by the German Government on this subject needs to be understood in the context of the continuous search for consensus among the social partners on all major economic issues and European integration in particular which has characterized German politics: social policies and the regulation of the labour-market also being among the keystones of the German 'social market economy'. The Nationale Europa-Konferenz was set up to discuss, on a regular basis, the impact of 1992 on the social partners, and this called for a Community-wide formulation of minimum social standards. On the other hand, the arguments used on this subject have not always been consistent with the liberal economic views generally expressed by German representatives in other areas of EC policy. But consistency is not, after all, one of the main features of the political debate in any country.

Support in the same direction has also been provided by the Socialist governments in France and some of the southern European countries, basically for ideological reasons while also serving as a means of satisfying their working-class constituencies back home. Governments in less developed economies have been, in fact, faced with a serious dilemma to the extent that the social dimension of the internal market would consist mainly of a harmonization upwards of wages, social provisions and working standards, without very much in terms of budgetary transfers across frontiers. If this process were taken too far, it would risk wiping out the only factor which could compensate for the large differences in productivity levels between members of the EC, unless the increase in labour costs were to be accompanied by similar increases in productivity, which is not usually the case. The way to reconcile political ideology with economic constraints has been sometimes to talk much and do little, which also partly explains the slow progress registered in this area.

At a general level, there is broad political consensus in Western Europe ranging from right to left of centre regarding welfare policies and employment protection legislation. Most Christian Democratic parties, with close links with the trade-union movement in their respective countries, form part of this consensus. One notable exception among the major political parties are the British Conservatives who have acted here again as champions of the opposition against the extension of EC action in the social field. The strong ideological dislike for measures that, according to Mrs Thatcher, smacked of 'social engineering', with socialist and Marxist connotations, has its source in the very nature of this party and in the Conservative interpretation of recent British history in which market rigidities and powerful trade unions are identified as the main culprits for the long period of economic decline. Ideological opposition has been reinforced by hostility to any further weakening of national sovereignty and the transfer of powers to European institutions in the field of social policy. The UK Government has found support for its stance on this issue among European business federations, and especially among its own businessmen.

The EC Commission has played a very active part in the promotion of the European social space. The desire to secure a wider political acceptance of the 1992 programme, through the active participation and endorsement of the programme by trade unions, has been combined with the ideological preferences of President Delors, who has played a major role in the promotion of social action at the EC level, and the federalist aspirations of the Brussels executive. Pressure from the European Parliament has also been in the same direction. The left of centre majority of MEPs (Socialists, Greens, and Communists) has acted as an important pressure group in an area where ideological preferences and potential votes have been conveniently married with the Parliament's interest in the further transfer of powers to the centre.

... And Light on Substance

The discussion about the creation of a European social space has often taken place without reference to its actual contents, this being a characteristic of a debate where words have counted more than deeds and where symbols have been sometimes more important than reality. In his address to the congress of the European Trade Union Confederation (ETUC) in Stockholm in May 1988, Mr Delors identified three main areas for action in an attempt to give the European social space a concrete meaning. These were the adoption of a European charter of fundamental social rights; legislation on a European company statute with provisions for the participation of workers in the management of firms; and the promotion of dialogue between representatives of business and labour at the European level. The EC Commission has delivered on all three promises, although the same is not yet true of all the other interested parties.

The idea of a social charter was first introduced by a Belgian Minister, Mr Hansenne, during his country's Presidency of the EC in the first half of 1987. Belgium had recently adopted a similar charter and presumably the idea was that if the charter was good for Belgium, why should it not be for Europe? An alternative explanation could be that if the adoption of the charter implied any economic cost for Belgium, it would be better if the same applied to the other EC partners in order to avoid any loss of competitiveness. The idea was taken up by the Commission, with strong support from France and Germany and the fierce opposition of the UK. Successive Presidencies from countries with Socialist governments (Greece, Spain, and France) actively promoted the charter, and this finally led to its adoption by the European Council in Strasbourg in December 1989. It took the form of a solemn political declaration from which the UK abstained; and this became an important precedent for subsequent action taken by the Eleven.

The Community Charter of the Fundamental Rights of Workers built on the existing Social Charter of the Council of Europe, the conventions of the International Labour Organisation (ILO) and the various UN instruments. It is a political declaration of a non-binding character, which contains twenty-six provisions under twelve general headings, including freedom of movement, employment and remuneration, living and working conditions, freedom of association and collective bargaining, training, information, consultation and participation, and equal rights for men and women. In an attempt to reconcile economic efficiency with the promotion of the rights of workers, it was stated in the preamble that 'social consensus contributes to the strengthening of the competitiveness of undertakings', thus aptly summarizing the fundamental approach of the majority of Western European countries in the post-war period.

The Charter was followed by a social action programme intended to give legal substance to the social rights enshrined in the declaration. Many of the provisions in the Charter are not really controversial; and the same could be said of a good part of the draft legislation already submitted by the Commission on the basis of the action programme. In fact, the adoption of the proposed measures would require few changes in the existing legislation of member states (see, for example, Perez Amorós and Rojo, 1991, for the case of Spain). The Commission has been very careful in repeatedly stressing the principle of subsidiarity in an attempt to calm any fears about a federalist conspiracy to encroach upon the rights of member governments in the social field. The Commission has also frequently employed simple recommendations addressed to national legislators. However, with the benefit of hindsight, it appears that this policy of appeasement has proved hardly successful.

There have been, though, several controversial issues related to the Charter and the social action programme. They include the Commission's proposed legislation on 'atypical' forms of employment, on the maximum duration of work and holidays, and even on equal opportunities for women, not to mention provisions for the participation and consultation of workers in enterprises. An increasing percentage of the new jobs created in recent years belongs to the category of 'atypical' employment, namely temporary and part-time jobs. Fearing that this would be used to dilute the rights of workers in terms of social security, health and safety protection, holiday and training entitlements, the Commission has proposed legislation in this area, aiming basically at the establishment of minimum standards. But some governments have considered this legislation as another attack on flexible labour-markets, and its adoption has been, therefore, strongly resisted. Other examples where consensus among member governments has proved difficult to reach are statutory limits on overtime work, the protection of pregnant women etc.

Different approaches to labour-market regulation have been combined with differences regarding the desired role of the EC in the social policy area. Sometimes, the main problem can be found in different national traditions which are difficult to reconcile in a common legal framework, despite the fact that the results are often broadly similar. As an example, we may cite the subject of paid holidays: some countries have legal provisions for the minimum amount of paid holidays, while in others this is a matter which is dealt with through collective bargaining. Under such circumstances, Community legislation becomes extremely difficult, unless it is restricted to rather anodyne statements.

The Charter also referred to an 'equitable wage' which may be interpreted as an attempt to introduce minimum wages in all member countries (they already exist in several of them). This is a controversial subject among economists, since the creation of an artificial floor is seen as leading to the narrowing of wage dispersion, while also adding an upward pressure on wages and salaries and eventually leading to an increase in unemployment. Those who are most likely to suffer as a result are new entrants and people with very low skills who are gradually pushed out of the labour-market. The labour-market has its own 'outsiders' whose interests are not effectively represented by trade-union organizations. They become 'outsiders', it is argued, because of excessive regulation and the fact that wages do not reflect supply and demand conditions in the market. Is there a trade-off between equity and unemployment (Flanagan, 1987), and if so, is that a legitimate trade-off? Again, the question reverts to the optimum degree of regulation. In view of the large wage disparities in the EC, a European minimum wage would make absolutely no sense. This means that legislation in this area is bound to be limited to general declarations with little concrete effect. A minimum wage is not the same as a minimum income which already forms part of the social security system of the majority of member countries. But there can be serious doubts as to whether this is an appropriate area for EC legislation, especially in view of the inability of Community institutions to undertake at least part of the financial cost implied.

Information, consultation and participation of workers has been the subject of never-ending debates which have usually remained on the ideological stratosphere. Participation is, of course, the most controversial. This subject will be discussed in more detail in the context of the proposed European company statute. The idea of a European company statute has a long history; it was initially launched by French legal practitioners in 1959, while the first Commission proposal was submitted as early as 1972. The economic logic behind it has been to produce a legal framework which would facilitate cross-border mergers and co-operation between firms. The enormous differences which still exist between national company laws make such co-operation virtually impossible, thus leaving takeovers as the only practical option. The creation of European companies would also provide a strong fillip for political and economic integration. However, proposals for the creation of a European company statute, which in itself is quite uncontroversial, have always been linked by the Commission with provisions for workers' participation, this linkage being, arguably, forced upon the Commission by already existing national legislation in this area.

As with company law, the situation among different European countries varies enormously, ranging from extensive legislation on employee participation in the Federal Republic (legislation for *Mitbestimmung* (codetermination) was introduced in 1976) and the Netherlands to informal arrangements determined by collective bargaining and no arrangements at all (Pipkorn, 1986). This diversity explains the insurmountable difficulties experienced in the past in reaching a common position on this subject, despite repeated efforts made by the Commission and the Parliament to make the proposals more flexible in order to increase their acceptability by governments which had not yet been convinced of the merits of industrial democracy (the British Conservatives acting, of course, as champions of the opposition). But there were also limits to this flexibility: too much of it and the consequent dilution of the provisions for workers' participation risked the wrath of trade unions and countries with advanced legislation in this area, since this would allow companies to bypass more stringent national rules. The result had been a long stalemate within the Council of Ministers.

Earlier experience did not prove sufficiently discouraging for the Commission which submitted new proposals for the adoption of a European company statute in August 1989, presumably trying to take advantage of the more favourable climate of the late 1980s, while also trying to restore the political balance by pushing forward measures with a strong social content. Legislation on industrial democracy was considered as a central plank of this strategy. Once again, provisions for workers' participation were tacked on to the proposed company statute which contained economic sweeteners for business in the form of a favourable tax regime: companies registered under the new legal regime would be able to offset losses in one country against profits in another. Participation was meant to offer workers a role in supervising the development and strategies of the company registered under the new European statute; participation did not, however, extend to day-to-day executive decisions. Information and consultation of workers would cover a wide range of issues, including the closure and transfer of plants, important changes in the organization of firms, and the substantial reduction or alteration of their activities.

Three alternative models were offered, and member countries would be allowed to restrict the choice among those models for companies with headquarters on their territory. The first model, following closely the German experience, envisaged the creation of a two-tier system, with a 33–50 per cent participation of workers' elected representatives in company supervisory boards. The second model, based on actual French and Italian practices, made provisions for the creation of works' councils, consisting of workers' representatives, which would remain outside the management structure. As for the third option, this was intended to offer more flexible arrangements which could be best classified in a separate category named 'other'. It referred to any other model jointly agreed by management and labour or any alternative scheme based on the most advanced national practices. The Commission proposal, with the three alternative models for workers' participation, constituted a delicate balancing act on a tightrope in an attempt to reconcile the maximalist positions of Germany and the Netherlands, backed by labour organizations, on the one hand, and the minimalist stance of the UK, with the strong support of employers on the other.

But again, there has been no success in breaking the deadlock. Although the proposed legislation would require few practical changes in most member countries, the opposition to it needs to be explained largely in ideological terms. Another factor should also be added, namely the fear of the opponents that developments at the European level could open the Pandora's box in this area, leading to more substantial changes in individual countries. The proposal for a European company statute, with provisions for workers' participation, is an example of the package deal method frequently employed in EC decision-making. It has not so far worked in this case, perhaps because there is not much in the statute for anybody, including the companies themselves which seem to have found ways of dealing with different national legislations.

Proposals concerning the information and consultation of employees in multinational firms have had a similar fate. It all started with the so-called Vredeling directive in 1980, named after the Dutch Commissioner in charge. The Commission's proposals had caused then a strong reaction from UNICE (the confederation of European industrialists) and several US multinationals, which found themselves under the political umbrella of the US and the UK Governments; and they were consequently buried some years later. They have been revived more recently in the context of the social action programme. The Commission's proposal for a European works' council for European-scale companies is a diluted version of the Vredeling proposals (Rhodes, 1992). Works' councils were meant to provide a channel for the information and consultation of workers of multinational firms on job reductions, new working practices and the introduction of new technology. But the views of traditional opponents have not changed in the meantime. The result has been a political stalemate.

The third area for action, identified by Mr Delors in his address to the ETUC in 1988, was the promotion of a dialogue between representatives of business and labour at the European level. The EC Commission has always been active in trying to associate representatives of business and labour in Community decision-making and in encouraging the creation of European pressure groups as a means of strengthening the political constituency on which the process of European integration is based. The Treaty of Rome had provided for the setting up of the Economic and Social Committee (ESC), consisting of representatives of the social partners, who would act in a consultative capacity in the Community decision-making process. This institution was in the good corporatist tradition of some member countries. But it has never taken off, partly because the various interest groups seem to have preferred national and informal channels instead of the ESC.

Consultation is one thing, and collective bargaining another. The European social dialogue has been in operation for some time, bringing together, under the presidency of Mr Delors, representatives of UNICE and the ETUC. Although the earlier experience had brought few tangible effects, the Commission sought to revive the dialogue in 1989 mainly as a means of producing common positions which could form the basis for EC legislation in the employment and social fields. The two organizations are loose confederations which represent little more than the sum of their parts, that is when those parts decide to act together, which is not always the case. Their membership extends beyond the EC, comprising virtually the whole of Western Europe.

In a leading article, *Le Monde* (10 Sept. 1988) wrote: 'L' Europe des syndicats avance mais, aura-t-elle des partenaires?' In fact, even the first part of the question appears doubtful. The capacities of trade unions for collective action at the European level are little developed, and the persisting diversity in terms of economic fundamentals acts as another constraining factor for the ability of organized labour to act effectively at the supranational level. Recent years have been characterized by the weakening of trade-union power in most European countries. More importantly, Europe is not as yet sufficiently integrated for effective negotiations to take place between representatives of business and labour at this level. On the other hand, the representatives of employers' federations have always resisted a strong EC role in the social field, and the last thing that UNICE wants is European collective agreements. As for the Commission, it does not have much to deliver in terms of dowry in order to bring this difficult couple together.

What can we then make of the agreement reached in the autumn of 1991 on a new procedure for greater involvement of the social partners in shaping EC policy in the social field. The social partners will be consulted prior to and during the drafting of Commission proposals. In addition, they can together decide to draft a joint position which can then act as the basis of EC legislation. This represented a significant extension of the role of social partners, and it later was incorporated in the social protocol adopted by eleven countries, with the exception of the UK, at Maastricht. It marked a shift in the employers' stance, which can be at least partly explained by the growing unease inside UNICE with the hardline policy pursued until then and identified mainly with the British members of the organization. But it could also be seen as a damage limitation exercise: given the strong political pressure for EC legislation in the social field on the eve of Maastricht, the employers' federation decided to become more actively involved in the legislative process. The results of this breakthrough remain to be seen.

The Padoa–Schioppa group argued in 1987 that '[the] principle of subsidiarity recommends minimal responsibility on the part of the Community on many aspects of social policy' (Padoa–Schioppa *et al.*, 1987: 43). This had remained true until the adoption of the social protocol at Maastricht, not so much because of a conscious collective decision to opt for a highly decentralized system but more because of the inability to agree on the contents of EC policy in the social field. With few exceptions, most notably with respect to health and safety standards where the SEA had provided for qualified majority voting, progress in the implementation of the social action programme had been minimal until then. Hence the frustration of proponents for further progress in this area and the prominent role occupied by social policy during the intergovernmental conference on political union. Given the negative experience until then, treaty revisions were considered as the only way of breaking the political deadlock.

The compromise reached in the end has no precedent in the history of the EC. Faced with the adamant opposition of the UK to any extension of EC legal competences in the social policy field, the other eleven countries decided to proceed separately by signing a social protocol. Although this formally lies outside the new treaty on European Union and future legislation based on the social protocol will not form part of the *acquis communautaire*, EC institutions will be used to adopt new legislation in this area which will only apply to the eleven signatories, as long as the UK continues to opt out. Thus, qualified majority voting has been extended to working conditions, the information and consultation of workers (but not participation which will remain subject to unanimity), equality between men and women, and the integration of unemployed persons into the labour market. Several other important aspects of social policy, such as social security and social protection of workers, have also been included in the social protocol, although in this respect the unanimity principle has been retained. Provision has been made for agreements between the social partners to be translated into legal acts, as mentioned above.

The social protocol should help to unblock several Commission proposals which have been lying for years on the Council's table. At the same time, it will inevitably constitute a legal minefield as the eleven countries try to reconcile the operation of an inter-governmental agreement with the EC decision-making process, even though special provision has been made for voting procedures in the Council from which the UK representatives will be excluded. It is difficult to imagine this complex procedure lasting for long without serious harmful effects on the internal cohesion of the Community; unless this is seen as another, but very important, step towards the institutionalization of a two- or multi-tier Community. If further harmonization of social rules forms part of the evolving package deal, can free trade, for example, be separated in the case of the UK and would this not constitute for the others an unacceptable form of unfair competition?

The EC layer in terms of social policy has remained thin until now. Its most important dimension has been the transfer of funds, limited and not always effective though it may have been, operated mainly through the ESF. Community expenditure has been mainly targeted at the reduction of inequalities, the compensation of losers from the integration process and the promotion of economic adjustment. Pressure has grown for a substantial increase of Community expenditure in the social field in order to tackle the growing problem of poverty and social exclusion in prosperous Europe and also to play a more important role in areas such as education and health. But whatever the progress expected in this respect, the big bulk of social expenditure will continue to be provided by national governments for a long time to come.

Despite the progressive elimination of barriers, intra-EC labour mobility has been limited. Linguistic and cultural factors, combined with shortages of housing and the remaining NTBs, will continue to prevent large population movements across national frontiers. Labour mobility is, therefore, unlikely to play a significant role as a means of economic adjustment in the context of the internal market. In fact, there is almost a consensus view that this would be politically and socially undesirable.

The nature of human flows inside the Community and the whole of Western Europe in general has been gradually changing. Instead of mass labour movements, European countries have been witnessing the rapid growth of a different kind of human flows, including student exchanges, movements of professionals, changes of residence after retirement, not to mention the explosion of tourism (Romero, 1990). Unlike earlier migratory movements, those flows are two-way in direction and also less permanent. However, the size of cross-border flows is still relatively small in size and this will allow national labour-markets to preserve for long their own separate identity. The creation of a European labour-market is not for tomorrow. The combination of low labour mobility, wide economic diversity, and different traditions and institutions will ensure that the European dimension of industrial relations also remains underdeveloped for a long time. Labour is much less mobile than capital, and its organizations stand to lose as a result. On the other hand, labour migratory pressures are likely to increase from outside. Given the low fertility rates in virtually all Western European countries, this might not be an unmitigated disaster for them. But economic considerations are likely to clash with domestic social pressures. This promises to become one of the most difficult issues for the prosperous countries of Europe (see also Chapter 9).

The other important dimension of EC social policy has been in terms of the harmonization of national rules and regulations. The attempt has been to establish minimum standards and thus create 'even playing fields' (see arguments on social dumping), while often also aiming at the raising of those standards in the laggard countries. With some important exceptions, such as the equality of men and women and health and safety regulations, progress in this respect has so far been very limited. National realities have survived for long the impact of economic integration, and the process of convergence has been rather slow. Large differences in productivity levels and economic conditions more generally continue to compensate in most cases for the large disparities in wages and working conditions. If harmonization and the establishment of minimum standards were to proceed irrespective of economic realities, the cost for the less-developed countries and regions could be very high. Social regulation cannot be determined in complete defiance of the market.

7. Economic and Monetary Union

The process of 1992 had little, if anything, to do directly with macroeconomic policies. It was basically a supply-side programme intended to eliminate the remaining NTBs in the intra-EC movement of goods, services and factors of production. However, in the Commission's analysis and strategy on the internal market, macroeconomic policies occupied a central role. A co-ordinated expansion had been originally advocated in order to take advantage of the new margin of manœuvre expected to be created by the completion of the internal market through reduced inflation and public sector deficits (Commission of the EC, 1988*a*; see also Chapter 4). Soon, as the internal market programme gathered momentum, the Commission appeared ready to go much further. It was no longer a question of co-ordination of macroeconomic policies; instead, economic and monetary union (EMU) was presented as the next logical step after the completion of the internal market.

Thus EMU has come to be seen as the post-1992 stage of economic integration, and this was ceremoniously confirmed with the treaty revision agreed at Maastricht in December 1991. Preliminary discussions, which led to the intergovernmental conference, had started earlier in 1988; yet another indication of the sea change that had taken place in Western Europe in a relatively short period of time. By then, the internal market was already taken for granted and the debate became centred on the next step; a step which would, however, have even wider political and economic ramifications.

European integration has already extended beyond trade; and the 1992 process represents both a quantitative and qualitative shift towards the creation of a regional economic system. Yet EMU is something very different. In terms of national sovereignty, the stakes are infinitely higher, and this is clearly recognized by all participants. Unlike the White Paper of 1985, the plans for the creation of EMU could not be camouflaged as a technical matter with limited political consequences. Money is, after all, at the heart of national sovereignty.

EMU already has a long and chequered history which may teach us a few lessons about its prospects for the future. Yet, history is not usually repeated, and the economic and political environment of the 1990s in Europe is very different from that prevailing when earlier attempts were made towards this goal. The decision to proceed towards a complete EMU before the end of the decade, backed up by a host of institutional and other measures which should be introduced in the meantime, guarantees that this issue will occupy a prominent, if not the most prominent, place on the European agenda for several years to come. Developments in this area will also largely determine the shape of the European economic system.

Werner, Emu, and the Snake

Although the main emphasis in the Rome Treaty was on the creation of a common market, with the progressive elimination of barriers for the free movement of goods, services, persons, and capital, there was at least some, albeit rather hesitant, recognition of the importance of macroeconomic policies, and monetary policy in particular, in the context of a common market. Article 2 referred to the task of 'progressively approximating the economic policies of Member States'. In Title II of the original treaty, we find references to the main objectives of macroeconomic policy, and the balance of payments in particular. The achievement of those objectives would be facilitated by the co-ordination of economic policies, the ways and means of which remained to be defined. Specific provisions were made only as regards monetary policy; hence the creation of the Monetary Committee. The treaty also referred to the exchange rate as a matter of common concern. The possibility of mutual aid in the case of balance of payments difficulties was envisaged as well as the progressive elimination of exchange controls to the extent necessary for the proper functioning of the common market.

There was clearly no intention to set up a regional currency bloc. The Bretton Woods system provided the international framework and the US dollar the undisputed monetary standard. On the other hand, limited capital mobility allowed European governments a reasonably wide margin of manœuvre in terms of monetary policy (hence the caution expressed in the articles of the treaty regarding capital movements). European regional integration was basically about trade in goods. Macroeconomic policies, and monetary policy in particular, remained the concern of national governments subject to the constraints imposed by the international monetary system.

During the 1960s, there was much talk about regional co-operation, but little concrete action. The central country of the Bretton Woods system, namely the United States, proved no longer willing or able to provide the public good of monetary stability, and this led to some discussions between the, by now, much more confident Europeans about alternatives, including closer regional co-operation. There was also some resentment about the asymmetrical nature of the international system, which was, however, combined with different degrees of economic and political vulnerability to US pressure. At best, the intra-European discussions of the 1960s helped to prepare the ground for subsequent plans for regional co-operation, while also leading to the expansion of the small infrastructure of committees at the EEC level. The creation of the CAP, based on common prices, was closely related to the perceived need for exchange rate stability. But this was in little doubt until 1968.

EMU was part of the new package deal agreed at the Hague summit of 1969. Growing trade interpenetration, largely the result of tariff liberalization measures, had reduced the effectiveness of autonomous economic policies; hence the new attraction of co-ordination procedures and joint action. The Six had proved unable both to insulate themselves from international monetary instability and to pursue a common policy in international fora. The events of 1968–9 provided clear evidence of this failure. This in turn endangered one of the main pillars of the EEC construction, namely the CAP. Thus the political decision to extend and deepen the process of integration led the Six, almost naturally, into the area of macroeconomic activity. And they decided to go all the way, committing themselves to the creation of an EMU which would replace the customs union as the main goal of the new decade. Money was also a means to an end, and the end was political union. This function of monetary union was widely recognized and aptly summarized by the French: 'la voie royale vers l'union politique'.

However, important divisions soon became apparent as regards the priorities and the strategy to be employed in order to achieve the final goal. Those divisions, which had their origins in an earlier debate provoked by the Commission memorandum of 1968 (better known as the first Barre plan) and the alternative proposals put forward at the time by the German Minister of Economics, Mr Schiller, dominated the discussions inside the high-level group which was entrusted with the preparation of a report on the establishment of EMU. This group was chaired by the Prime Minister of Luxembourg, Mr Werner.

The main conflict was between 'economists' and 'monetarists' and was based on the strategy to be adopted during the transitional period in order to achieve a sufficient harmonization of national economic policies which were seen as the main precondition for the elimination of payments imbalances (Tsoukalis, 1977a). More specifically, the crucial difference was whether the Community would move towards the irrevocable fixity of parities and the elimination of margins of fluctuation before the system of economic policy co-ordination had proved its effectiveness.

The 'monetarists', represented by France, Belgium, and Luxembourg, stressed the importance of the exchange rate discipline and the need to strengthen the 'monetary personality' of the EC in international fora. They also, presumably, would have liked to pass the adjustment burden on to surplus countries and thus face them with the choice of either financing the deficits of others or accepting a higher rate of inflation. This would in turn have largely depended on provisions for the financing of payments imbalances. It was precisely this choice that the potential surplus countries, namely the Federal Republic of Germany and the Netherlands, would have liked to avoid. Hence the insistence of the 'economists' on policy co-ordination, the results of which were, apparently, expected to be close to their own set of policy preferences, as a necessary condition for progress on the exchange rate front. What the 'economists' in fact implied was the convergence of the inflation rates of other countries to their own. There was a certain degree of ambivalence and confusion on both sides, and also some double-talk.

The debate between 'economists' and 'monetarists' did not touch upon the big question of the economic feasibility of EMU within the relatively short time-scale envisaged. The harmonization of policy preferences was considered as the main precondition for the elimination of payments imbalances, thus ignoring basically the various factors behind different wage and price trends as well as productivity levels between member countries. Turning the realization of EMU into a question of political will, the Community ran the risk of neglecting the possible economic costs associated with the abandonment of a major policy instrument such as the exchange rate. Furthermore, the difference between 'economists' and 'monetarists' was only partly a reflection of fundamentally different approaches to the establishment of EMU. It also served as an ideological cloak for different short-term interests of individual countries. The 'short-termism' of some governments became clearer through their subsequent actions.

The final report of the Werner group was submitted in October 1970 (Werner, 1970). A complete EMU was to be achieved in three stages within an overall period of ten years. The final objective was defined in terms of an irrevocable fixity of exchange rates, the elimination of margins of fluctuation, and the free circulation of goods, services, persons, and capital. The creation of EMU would require the transfer of a wide range of decision-making powers from the national to the EC level. All principal decisions on monetary policy, ranging from questions of internal liquidity and interest rates to exchange rates and the management of reserves, would have to be centralized. Quantitative medium-term objectives would be jointly fixed and projections would be revised periodically.

With respect to fiscal policy, the Werner group argued that an agreement would have to be reached on the margins within which the main national budget aggregates would be held and on the method of financing deficits or utilizing surpluses. Fiscal harmonization and co-operation in structural and regional policies were also mentioned as objectives. The creation of two main institutions of the Community was envisaged, namely 'the centre of decision for economic policy' and the 'Community system for the central banks'.

The final report was based on a consensus among its members, regarding the ultimate objective, and a compromise, couched in somewhat vague terms, between 'economists' and 'monetarists' about the intermediate stages. The compromise was embodied in the strategy of parallelism between economic policy co-ordination and monetary integration, with the Werner group concentrating mainly on the measures to be adopted during the first of three stages before reaching the 'Elysian harmony'[1] of a complete EMU.

The fragility of the compromise and the political commitment of several countries to the final objective was soon to be exposed by the dramatic deterioration of the international economic environment. One of the few concrete decisions was to narrow the intra-EC margins of fluctuation from 1.5 to 1.2 per cent on either side of the parity; a decision, however, which was never implemented. This small, first step on the road towards monetary union took Bretton Woods and fixed exchange rates against the dollar for granted. This proved to be one of the biggest weaknesses of the European strategy. Massive capital inflows and the US policy of 'benign neglect' ('unbenign' would, perhaps, be a more accurate term) showed once again the extreme fragility of EC unity and of only recently concluded agreements. Faced with the Nixon measures of August 1971, the Six suddenly forgot the objective of EMU and produced instead an impressive variety of national exchange rate regimes ranging from independent floating to 'pivot rates' and a two-tier market.

The Smithsonian agreement of December 1971 created only a short-lived illusion of a new international order. The Six hastened to build their regional system on its foundations; and, not very surprisingly, it soon crumbled. The 'snake in the tunnel', created in March 1972, represented a new attempt to narrow intra-EC margins of fluctuation to 2.25 per cent, instead of the 4.5 per cent resulting from the application of the Smithsonian agreement.

---

[1] This was a term used at the time by Mr Schiller, the German Minister of Economics.

The snake would therefore consist of the EC currencies jointly moving inside the dollar tunnel, that is the 2.25 per cent bands on either side of their parity against the US currency. Wide margins of fluctuation of intra-EC exchange rates were generally considered as incompatible with the functioning of the common market and the CAP. The birth of the snake was accompanied by rules for joint intervention in the exchange markets and provisions for very short-term credit between central banks for the financing of those interventions. As part of the strategy of parallelism, a steering committee was also set up for the more effective co-ordination of economic policies.

The continued instability in exchange markets caused the progressive mutilation of the snake, thus creating further unhappiness in the European monetary zoo. Sterling and the punt left the EC exchange rate arrangement in July 1972, soon to be followed by the Danish krone which subsequently returned to the fold. Only a few months after the Paris summit of October 1972, which reiterated the political commitment to a complete EMU by the end of the decade, the lira was floated and then the remaining currencies of the snake decided to make virtue out of necessity by accepting a joint float against the dollar. The snake had, therefore, lost its tunnel and would from then on have to wriggle its way into the open space. The final exit scene was reserved for the French franc which left the snake in January 1974, while a subsequent attempt to return did not last for very long (July 1975–March 1976).

In the following years, the snake was little more than a Deutschmark (DM) zone. The Benelux countries and Denmark remained in it recognizing the importance of stable exchange rates for their small and open economies and the relative weight of their trade relations with the Federal Republic. It was rather ironical (or was it just an illustration of the short-term considerations which had prevailed in national attitudes towards EMU?) that the core countries of the 'economist' group had chosen to remain in the regional exchange rate system, while France, the leader of the 'monetarists', had been forced to leave. The snake was hardly a Community system, with almost half of the EC members staying outside it, while a number of other Western European countries (Austria, Norway and Sweden) became associate members of the DM-zone.

The most common criticism made against any system of fixed exchange rates, with limited amounts of liquidity to finance payments deficits, used to be that the burden of adjustment is likely to fall on the deficit countries, thus creating a deflationary bias in the system. For Germany, the snake contributed towards a certain undervaluation of the DM, while at the same time it did not impose any additional constraints in the conduct of German monetary policy. For the smaller countries, the corresponding appreciation of their currencies against the DM seemed to be offset by the greater stability of the exchange rate and the increased credibility of their economic policies. Concentrating on the Danish experience, Thygesen (1979) did not find any convincing evidence of deflationary bias. On the other hand, the snake did not lead to any serious co-ordination on the external monetary front, except for one-way co-ordination on the basis of German policies. It was, undoubtedly, an asymmetrical system.

In the attempt to preserve stable intra-EC exchange rates, in the midst of the general upheaval which characterized the early 1970s, Community countries increasingly resorted to capital controls, despite the declared objective of a complete liberalization of capital movements which was supposed to be an integral part of EMU. Furthermore, there was neither uniform action nor even some form of broad agreement on this subject. The question of capital controls produced very different responses on behalf of individual member countries. This continued to be true until the end of the decade when Germany, Britain, and the Netherlands were the only countries of the enlarged EC with virtually no restrictions imposed on international capital flows. On the other hand, no attempt was made to discriminate in favour of intra-EC movements, probably recognizing the futility of such an attempt in view of the permeability of national frontiers.

The combination of an unfavourable international environment, divergent national policies, a half-baked economic strategy, and a very weak political commitment ensured the quick death of EMU. The latter became the biggest non-event of the 1970s. With the benefit of hindsight, it can be argued that the ambitious initiative, originally intended to transform radically the economic and political map of Western Europe, had been taken at the highest level without much thought of its wider implications. It certainly did not survive the test of time and economic adversity.

Monetary Stability and the EMS

Despite the serious setbacks suffered in the attempt to move towards an EMU in the 1970s, interest in the subject never disappeared. The mini-snake was generally considered as only a temporary arrangement which would be improved and extended when the economic conditions became more favourable. Various plans were put forward which served to keep interest alive and also prepare the ground for a new political initiative. The aim of official plans was usually to design a more flexible exchange rate system which would incorporate all EC currencies. On the other hand, many proposals originating from professional economists and academics concentrated on the creation and develop- ment of a European parallel currency (Fratianni and Peeters, 1978). Finally, a proposal put forward by the President of the Commission, Mr Jenkins, in October 1977 acted as the catalyst for the relaunching of monetary integration. And this in turn led to the establishment of the EMS in March 1979.

The EMS was the product of an initiative taken by Chancellor Schmidt, against the advice, if not the outright opposition, of his central bank. This was later presented as a joint Franco-German initiative, something which was becoming increasingly a regular pattern in EC affairs. It could have been an arrangement among the Big Three to which the other EC countries would have been invited to join. But Britain, once again, decided to stay out (Ludlow, 1982).

The creation of the EMS was seen, first and foremost, as a means of reducing exchange rate instability among EC currencies. Despite the agnosticism expressed by many academic economists on the subject, exchange rate instability was believed by most political leaders and businessmen in Western Europe to have deleterious effects on the real economy, and more precisely on trade, investment, and growth. Our understanding of those effects has improved considerably since then, even though econometric evidence still remains inadequate (Krugman, 1989).

Concern about the proper functioning of the common market was combined with the desire to preserve the system of common agricultural prices. On the other hand, the initiative for the creation of the EMS was linked to the expectation that there would be no substantial reform of the international monetary system and hence no prospect of a return to some form of exchange rate stability in the near future. The construction of a regional system was, therefore, seen as a second- or third-best solution.

Exchange rate stability was to be backed by an increased convergence between national economies, with the emphasis clearly placed on inflation rates. The EMS was considered as an important instrument in the fight against inflation, and its creation meant an implicit acceptance of German policy priorities by the other EC countries. The experience of the 1970s was seen as validating the uncompromising anti-inflationary stance combined with the strong currency option adopted by the Federal Republic. The EMS was also intended as a European defensive mechanism against US 'benign neglect' as regards the dollar, and, more generally, what was perceived to be a political vacuum in Washington at the time of the Carter Administration; even though it was never made very clear how the EMS could perform such a role. It was also seen as a means of strengthening Europe economically and politically through closer co-operation at a time when US leadership was seen as waning. Once again, monetary integration was partly used as an instrument for political ends.

The EMS was intended as a system of fixed but adjustable exchange rates. One of the novelties of the system was the European Currency Unit (ECU) consisting of fixed amounts of each EC currency, including those not participating in the exchange rate mechanism (ERM). Provision was made for the revision of these amounts every five years. Two revisions have already taken place in 1984 and 1989, and they also led to the inclusion of the currencies of the three new EC members (Table 7.1). The relative weights of each currency are a function of the economic and trade weight of the country concerned, with a clear tilt towards the stronger currencies (see relative weight of the Deutschmark and the Dutch guilder).

Each EC currency has a central rate defined in ECUs. Central rates in ECUs are then used to establish a grid of bilateral exchange rates. The margins of fluctuation around those bilateral rates were set at 2.25 per cent, with the exception of the lira which was allowed to operate within wider margins of 6 per cent and sterling which stayed completely out of the ERM. Central bank interventions were compulsory and unlimited, when currencies reached the limit of their permitted margins of fluctuation. Central rates could be changed only by common consent.

Table 7.1. Composition of the ECU

Currency	13 March 1979		17 September 1984		21 September 1989	
	1	2	1	2	1	2
Deutschmark	0.828	33.00	0.719	32.00	0.6242	30.53
French franc	1.15	19.80	1.31	19.00	1.332	19.43
Netherlands guilder	0.286	10.50	0.256	10.10	0.2198	9.54
Belgian and Luxembourg franc	3.80	9.50	3.85	8.50	3.431	7.83
Italian lira	109.0	9.50	140.00	10.20	151.8	9.92
Danish krone	0.217	3.00	0.219	2.70	0.1976	2.53
Irish punt	0.00759	1.10	0.008781	1.20	0.008552	1.12
Pound sterling	0.0885	13.60	0.0878	15.00	0.08784	12.06
Greek drachma	—	—	1.15	1.30	1.44	0.77
Spanish peseta	—	—	—	—	6.885	5.18
Portuguese escudo	—	—	—	—	1.393	0.78

Note: Column 1 indicates the number of national currency units in each ECU while column 2 gives the percentage weight of each currency in the ECU basket.

Source: Eurostat.

Against the deposit of 20 per cent of gold and dollar reserves held by participating central banks, which took the form of three-month revolving swaps, the European Monetary Co-operation Fund (EMCF) issued ECUs in return. Those ECUs were therefore intended to serve as an official reserve asset, although subject to many restrictions which could qualify them basically as a non-negotiable instrument of credit. ECUs would also serve as a denominator for market interventions arising from the operation of the ERM and as a means of settlement between the monetary authorities of the EC. This in turn meant a sharing of the exchange risk between creditor and debtor countries. Exactly the same applied to the credit mechanisms of the EMS.

Another novelty of the EMS was the so-called divergence indicator intended to provide a certain degree of symmetry in the adjustment burden between appreciating and depreciating currencies and an automatic mechanism for triggering consultations before the intervention limits were reached. The device would make it possible to locate the position and the movement of an EMS currency relative to the EC average represented by the ECU. There was a so-called 'presumption to act', when the divergence threshold was reached. The creation of the divergence indicator also suggested that, at least initially, policy convergence (including the convergence of inflation rates) was expected to be towards the EC average, instead of the best performance which was later adopted as the target.

Very short-term credit facilities, in unlimited amounts, were to be granted to each other by participating central banks, through the EMCF, in order to permit intervention in EC currencies. Provisions were also made for other credit facilities, building on the already existing short-term monetary support and the medium-term financial assistance mechanism of the snake. In principle as a means of fostering economic convergence and in practice as a carrot to lure the economically weaker countries into participating in the ERM, provision was made for the granting of subsidized loans by EC institutions and the European Investment Bank.

Thus the EMS was built on the existing snake with some important novel features intended to ensure the enlargement of its membership and the smoother functioning of the exchange-rate mechanism. It was based on a political compromise between

Germany which feared the effects of prolonged international monetary instability and an excessive revaluation of the DM, resulting from the continuous sinking of the dollar in exchange markets, and France and Italy which saw their participation in the EMS as an integral part of an anti-inflation strategy. All three also shared the broader political objectives associated with the EMS, namely support for European unification and the strengthening of Europe's identity in relations with the United States. It was clearly a decision of high politics. Writing about Italy, de Cecco (1989: 90) argued, with some element of exaggeration, that '[to] be in favour of the EMS meant to be in favour of freedom and of Western civilization'!

The other members of the old snake, countries with small and highly open economies and hence little prospect of independent monetary policies, were only too happy to see an extension of the area in which stable exchange relations applied. As for Ireland, the decision to join the ERM was partly a function of the side-payments offered and the attraction of the external discipline on monetary policy and partly an expression of political independence against Britain. The latter was a totally different case. The fear of deflationary pressures, stemming largely from the traumatic experience with earlier attempts under the Bretton Woods system to keep the exchange rate of sterling fixed, were combined with strong opposition to the political objectives behind the EMS. Thus, Britain stayed out of the ERM for more than eleven years, limiting its participation to the other, much less constraining manifestations of the new system.

At the time of writing, after the thirteenth birthday of the EMS, there is little doubt about the success of the system. It has survived against the odds and the expectations of most professional economists and central bankers who had greeted its birth with considerable scepticism, if not sheer cynicism, drawing largely from the earlier experience in regional monetary integration. Furthermore, it has expanded its membership, acting as a major source of attraction for non-EC members as well, and it has also acted as a launching board for the renewed attempts to create a complete EMU.

The EMS has been generally described as a zone of monetary stability, and this is linked to both exchange rates and inflation rates. Short-term volatility of bilateral exchange rates has been substantially reduced. This is true when a comparison is made with the pre-EMS experience of participating currencies or with the experience of other major currencies, including sterling for the period it stayed outside the system. The greater stability in nominal exchange rates has been achieved through a convergence of inflation rates, the gearing of the interest rate towards the exchange rate target, joint interventions in the exchange markets, capital controls and the increased credibility of the system; the mix of all those factors having changed considerably over the years. Stability does not, however, mean rigidity. During the first thirteen years of the EMS, there were twelve realignments, five of which involved more than two currencies (Table 7.2). They soon became a matter of genuinely collective decisions, while the element of drama, which had often accompanied the negotiations leading to the early realignments, gradually disappeared.

It is very difficult to generalize for the whole period, since the EMS has experienced major changes both in terms of the general macroeconomic environment and its own operating rules. Three different phases in the history of the EMS can be distinguished (see also Ungerer et al., 1990; Gros and Thygesen, 1992). The first phase ended with the realignment of March 1983. It was a turbulent period, marked by frequent realignments of exchange rates and wide policy divergence (also manifested in terms of inflation rates — see Fig. 7.1). This suggested that the consensus on which the creation of the EMS had been based was rather flimsy. But after all, both parents of the EMS (Chancellor Schmidt and President Giscard d'Estaing) had left the political stage before the end of 1982.

The second phase was one of consolidation, and it ended with the realignment of January 1987. There were few realignments during this period, usually involving small changes in the central rates and only a few currencies. It was also the period of increasing price convergence downwards. Following the ill-fated attempt by the French socialists to apply Keynesianism in one country in 1982–3, exchange rate stability was based on a convergence of economic preferences, which was in turn reinforced by the operation of the system. Most countries participating in the ERM reached the lowest levels in terms of inflation rates between 1986 and 1987; since then there has been a shift upwards which was initially linked to the economic boom.

The third phase has been characterized by a remarkable stability

Table 7.2. EMS Realignments: Changes in Central Rates

(% change: minus sign (−) denotes a devaluation)

Currency	24 Sept. 1979	30 Nov. 1979	23 Mar. 1981	5 Oct. 1981	22 Feb. 1982	14 June 1982	21 Mar. 1983	22 July 1985	7 Apr. 1986	4 Aug. 1986	12 Jan. 1987	8 Jan.[a] 1990
Deutschmark	2.0			5.5		4.25	5.5	2.0	3.0		3.0	
French franc				−3.0		−5.75	−2.5	2.0	−3.0			
Netherlands guilder				5.5		4.25	3.5	2.0	3.0		3.0	
Belgian and Luxembourg franc					−8.5		1.5	2.0	1.0		2.0	
Italian lira			−6.0			−2.75	−2.5	−6.2				−3.0
Danish krone	−2.9	−4.8		−3.0			2.5	2.0	1.0			
Irish punt					−3.0		−3.5	2.0		−8.0		
Spanish peseta[b]												
Pound sterling[c]												
Portuguese escudo[d]												

[a] On this date the Italian lira moved from fluctuation bands of ±6% around its central rate to narrow fluctuation bands of ±2.25%.
[b] The peseta joined the ERM on 19 June 1989 with a fluctuation band of ±6% around its central rates.
[c] The pound entered the ERM on 8 Oct. 1990 with a fluctuation band of ±6% around its central rates.
[d] The escudo entered the ERM on 6 April 1992 with a fluctuation band of ±6% around its central rates.

Source: Eurostat

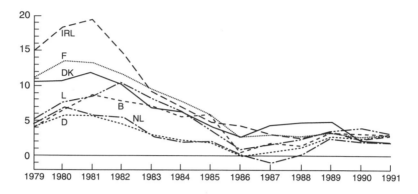
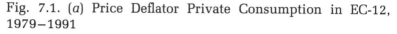

Fig. 7.1. (*a*) Price Deflator Private Consumption in EC-12, 1979–1991

Source: Commission of the EC, 1991*a*.

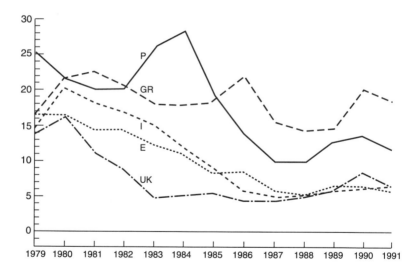

Fig. 7.1. (*b*) Price Deflator Private Consumption in EC-12, 1979–1991

Source: Commission of the EC, 1991*a*.

of exchange rates, which could be, perhaps, also described as rigidity. For more than five years, there has been no realignment of exchange rates, with the exception of a small repositioning of the lira, which was announced in January 1990, together with the decision to reduce the margins of fluctuation to 2.25 per cent. It had been preceded by the most important so far modification of the operating rules of the system (see below, for a discussion of the Basle–Nyborg agreement of 1987). The stability of exchange rates has been based essentially on the continued convergence of inflation rates, even though there has been an upward shift in recent years, and the increased credibility of the system in exchange markets.

During the third phase, there has also been a gradual extension of ERM membership, a sure sign of the success of the system.[1] Having brought down the rate of inflation and in search of a credible exchange rate target and an external discipline, Spain joined in June 1989, with a 6 per cent margin of fluctuation. Later, it was the turn of the real heretics to take the oath of allegiance to the true faith. After a long internal debate, in which arguments about the petrocurrency status of sterling and the role of London as an international financial centre had been used interchangeably with political arguments about the loss of national sovereignty, the UK finally decided to join the ERM in October 1990, also with a 6 per cent margin of fluctuation. It was followed by Portugal in April 1992, thus leaving only the Greek drachma outside the ERM. In all cases, the decision to join was based on a combination of economic and political considerations: the search for a stable anchor for the exchange rate and an external discipline for monetary policy, offering further proof of one's commitment to Europe, and a desire not to be left out of an increasingly important part of regional integration. High inflation rates still prevented Greece from joining the club.

There were three other important events which marked the latter part of the third phase in the history of the EMS. One was the liberalization of capital movements; the second was the decision to proceed to a complete EMU, with the first stage starting in July 1990 and linked precisely to capital liberalization;

---

[1] For an analysis of the three southern European countries and Britain in relation to the EMS, see relevant chapters in de Grauwe and Papademos (1990). For a very critical approach, see also Walters (1990).

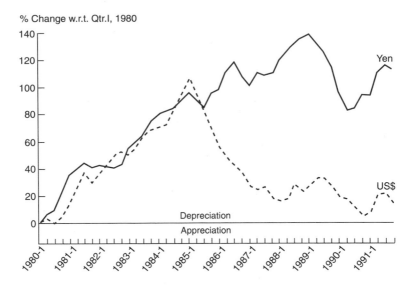

Fig. 7.2. Bilateral Exchange Rates of the US Dollar and the Yen against the ECU, 1980–1991

Source: Eurostat.

and the third was the unification of Germany, with profound effects on the German economy and also indirectly on the other members of the system. The combination of all three may suggest that the EMS had already entered a new phase; but it is still early to judge. This subject will be discussed further below.

The reduction of short-term volatility and the avoidance of substantial long-term misalignment have not applied to exchange relations with major currencies outside the ERM. This is particularly true of the exchange rate of the ECU with the dollar (Fig. 7.2). Thus, the EMS did not provide the insulation against external instability which some Europeans had hoped for, although the centrifugal effects of the dollar gyrations on intra-ERM exchange rates seem to have been reduced in recent years. Until the Plaza agreement of 1985, there had been no serious attempt at policy co-ordination at the international level. The gross overvaluation of the dollar during the first half of the 1980s, seen in the beginning as a sign of national virility by President

Reagan and several members of his Administration, and its disastrous effects on the US external trade balance finally brought about a change in US attitudes. The experience since then suggests that the conversion to the virtues of international economic co-operation and the joint management of exchange rates was neither deep nor long-lasting; but this applies virtually to all parties concerned, which is a further sign of the limitations of international policy co-ordination. Nevertheless, the Plaza agreement marked a turning-point at least in terms of the prevailing exchange rate fashion, since even the Anglo-Americans appeared ready then to abandon their earlier theological attachment to free floating (Funabashi, 1988).

Intra-ERM exchange rate stability is closely linked to price convergence. This convergence started with a time-lag in 1983 and it was downwards until 1986–7, when there was again some acceleration in inflation rates (Fig. 7.1). On the basis of the experience of recent years, there are three distinct groups of countries inside the EC. The first group comprises the seven original members of the narrow band of the ERM, which have experienced the highest degree of convergence and the lowest rates of inflation. The second group consists of the countries with the longest history in the wider band of the ERM, namely Spain, the UK and Italy which accepted the discipline of the narrower margins of fluctuation in 1990 despite the continued divergence in terms of inflation rates. The remaining two countries, namely Portugal and Greece, have continued all along with double digit rates of inflation.

To the extent that exchange rate changes have not compensated fully for price and/or wage differentials during this period, this has had inevitably an effect on the external competitiveness of the countries concerned. Although major misalignments have been avoided in most cases, exchange rates have been used, especially since 1983, as an important anti-inflationary instrument, with realignments compensating only partially for inflation differentials. This has, therefore, led to the progressive overvaluation in real terms of the more rapidly inflating currencies (Fig. 7.3). Measurements of competitiveness of the different ERM currencies vary depending on the indicator used (consumer prices, unit labour costs etc; see also Ungerer *et al.*, 1990; de Grauwe and Gros, 1991). However, whatever the indicator used,

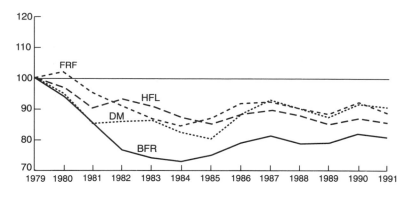

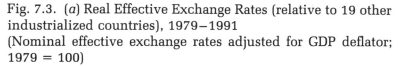

Fig. 7.3. (*a*) Real Effective Exchange Rates (relative to 19 other industrialized countries), 1979–1991
(Nominal effective exchange rates adjusted for GDP deflator; 1979 = 100)

Source: Commission of the EC.

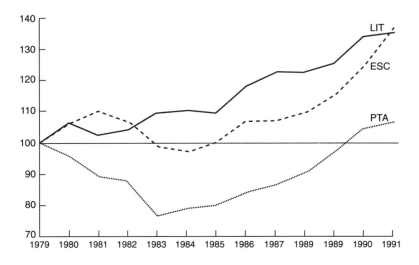

Fig. 7.3. (*b*) Real Effective Exchange Rates (relative to 19 other industrialized countries), 1979–1991
(Nominal effective exchange rates adjusted for GDP deflator; 1979 = 100)

Source: Commission of the EC.

the lira stands out in terms of its cumulative overvaluation which in turn raises questions about the sustainability of a policy which has relied largely on the exchange rate as an anti-inflation instrument. But this is, to a large extent, also true of the currencies of other southern European countries, such as the peseta and the escudo which had stayed for long outside the ERM. This in turn suggests that the exchange rate has been used more widely as an anti-inflation instrument. The counterpart to this overvaluation has been the gain in competitiveness of other countries, and especially the Benelux countries (Fig. 7.3).

There has been much discussion about the link between membership of the ERM and price convergence. After all, inflation rates of most industrialized countries, including non-EC countries of Western Europe, followed a very similar path during the same period. How much of the inflation experience inside the EC can be attributed to the operation of the ERM? The EC Commission (1991a: 199), among many others, has argued that '[D]uring the 1980s, the EMS was a powerful engine for disinflation in the Community.' This statement is very difficult to prove econometrically, although there is more than circumstantial evidence to indicate that participation in the exchange rate mechanism served, at least for a large part of the period under consideration, as an important additional instrument in the fight against inflation; and this seems to be particularly true of France, Italy and Ireland. Participation in the ERM acted as an external constraint on domestic monetary policies, while the exchange rate has also been used, as shown above, as an instrument of disinflation. With growing price convergence and the increased credibility of the system, there has been in recent years a significant convergence of both short- and long-term interest rates in the countries of the original narrow band of the ERM (for long-term interest rates, see Fig. 7.4). This subject will be discussed in more detail below in connection with the asymmetrical nature of the system.

On the other hand, there is little evidence of convergence in the case of budgetary policies. To the extent that such a convergence has taken place, it has been the result of autonomous decisions leading to a reduction of public sector deficits in several member countries for a large part of the 1980s and not the product of an effective co-ordination of national fiscal policies within the EC. In fact, the mechanism set up for the co-ordination

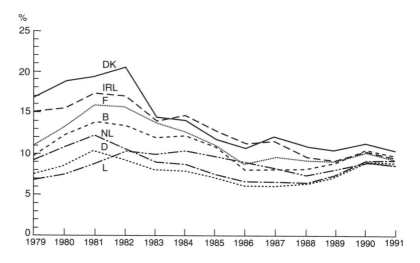

Fig. 7.4. (*a*) Nominal Long-term Interest Rates in the EC, 1979–1991

Source: Commission of the EC, 1991*a*.

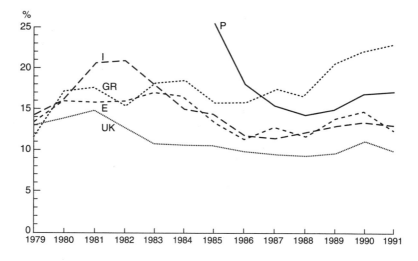

Fig. 7.4. (*b*) Nominal Long-term Interest Rates in the EC, 1979–1991

Source: Commission of the EC, 1991*a*.

of policies and based on the 1974 Council Decision on convergence, itself a product of the old days of Keynesian 'fine tuning' (see also Mortensen, 1990), never worked properly: it was strong on procedure and weak on implementation. Council deliberations often resembled the dialogue of the deaf, with each minister expounding at length on the virtues of his country's policies, without the slightest inclination of modifying them for the 'general good'. Linked to the first stage of EMU, a new co-ordination framework was created with the Council Decision of April 1990. Multilateral surveillance is likely to acquire new teeth, mainly because of the various criteria established at Maastricht for the participation of member countries in the final stage of EMU.

Although the objective of monetary stability, both external and internal, has been largely achieved, even though fiscal policies have hardly converged, other objectives initially associated with the creation of the EMS have been only partially or not at all fulfilled. For example, the balance sheet on the development of a common currency has been mixed, at best. The official ECU has remained underdeveloped and under-utilized. The problem partly lies with the various restrictions imposed on its use and particularly with its non-permanent character, being still based on the three-month revolving swaps of foreign reserves held by EC central banks. On the other hand, the role of the ECU as a basket of currencies has led to a somewhat unexpected development, namely the creation of ECU-denominated instruments in the financial markets. There was rapid growth during the first half of the 1980s, which later slowed down as the uncertainty in terms of intra-ERM exchange rates was reduced and hence also the attraction of the basket.

The transition to the second stage of the EMS, which had been envisaged in the original agreement, never materialized. The system continued to operate for many years on the basis of transitional structures and arrangements, although it was legitimized as an integral part of the Community system through the SEA. A few years later, it provided, however, the basis for plans and eventually for the new articles of the treaty referring to the establishment of EMU. The first stage of the latter started in July 1990, and the EMS constituted, without any doubt, the central pillar around which the new edifice started to be constructed.

Changing Asymmetry

Despite special provisions, such as the creation of the divergence indicator, the EMS has operated in an asymmetrical fashion, thus following the earlier example of the snake. The degree and nature of the asymmetry have changed over time and so has the assessment of its effects, which has also been a function of the economic paradigm used to analyse those effects. After all, this is hardly a topic on which economists would be expected to agree.

As with the snake, the asymmetry inside the system, and more precisely the ERM, relates to the central role of the Deutschmark. The Delors report on EMU (Committee for the Study of Economic and Monetary Union, 1989: 12) explicitly referred to the German currency as the 'anchor' of the system, to which other currencies were pegged. The source of the asymmetry is dual: the German low propensity to inflate, at least before unification, and the international role of the DM. The former, combined with the economic weight of the country and the priority attached by the other EC partners to exchange rate stability and the fight against inflation, enabled Germany for many years to set the monetary standard for the other countries. On the other hand, the increasingly important role of the DM as an international reserve currency has placed the German central bank in a key position with respect to the external monetary policy of the EMS as a whole.

No realignment of ERM central parities in the first thirteen years of the system ever involved a depreciation of the DM in relation to any other currency (see Fig. 7.5 for the evolution of bilateral DM rates of the original member currencies of the ERM). Fig. 7.5 also gives some indication of the margin of manœuvre used by other countries *vis-à-vis* Germany, ranging from the Netherlands at one extreme, with an almost complete alignment on German monetary policy, to Italy and France which for several years made repeated use of currency realignments as well as foreign exchange interventions and capital controls, at the other.

Asymmetry in a system of fixed (even if periodically adjustable) exchange rates is linked with the unequal distribution of the burden of intervention and adjustment and also of influence in the setting of policy priorities. The asymmetry in the EMS has manifested itself in different ways, and it reached the highest

% change w.r.t. Jan. 1980

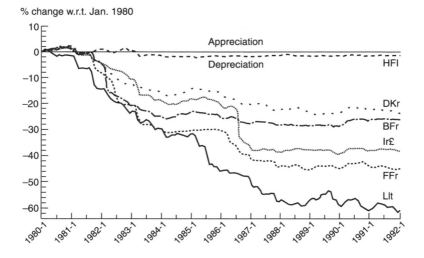

Fig. 7.5. Bilateral DM Rates of 6 ERM Currencies, 1980–1992 (Price of DM in terms of each of the 6 ERM currencies; the exchange rates are monthly averages)

Source: Deutsche Bundesbank.

point during the second phase between 1983 and 1987. Before that, there had been little policy convergence and frequent currency realignments; hence relative symmetry being combined with monetary instability. We shall concentrate here on the experience of the second phase, and then we shall go on to examine the main changes which have happened since then.

Especially after the March 1983 realignment, most central bank interventions in the exchange markets were taking place inside the permitted margins of fluctuation. The aim of intra-marginal interventions was to strengthen market confidence in the stability of existing bilateral rates and thus avoid the development of crisis situations. The heavy reliance on intra-marginal interventions had a number of important consequences. First of all, the burden of supporting the exchange rate system rested disproportionately on the shoulders of the countries with the weaker currencies. Furthermore, the very short-term credit facility, which could be triggered off automatically only as a result of marginal interventions, fell increasingly into disuse; and this also had a negative effect on the use of the ECU in the credit mechanism. The recourse to intra-marginal interventions also meant that the threshold of the divergence indicator was rarely reached, thus contributing to what had already appeared as the inoperational nature of this new mechanism. Therefore, the attempt made with the EMS to design rules which would guarantee a certain degree of symmetry between strong and weak currencies proved almost totally ineffective.

The asymmetry in foreign exchange interventions was also manifested in the implicit division of labour between the German and the other central banks in the system, with the latter intervening mainly in order to support intra-ERM exchange rates, while the former concentrated on exchange relations with third currencies, and most notably the US dollar. This division of labour was also reflected in the holding of foreign currencies by EC central banks; the Bundesbank holding only relatively small amounts of other EC currencies.

Another manifestation of asymmetry in the EMS was through the interdependence of interest rates in different national markets. Short-term interest rates have been consistently used as a key instrument in the exchange rate policy of participating countries. There is strong evidence to suggest that in times of crisis in the past, usually preceding realignments, the burden of adjustment was borne mainly by France, Italy and the smaller countries, with capital controls being used to insulate domestic interest rates from the effects of foreign exchange speculation. According to Giavazzi and Giovannini (1989: 75): 'the data on interest rates suggest that only Germany sets monetary policy independently. Italy and France can either accommodate German monetary policies perfectly or decouple domestic and foreign interest rates, at least temporarily, by resorting to capital controls.'

This is also confirmed by the fact that, more often than not, interest rate changes initiated by the Bundesbank for domestic stabilization purposes were followed almost immediately by many other European central banks, including those of countries not even participating in the ERM. There was little doubt among central bankers as to who was the leader in the European game. Although there is evidence to suggest that, even at the peak of the asymmetry inside the EMS, German monetary policy did not remain completely unaffected by policies pursued in the other countries, the distribution of influence was undoubtedly highly unequal (Gros and Thygesen, 1992).

Is asymmetry necessarily bad, at least from an economic point of view? Giavazzi and Giovannini (1989: 63), for example, argued that 'the EMS reproduces the historical experiences of fixed exchange rate regimes'. The literature on fixed exchange rates refers to the so-called N-1 problem which means that in a group of N countries (and currencies) there can only be N-1 policies that can be set independently. In the EMS, as with the Bretton Woods system earlier, the centre country retains the ability to set its monetary policy independently, while the other countries peg to its currency.

There are two sides to the asymmetrical nature of the EMS. On the one hand, it can be argued that countries with a higher propensity to inflation, such as France, Italy, and Ireland, borrowed credibility by pegging their currencies to the DM and they consequently reduced the output loss resulting from dis-inflationary policies. This is the alleged advantage of 'tying one's hands' to the DM anchor (Giavazzi and Pagano, 1988). Participation in the ERM is also supposed to have strengthened the commitment of national authorities to non-inflationary policies. The latter, softer version of the argument is perhaps more plausible. Participation in the ERM introduced an external discipline and thus reinforced the hand of institutions and interest groups inside a country fighting for less inflationary policies. This largely explains the popularity of the system with most central bankers, which is contrary to their earlier expectations. The Banca d'Italia is the best example in this respect. However, faced with the profligacy of the domestic political class and an inflexible labour market, the Italian central bank has also experienced the serious limitations of an anti-inflation strategy based almost entirely on the exchange rate and monetary policy.

The advantage of 'tying one's hands': as with most arguments in economics, there is also the other hand. The asymmetry of the EMS has also been seen as leading to a deflationary bias in the system (de Grauwe, 1987; Wyplosz, 1990). The early years of the EMS coincided with the worst and longest recession of the post-war period in Western Europe, which saw unemployment rates reach unprecedented levels. Until the mid-1980s, growth rates of the participating countries were signfcantly below those enjoyed by their main outside competitors, and intra-EMS trade remained stagnant, thus raising doubts about the beneficial effects of exchange rate stability on trade.

The criticism referring to the deflationary bias of the EMS relates to the more general debate of the early and mid-1980s in Western Europe regarding the causes of low growth. The latter was usually attributed either to supply-side factors, as part of the notorious 'Euro-sclerosis', or to the restictive policies pursued by European governments during this period. In different terms, the debate was largely about how much of the unemployment in Western Europe could be attributed to classical or Keynesian factors and thus whether there was room for macroeconomic expansion.

The combination of low inflation and a large trade surplus, coupled with its central position in the EMS and the economic weight of the country, made Germany the natural candidate for the adoption of reflationary measures. As the progressive devaluation of the US dollar after 1985 started eliminating one of the main factors behind Western Europe's modest export-led growth, the pressures for the so-called two-handed approach mounted (Commission, 1985*b*; Drèze *et al.*, 1987). This piece of Euro-jargon meant that supply-side measures aiming at greater flexibility of labour and product markets and the attempt to bring down real wages and salaries should be complemented by a co-ordinated expansion of fiscal policies; a unilateral expansion being excluded because of large import leakages in a system of fixed exchange rates. With inflation rates having already reached historically low levels, some of the participating countries started then to adopt a less benign view of German leadership. But the crisis in the EMS was at least temporarily averted as a result of changes in the rules and the dramatic improvement in the economic environment.

Following strong pressures, mainly from France and Italy, an agreement was reached in September 1987 (the so-called Basle–Nyborg agreement) aiming at a partial correction of the asymmetrical nature of the system. It referred to a 'presumption' that loans of EC currencies would be available through the EMCF for the financing of intra-marginal interventions. It also announced the extension of the very short-term credit facility and the use of ECUs for the settlement of debts as well as the intention of central bankers to achieve a closer co-ordination of interest rates. Although they have not completely eliminated the asymmetry, those measures have led to a more equitable distribution of the burden of intervention between different countries. As a counterpart to the extension of credit facilities, participating countries also expressed the intention to make realignments of central parities as infrequent and as small as possible, while also making explicit their intention not to compensate fully for inflation differentials. This intention has been confirmed by the development and the further strengthening of the EMS in the third phase.

The change of operating rules in September 1987 was accompanied by a largely unexpected improvement of the economic *conjoncture*. This was due to exogenous factors such as the decline in oil prices, the relaxation of monetary policies in the aftermath of the 1987 stock exchange crash (the effects of which on consumption were grossly exaggerated), and the 1992 effect. This led to the relaxation of tension inside the EMS, which lasted for some time. But other important developments have happened since then, which have had a major impact on the functioning of the system. One such development has been the liberalization of capital movements, as part of the internal market programme. Higher capital mobility, combined with no currency realignments, has imposed much tighter constraints on the monetary policy of individual countries, while at the same time loosening further the link between the exchange rate and the current account. This is exemplified in the case of countries such as Italy and Spain which have been able to attract for years large inflows of capital through high nominal interest rates; at least as long as ERM central parities remained credible. One consequence of the large capital inflows was that the peseta stayed for long at the upper end of its permitted margin of fluctuation, leading to complaints about the upward pressure exerted on the interest rates of partner countries. Giavazzi and Spaventa (1990) wrote about the new EMS in which the monetary standard would be effectively set by the weaker members.

And then came German unification, the early effects of which preceded the actual event. Rapidly growing public expenditure in the new *Länder* led to high budget deficits in Germany and the acceleration of inflation. The burden of the stabilization effort fell then almost entirely on monetary policy, and this led to higher interest rates in Germany and also indirectly in the other countries of the narrow band of the ERM. This coincided with

Mrd US$

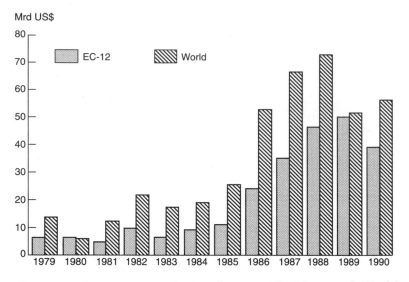

Fig. 7.6. Germany's Trade Surpluses with EC-12 and World, 1979–90

Note: excluding Eastern Germany.

Source: OECD.

the end of the economic boom, when the anticipation effect of 1992 also started petering out. In times of recession and growing unemployment, high interest rates, seen as imposed by the German policy-mix, were strongly resented by Germany's partners. And this led to renewed tension in the system. The effect of the German policy-mix was, however, attenuated (for how long?), because the weakening of the DM in exchange markets, coupled with the increased credibility of the system, allowed the other members of the ERM to reduce progressively their nominal interest rate differentials with respect to the Deutschmark.

On the other hand, the rapid expansion of domestic demand in unified Germany brought about a large increase in imports, which helped to reverse the previous trend of steadily growing trade surpluses for Germany (Fig. 7.6). Those surpluses had been seen in the past as one of the prizes gained by the leading country, linked at least partially with the overvaluation of some currencies inside the EMS. According to the EC Commission (1991*a*: 8), the positive impact of German unification through

trade on the rate of growth of the other EC countries was estimated to be around 0.5 percentage points for 1990 and 1991. This
should be, therefore, set against the negative impact of high
interest rates (see also Siebert, 1991; Atkinson *et al.*, 1992).
The overall economic effect of German unification on different
EC countries varied, depending among other things on their
competitiveness and the size of their public debt.

The weakness of the DM, linked to the acceleration of German
inflation which by 1991 was higher than that of several members
of the ERM, raised questions about the continued role of the
DM as the anchor of the system, and hence also about the
continued asymmetry inside the EMS. Would this prove to be
a short-term aberration or would it, instead, lead to a more
symmetrical system based on a 'more explicit co-ordination of
policies with Community-wide considerations' (EC Commission,
1991*a*: 198)?

The success of the EMS has been based on a consensus about
exchange rate stability and disinflation, which was centred for
several years around German monetary policy. Is asymmetry the
inevitable price to pay for monetary stability? Views have varied
on this subject, and the leading role of Germany has not been
always accepted with equanimity by the other members of the
system, especially in times of low growth. The asymmetrical
nature of the EMS has been both modified and attenuated in
recent years. In fact, there may be question of a more fundamental change being brought about by a combination of factors,
such as the progressive convergence of inflation rates, the increased credibility of the system, the liberalization of capital
movements and the destabilizing effects of German unification.
In the intervening period leading to the final stage of EMU, the
system may come under further strain as markets anticipate the
last realignment before the irrevocable fixing of parities.

EMU Is Back on the Stage: The Politics of Economic Fundamentals

In the late 1980s, EMU came back with a vengeance on the
European agenda, and it very quickly reached the most prominent
place. It was in many respects a repetition of what had happened
twenty years earlier, although this time there is a reasonable

chance that the final outcome will be different. EMU was generally presented as the logical continuation of the internal market programme, the next target to aim for in the long process of economic and political unification of Europe. The decision to proceed with monetary integration would help to sustain business confidence, a major factor behind the boom of the late 1980s, which started showing signs of petering out; hence also the urgency for a political commitment to the next stage of integration while the 'Euro-euphoria' still lasted. Monetary union would be the final and irrevocable confirmation of the reality of the single European market and the European economy. A common currency was seen as the means of welding national economies together, and also the means of accelerating the movement towards political union. The developments in Eastern Europe in 1989 and German reunification provided a powerful catalyst which rapidly brought EMU back to the centre of the stage. Those developments called for a stronger Community on the European scene and also a Community which would provide a stable and secure framework for larger Germany; and money once again provided the instrument.

Proposals for EMU were directly linked to the 1992 process. Although mentioned in the SEA, no binding commitment had been undertaken in the latest revision of the treaties for the realization of EMU. Yet the increase in trade and economic interdependence, which would inevitably result from the completion of the internal market, should be expected to reduce further the effectiveness of the exchange rate as an instrument for the correction of payments imbalances. Furthermore, the very concept of the internal market was, arguably, difficult to reconcile with the considerable transaction costs involved in converting between different currencies and the uncertainty associated with changes in exchange rates. Thus exchange rates were presented as another NTB to be eliminated. On the other hand, unlike previous occasions, CAP considerations did not seem to play an important role. The exchange rate stability already achieved through the EMS and the diminishing importance of agriculture in European integration had helped to sever the link between EMU and the CAP.

The positive balance sheet of the EMS and the considerable experience and knowledge which central banks had acquired in the joint management of exchange rates were important additional factors. Gone was the old enthusiasm for independent floating, as a means of achieving national monetary independence, and also the belief in the 'efficiency' of exchange markets (Allsopp and Chrystal, 1989); although, with the exception of Britain and, to a lesser extent, Germany, such enthusiasm had never been very pronounced in the open economies of Western Europe. The transition from the EMS to EMU also had a major significance for countries such as France and Italy, because the creation of common institutions would lead to a shift from the existing German-centred system to new forms of collective management. Proposals for EMU were at least partly driven by the search for greater symmetry. Hence the considerable scepticism shown by the central bankers in Frankfurt.

As far as interest in EMU is concerned, the most important aspect of the internal market was, undoubtedly, the decision to liberalize capital movements. Padoa-Schioppa (1988) talked about the 'inconsistent quartet' of economic objectives. It includes free trade, free capital movements, fixed exchange rates, and monetary autonomy. His conclusion was that since the EC countries have committed themselves to the first three objectives, through the internal market, the EMS, and the decision to liberalize capital movements, monetary autonomy would also have to give way, moving gradually from an effective co-ordination of national policies to the centralization of monetary policy at EC level. This is also the main argument on which the EC Commission strategy was later based; another example of the Commission trying to make full use of functional spill-over.

Padoa-Schioppa's argument assumes, first of all, that EC countries are fully committed to the other three objectives, which has not always been true in the past. Thus, while Britain opted for most of the 1980s for free capital movements and monetary autonomy, at the expense of exchange rate stability, France and Italy concentrated during this period on the latter objective, while sacrificing free capital movements and part of their monetary sovereignty in the process. Wider margins of fluctuation and more frequent realignments inside the ERM could have been a possible compromise in the search for politically acceptable trade-offs between the four above-mentioned objectives. On the other hand, the fear of destabilizing speculation was clearly the most important factor behind the safeguard clause included in the directive on the liberalization of capital movements, although it remains to be seen whether capital controls could be effectively re-applied in cases of emergency.

Earlier initiatives in the field of European monetary integration had been largely motivated by external preoccupations: the instability of the dollar and US policies of 'unbenign neglect' serving as powerful federalizing factors in Europe. This was not true of the various initiatives which finally led to the new treaty provisions for the establishment of EMU; or, at least, not to the same extent as in the past. True, the reform of the international monetary system was not on the cards and the lack of unity of European countries remained an important factor behind the continuing asymmetry in the international system. But this asymmetry was now less evident, the Bush Administration did not adopt the aggressive stance of its predecessor, and intra-ERM exchange rates appeared to be less vulnerable to the gyrations of the dollar. Perhaps less preoccupation with external factors was also a sign of the new collective confidence of the Europeans.

The main driving force for the relaunching of monetary union came from Brussels and Paris, with EMU representing the flagship of the European strategy of both the EC Commission and the French Government for the 1990s. There was also strong support from Italy, Belgium, and, with some qualification, from Spain. Initially, Germany showed little enthusiasm; the Government and the central bank were happy with the *status quo* and any move towards monetary union was perceived, quite rightly, as leading to an erosion of Germany's independence in the monetary field. What later tipped the balance was the perceived need to reaffirm the country's commitment to European integration in the wake of reunification. This is how the matter was presented in Paris and Brussels. Thus the German decision to proceed with EMU was fundamentally political, as it was, indeed, in most other member countries. Once a Franco-German agreement had been reached on the subject of EMU, the process appeared almost unstoppable; thus becoming a repetition of earlier patterns of European decision-making. The Dutch shared much of the economic scepticism of the Germans, but their margin of manœuvre was extremely limited. The main concern of the small and less developed economies was to link EMU to more substantial budgetary transfers and to avoid an institutionalization of two or more tiers in the Community. As for

Britain, it remained the only country ready to question in public the desirability and feasibility of EMU, both on economic and political grounds. However, drawing some lessons from the earlier stages of European integration, the British Government, even under Mrs Thatcher, made a conscious effort to remain in the negotiating game.

In terms of decision-making, the approach to EMU bears considerable resemblance to earlier European initiatives and especially the one which had led some years ago to the adoption of the internal market programme. The gradual build-up of momentum, the steady expansion of the political support base through coalition building, and the isolation of opponents have been combined with an effective marketing campaign addressed primarily to opinion leaders and the business community. Functional spill-over was also successfully mixed with high politics and the appeal to 'Euro-sentiment'; a recipe which had proved quite powerful in the past.

In June 1988 the European Council of Hanover set up the 'Committee for the Study of Economic and Monetary Union', under the chairmanship of the President of the Commission, Mr Delors. This decision was taken only a few months after the adoption of the directive on capital liberalization, and the two were directly linked. The committee included all governors or presidents of EC central banks plus a few independent experts; its unanimous report was submitted in April 1989 (Committee for the Study of Economic and Monetary Union, 1989). There were many similarities with the Werner report which had appeared almost twenty years earlier, something which should come as no surprise since the briefs of the two committees were virtually identical. The final objective in terms of monetary union remained the same, namely complete liberalization of capital movements, the irrevocable fixity of intra-EC exchange rates, coupled with the elimination of margins of fluctuation (and possibly the replacement of national currencies by a single currency), and the centralization of monetary policy. The Delors report was, however, more explicit about the necessary transfer of powers to the level of the union and the institutional changes required.

In fact, the central bankers, constituting the large majority in the committee, appeared only too keen on stressing the full economic and institutional implications of EMU to their political masters. They called for a system of binding rules governing the size and the financing of national budget deficits and referred to the need to determine the overall stance of fiscal policy at the EC level, with decisions taken on a majority basis. The disciplinary influence of market forces on national budgetary policies was not deemed to be sufficient on its own. Much emphasis was also placed on the independence of the new institution which would be in charge of monetary policy for the union. This consensus on the question of independence of the future European central bank reflected not only the composition of the committee but also the leadership role and the prestige enjoyed by the Deutsche Bundesbank which had provided the role model.

As for the intermediate stages, the strategy of parallelism seemed to have survived the long time separating the publication of the Werner and the Delors reports. In the latter, the major institutional changes and the application of the new treaty rules were reserved for the second stage of EMU. On the other hand, it stressed that 'the decision to enter upon the first stage should be a decision to embark on the entire process' (p. 31); certainly, not the right message for the faint-hearted.

The Delors report set the EMU ball rolling; and it did roll very fast indeed. On the basis of the report, a decision was taken at the Madrid European Council of June 1989 to proceed to the first stage of EMU on 1 July 1990 which coincided with the complete liberalization of capital movements in eight members of the Community. This was followed by the decision reached at the European Council in Strasbourg in December of the same year to call for a new intergovernmental conference to prepare the necessary treaty revisions for a complete EMU. Both decisions were taken unanimously, despite the expressed opposition of the UK and the persisting differences on important aspects of EMU and the nature of the transitional period among the other members. In a subsequent meeting of the European Council, the date for the second stage of EMU was fixed for 1994, thus following one year after the expected completion of the internal market. Before the official opening of the new intergovernmental conference in December 1990, a great deal of the preparatory work had already been done in the context of the Committee of the Governors of Central Banks and the Monetary Committee. This included the draft statutes of the European Central Bank. The Commission had also published a major study of the potential costs and benefits of EMU, very much along the lines of the Cecchini report (Commission of the EC, 1990e). Furthermore, between the publication of the Delors report and the opening of the intergovernmental conference, important changes had taken place inside the ERM, with two more currencies joining (the peseta and sterling) with a 6 per cent margin of fluctuation, while the margin for the lira had been reduced to 2.25 per cent.

Thus, money was once again at the centre of European high politics and, even more so than with the ERM, commitment to monetary union was almost indistinguishable from the more general commitment to European unification. It was certainly not a coincidence that the other intergovernmental conference on political union was running concurrently with that on EMU. But economic fundamentals do not always conform to high political priorities. From an economic point of view, the discussion on the desirability of the final objective essentially boils down to a comparison of the benefits of monetary union with the costs of forgoing the exchange rate as an instrument of economic policy (Commission of the EC, 1990e; see also Allsopp and Chrystal, 1989). A corollary to this question would be to compare the effectiveness of the exchange rate to other alternative instruments of policy geared towards economic adjustment and changes in the real competitiveness of national economies which, presumably, will continue to be needed for some time.

The benefits of monetary union are linked to the elimination of foreign exchange transaction costs and the factor of uncertainty associated with exchange rate fluctuations. According to the Commission's study, these direct benefits would be relatively small, although, as with the earlier work on the internal market, the dynamic gains were expected to be much larger and more difficult to quantify. The creation of a single currency (the emphasis had already shifted towards the creation of a single currency as the final objective as opposed to the retention of national currencies linked through irrevocably fixed exchange rates) would secure further the credibility of the internal market and the gains associated with its completion; 'one market' and now 'one money'. The Commission's study also talked in terms of the benefits from a lower average rate of inflation which would be expected to result from EMU. However, this would obviously depend on the policy stance and the credibility of the new European Central Bank, which would be, of course, one of the crucial questions for the future. On the other hand, the adoption of the ECU as an international currency, thus replacing and eventually extending the international role of the Deutschmark, was supposed to bring seigniorage gains resulting from the readiness of outsiders to hold the new currency. An international currency would also be accompanied with gains in prestige and political power, associated with a common monetary policy *vis-à-vis* the rest of the world. However, the Commission's expectations regarding the gains from the future international role of the ECU appeared to be somewhat vague and also exaggerated (Goodhart, 1993).

The relative effectiveness of the exchange rate instrument at the national level and the possible costs resulting from an irrevocable fixity of intra-EC exchange rates (and/or the adoption of a common currency) remain big questions on which there are as many views as there are economists. Many would agree about the limited effectiveness of the exchange rate because of large trade interdependence, the decisive importance of financial transactions in determining the price in exchange markets, and the limited 'efficiency' of those markets. The experience with floating exchange rates has not been particularly encouraging, in view of frequent 'overshooting' and the persistent misalignment of currencies. An increasingly fashionable paradigm in the economics profession stresses the relative ineffectiveness of the exchange rate and links this to inflation which is basically seen as a credibility problem (Mortensen, 1990; Gros and Thygesen, 1992). However, still only a minority of economists would go as far as arguing the complete ineffectiveness of the exchange rate, even for a small and open economy such as Belgium which has, for example, made a limited but apparently successful use of this instrument in the context of the ERM (the franc was devalued by 8.5 per cent in February 1982).

Very few would also argue that the EC is already an optimum currency area. This is a concept which became very popular in the literature during the 1960s when economists looked for factors which could act as near substitutes for the exchange rate instrument (Mundell, 1961; McKinnon, 1963; see also Krugman, 1990); such factors also being, usually, good indicators of the degree of economic integration inside the potential currency area. That means that economic integration inside the EC has not yet reached the level where capital and especially labour mobility could act as near substitutes for changes in the exchange rate; or that wage and price movements in different countries correspond to changes in productivity rates so as to make exchange rate realignments redundant; or even that the EC economy is sufficiently homogeneous so that different countries and regions are not frequently subject to asymmetric external shocks. But is this true of the United States or even Italy, and if not, are those countries optimum currency areas?

There is still a long distance to cover in the Community in terms of economic convergence which, interestingly enough, means very different things to different people. Thus, when the Germans talk about convergence, they invariably mean convergence of inflation rates, while the Greeks, the Irish, and the Portuguese are more interested in the convergence of income levels. The experience of the EMS in terms of both internal and external monetary stability has been positive. However, in terms of real economic variables, the experience has been mixed. The EC could be approaching fast the uncomfortable state in which, while national wages and prices still have a life of their own, being a function of history, political institutions, social traditions, and labour power among other factors, and while the need for adjustment differs considerably from one country to the other, the effectiveness of the exchange rate becomes increasingly curtailed, even before the instrument is politically banished. This could become even more uncomfortable if the move towards monetary union is not accompanied by a corresponding convergence in some of the economic fundamentals. The loss of the exchange rate instrument, coupled with a continuing need for real adjustment, could turn countries into depressed regions; and other factors, such as labour mobility and interregional budgetary transfers, would remain too weak to act as effective substitutes.

A comparison has been sometimes drawn between EMU at the European level and monetary union between West and East Germany, although the differences between the two are quite substantial. German monetary union led, at least in the early years, to a major decline in production in the Eastern *Länder* and a large increase in unemployment. The shock of unification in the East was, however, attenuated by substantial budgetary transfers from the West (and only to a much lesser extent through private investment flows). Wage convergence proceeded much faster than convergence in productivity rates, and this threatened to jeopardize the economic prospects of the new *Länder*. At the same time, labour mobility was high, thus adding to the pressure for wage convergence (see also, Siebert, 1991). Thus, on the one hand, public transfers and labour mobility helped significantly to ease the burden of adjustment for what used to be the German Democratic Republic; and in this respect, German monetary union is likely to be very different from European EMU. On the other hand, the state of the former centrally planned economy and the artificially high exchange rate, adopted for political reasons, made the scale of economic adjustment very large. Though the German experience could serve as a warning sign of some of the problems lying ahead, the qualitative differences with European EMU should not be underestimated.

Of course, in the discussion about EMU, some differentiation is needed with respect to individual countries. The attachment to stable, although not necessarily fixed irrevocably, exchange rates is usually closely related to the openness of an economy to international trade. In the case of the EC, the readiness of individual countries to accept the implications of a monetary union should also be a function of the relative importance of intra-regional trade. A broad distinction can be drawn between small and big countries in the EC as regards their openness to international trade (Table 7.3). The Benelux countries and Ireland on the one hand and the Big Four on the other (the German economy — the data only refer to West Germany — being significantly more open than the other three) fit nicely into those two categories. By comparison, Greece and Spain are still relatively closed economies (at least in terms of trade in goods, since the inclusion of services would significantly modify this picture), while Denmark and Portugal find themselves in between the two groups. As for the importance of intra-EC trade, this is relatively more pronounced in the case of the Benelux countries and Ireland, although differences between member countries have gradually diminished over the years. Intra-EC trade now represents on average 60 per cent of total trade for member countries (ranging from 50 per cent for the UK to 72 per cent for Belgium – Luxembourg and Ireland).

On the other hand, the Community does not start from zero on the road to EMU. The use of the exchange rate is already severely curtailed in the context of the ERM. In fact, the lack of currency realignments for several years, combined with capital liberalization, has brought the system very close to a *de facto*

Table 7.3. Openness of EC Economies, 1960–1990

(Average of imports and exports of goods as a percentage of GDP;[a] intra-EC trade is given as a % of total trade)

	1960–7		1968–72		1973–9		1980–4		1985–90	
	World	Intra-EC	World	Intra-EC	World	Intra-EC	World	Intra-EC	World	Intra-EC
Belgium and Luxembourg	37.5	64.8	42.8	71.2	48.9	71.2	61.5	67.0	60.8	71.7
Denmark	27.0	52.3	23.4	46.7	25.3	48.2	29.1	49.1	26.7	50.8
Germany	15.9	44.8	17.6	50.9	20.4	50.3	24.8	50.5	24.9	53.1
Greece	12.7	50.6	12.7	53.4	17.6	47.3	19.5	48.1	20.8	60.5
Spain	8.1	47.8	9.2	44.4	11.2	41.3	14.4	39.3	14.4	56.0
France	11.0	45.8	12.5	57.2	16.6	55.0	18.9	53.9	18.6	61.9
Ireland	33.6	72.2	35.3	71.6	45.8	74.3	50.1	73.1	52.0	72.2
Italy	11.7	42.9	13.1	49.5	18.4	48.7	19.4	46.1	16.9	55.0
Netherlands	37.7	62.7	36.4	67.8	40.6	65.9	47.8	64.1	49.0	67.8
Portugal	19.8	48.3	20.3	49.6	22.2	50.2	30.1	50.1	32.5	65.1
United Kingdom	16.0	26.7	16.7	31.2	22.3	37.9	21.6	44.1	21.0	49.9
EC-12	8.8	45.0	8.3	51.9	10.2	52.6	11.5	52.2	9.8	59.8

[a] Figures for world trade as a % of GDP have been calculated by inserting country data for imports and exports of goods (SITC categories 0–9) in the formula:

$$\frac{\frac{1}{2}\Sigma(x + m)}{GDP} \times 100.$$

For the Member States, these figures include intra-Community trade; for EC-12, intra-Community trade has been excluded.

Source: Eurostat.

monetary union, without some of the benefits. The question, however, remains whether this situation can be sustainable for long, even though there has been significant convergence of inflation rates among most of its members. In the aftermath of the Maastricht decisions, the costs and risks associated with an irrevocable fixity of exchange rates would appear to be much bigger for Britain than for the Netherlands, and certainly even bigger for Greece and Portugal. We shall return to this subject when we discuss the transition to the final stage of EMU.

To the old familiar arguments about the flexibility of labour markets and trade union militancy, Dornbush (1988; see also Commission of the EC, 1990e) has added the argument about seigniorage as an important means of financing large public debts in some countries. The gains from seigniorage are mainly a function of the rate of inflation, the amount of cash transactions and the reserve requirements for commercial banks; and those gains could not be easily replaced by increased tax revenues because of political and institutional constraints. This argument is relevant for some EC members. According to the Commission's study on EMU, the revenue effects through seigniorage exceeded in 1988 2 per cent of GDP in both Greece and Portugal, and 1 per cent in Italy and Spain. Thus, the transition to low inflation rates for the southern European countries could be both painful and long.

Define Virtue and Achieve it Later

During the intergovernmental conference, the economic and political desirability of EMU was not seriously put in question. This matter was supposed to have been already settled. The political decision had been taken at the highest level, and only the British were ready to express their economic and political doubts in public. The other doubters, and they did seem to exist inside the governments and central banks of several member countries, preferred to concentrate on specific problems, instead of challenging the main principles and objectives. Thus, the negotiations revolved around other issues which could be grouped together into three main categories: the institutional framework of the union, and especially the role and statutes of the European central bank; the balance between the monetary and the fiscal arm of the union; and the nature and length of the transitional period leading to a complete EMU, including the criteria for participating in the final stage. The influence of Germany on the overall package agreed at Maastricht was decisive.

We shall start by examining the transitional arrangements before the final stage. The French wanted a firm commitment about a date for the final stage, which would tie the others, and especially the Germans, to the objective of a complete EMU. The Germans, from their side, wanted to delay any serious transfer of powers to central institutions which, once created, should be as close as possible to the German model. They also insisted on strict rules for the admission of countries to the final stage, consistent with the old 'economist' line that economic convergence should precede monetary union. The economically weaker members stressed the link between EMU and cohesion (read budgetary transfers), while also favouring loose criteria for admission to the final stage, which might allow them to sneak in. As for the British, they refused any firm commitment about the final stage, which they could not apparently sell at home.

The final compromise was essentially based on a French time-table and German conditions. The first stage, which had already started in July 1990, was meant as a consolidation of the *status quo*. This should include the liberalization of capital movements and the inclusion of all currencies in the narrow band of the ERM. Economic convergence should be promoted on the basis of multiannual programmes for each member country to be discussed in the context of the Council Decision of 1990 on multilateral surveillance. The second stage was planned to start on 1 January 1994. Although monetary policy will remain the exclusive responsibility of national authorities, a new institution will be created, the European Monetary Institute (EMI), which will prepare the way for the future European Central Bank (ECB) by further strengthening the co-ordination of national monetary policies. It remains, however, to be seen whether the setting up of the EMI, which is intended to have only a short spell of life, will constitute an important qualitative change to the co-ordination mechanisms already existing in the Community. During the second stage, member countries will also be expected to take the necessary legal steps to ensure the independence of national central banks, wherever this independence does not already exist.

The crucial part of the transitional arrangements referred to the conditions to be fulfilled for the beginning of the third and final stage which will include the irrevocable fixity of exchange rates and the adoption of a single currency as well the creation of a European System of Central Banks (ESCB), with the ECB in its centre. The first attempt will be made before the end of 1996, the European Council deciding by qualified majority and on the basis of reports from the Commission and the EMI whether a majority of member countries fulfil the conditions for being admitted to the third stage. A qualified majority decision about whether there is a majority fit to enjoy the fruits of paradise (and EMU): this will be a rather uncommon case in international and even domestic politics.

In fact, the authors of the new treaty went further. It was stipulated that the third stage will start on 1 January 1999 at the latest, irrespective of how many member countries are found to fulfil the necessary conditions; again the European Council deciding on each case and on the basis of qualified majority. Those failing the test will remain in derogation and, to all intents and purposes, they will be excluded from the new institutional framework. Their case will, however, be examined at least every two years. There were two further complications: the UK refused to commit itself in advance to participating in the final stage of EMU, with an 'opt out' protocol which left the decision for a future government and parliament, while Denmark chose a softer version of 'opting out', thus referring to the possibility of a referendum prior to its participation in the final stage.

The conditions for admission to the final stage, otherwise known as convergence criteria, were quite explicit and they concentrated exclusively on monetary variables. The first referred to a sustainable price performance, defined as a rate of inflation which should not exceed that of the best performing member countries by more than 1.5 percentage points. The second related to national budgets: actual or planned deficit should not exceed 3 per cent of GDP, while the accumulated public debt should not be above 60 per cent of GDP. With respect to this criterion, the wording of the new Article 104c of the treaty left some margin of manœuvre by allowing for exceptional and temporary deficits and also for government debts that may exceed the 60 per cent ratio as long as they have been declining 'substantially and continuously'. Exchange rate stability was the third criterion: the national currency must have remained within the narrow band of the ERM for at least two years prior to the decision about the final stage, without any devaluation and without any severe tension. The fourth criterion was meant to ensure that the exchange rate stability is not based on excessively high interest rates; thus, the average nominal interest rate on long-term government bonds should not exceed that of the three best performing member states by more than two percentage points.

Thus, the treaty on European Union and the attached protocols went into considerable technical detail on the subject of EMU. As the price for accepting a firm commitment about the date of the final stage, the Germans had insisted on the application of strict convergence criteria before any country were to be admitted to the final stage. In this respect, they also had the support of the Dutch. The crucial question is how much convergence in terms of inflation or budgetary policies is required before the irrevocable fixity of exchange rates; and there can be no single answer from economists. The criteria adopted in the treaty can be, perhaps, treated as too mechanistic and arbitrary; it remains to be seen how much convergence will be achieved in the intervening period and how flexible an interpretation is given to those criteria, when the European Council decision is taken.

It is also interesting that only monetary variables were used for the convergence criteria, which may be seen as a further confirmation of the prevailing economic paradigm that there is no real trade-off between price stability on the one hand and growth and employment on the other. Gros and Thygesen (1992) raise a pertinent question: can a temporary fall in inflation achieved at the expense of high unemployment be sustainable, since the disinflationary policy is apparently not credible in labour markets? On the other hand, a strict application of the convergence criteria, when the time of judgement comes, risks leaving outside the gates of paradise several member countries, including some who belong to the old 'core' group of the EC. Is that exclusion politically realistic, and if so, what would be the implications for the internal balance of the Community? Since a further enlargement of the EC is likely to have taken place before the beginning of the final stage of EMU, it is also worth noting that some of the present candidates for accession are closer to fulfilling the conditions for membership in the final stage than several of the existing members of the EC. This subject will be discussed further in Chapter 9.

In terms of inflation, we have already referred to three groups of countries in the EC on the basis of the experience of recent years (Fig. 7.1): the original members of the narrow band of the ERM on the one end, and Portugal and Greece on the other. In terms of budgetary policies, the situation appears somewhat differentiated. Belgium, Italy, Greece, and Ireland (although steadily improving its debt situation in recent years) are still way above the 60 per cent public debt target in terms of GDP (Fig. 7.7). As for current deficits, Greece and Italy form a group of their own. The budget deficits of those two countries have been persistently above 10 per cent of GDP for several years. The adjustment effort required for those countries in order to meet the two budgetary criteria set in the new treaty will therefore be more than considerable. For Greece and Italy, it may, indeed, necessitate major political and institutional reforms which are long overdue.

With respect to exchange rate stability, what will count will be the performance of each currency over a period of two years prior to the final decision. Judging from the experience so far in terms of internal and external monetary stability, the countries of the original narrow band of the ERM start with a distinct advantage. The situation is not very different in terms of long-term interest rates, the fourth criterion set out in the Maastricht treaty. As already explained before, convergence of inflation rates and exchange rate stability have been accompanied by a progressive narrowing of interest rate differentials (see Fig. 7.4). In this respect, the UK is closer to the countries of the original narrow band than either Italy or Spain.

In 1992 only France, Luxembourg and Denmark (with a small question mark in terms of its public debt) would have fulfilled all four convergence criteria, the irony being that Germany, the country which had insisted all along for raising the standards for admission, would have been excluded for failing both the inflation and the budget deficit criteria. Should reunification be treated as a temporary exogenous shock and what conclusions should be drawn about the future application of rather mechanistic rules? Both the snake and the EMS have contributed to the institutionalization of a variable speed Europe, although it could be argued that the EMS experience in particular suggests that

% of GDP

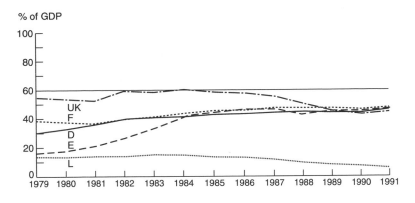

Fig. 7.7. (*a*) Gross Public Debt in EC Member States, 1979–1991
(Percentage of GDP at market prices)
Source: Commission of the EC.

% of GDP

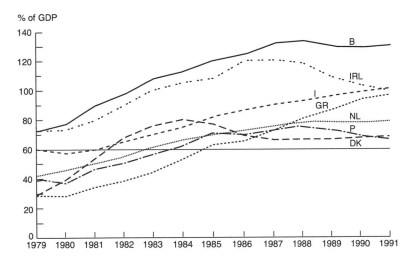

Fig. 7.7. (*b*) Gross Public Debt in EC Member States, 1979–1991
(Percentage of GDP at market prices)
Source: Commission of the EC.

an initial central core can act as a powerful pole of attraction towards which other countries try to converge. It is difficult to predict whether the same will happen with EMU. If the Community continues moving along in different speeds, this will inevitably have serious repercussions for the model of integration as a whole. Leaving aside the political problems which may be raised with respect to the future participation of the UK, even countries such as Belgium and Italy will have a hard time in meeting all criteria before the end of the decade. Will national political systems be able to deliver the goods under peer pressure and the threat of exclusion or will the final stage start with only a limited number of members? A third scenario would involve a more flexible application of rules.

On the other hand, EC negotiators have opted for a relatively long transitional period before the final stage, with no substantial institutional changes taking place in between. Considerations of economic convergence, as a pre-condition for monetary union, were happily married with the natural instinct of politicians and lesser mortals to postpone difficult decisions. 'God, give me virtue (and EMU), but not yet.' In the intermediate period, there may be uncertainty and instability in exchange markets, as agents try to anticipate the last realignment inside the ERM, which, according to the rules, cannot take place later than the end of 1996, at least for those who would like to be with the first group in entering the final stage (or 1994, if an early agreement were to be reached for the transition to the final stage). Will EMU, therefore, destabilize the EMS? And how certain is it that EMU will happen as planned? In view of the length of the transitional period and the importance of decisions which will need to be taken some years from now, economic and political realities prevailing at the time are likely to count more than any treaty obligations; hence another element of uncertainty.

The institutional framework of the union constituted a major part of the whole debate around EMU and the negotiations which took place in the context of the intergovernmental conference. In fact, the debate went much further by establishing a direct link between EMU and political union, the latter aiming among other things at the strengthening of central political institutions, although with relatively limited success in the end. Only the UK opposed the institutional approach to EMU, and proposed instead a market approach which was based on a plan for competing currencies and later on another one for the creation of a parallel currency. But those proposals had, in the end, little effect on the negotiations.

The new treaty created the ESCB which will be based on a federal structure, composed of the ECB and the national central banks. The latter will operate in accordance with the guidelines and instructions of the ECB. The primary objective of the new monetary institution will be to maintain price stability. It will be responsible for the conduct of monetary policy for the union as a whole and for foreign exchange operations. It will also hold and manage the foreign reserves of the member states; and provisions have been made for the transfer of part of those reserves to the ECB. The treaty and the attached protocol on the statute of the ESCB and the ECB refer explicitly to the political independence of the new institution, thus following the German model. This will be true with respect to domestic monetary policy. In terms of exchange rate policy *vis-à-vis* third currencies, the Council of Ministers will also be expected to play an important role in setting the guidelines.

The ECB, the seat of which is supposed to be decided before the end of 1992, will be governed by a six-member Executive Board, appointed for eight years by the heads of state or government of member countries, and a Governing Council consisting of the members of the Executive Board and the governors of the national central banks. To ensure the independence of the ESCB, national legislations regulating the operation of national central banks are also expected to be changed accordingly. Thus, the entry into the third stage of EMU will imply not only the transfer of monetary powers from the national to the European level, but also a significant change in relations between political authorities and central banks.

The emphasis on price stability as the primary objective of the ESCB and the political independence of the new monetary institution is consistent with the prevailing idea that there is no trade-off between inflation and growth, not to mention unemployment. The Commission's study took a clear stance on this issue: 'The old theory that there is a trade-off between high inflation and low unemployment is now unsupported as a matter of theory or empirical analysis, except for short-run periods' (Commission of the EC, 1990e: 22). Although the short-run may be a long time in politics, the statement did, in fact, represent a fairly broad political consensus at the time, which cannot, however, be treated as completely irreversible. But the problem goes much further: can there be a central bank without a corresponding political authority? There is hardly any historical precedent for that. Given the fragmentation and the relatively underdeveloped nature of political power in the nascent European economy, a reality which does not seem likely to change drastically in the foreseeable future, the independence of the ESCB and the ECB will be even greater in real terms.

This also brings us back to the old question of the link between economic and monetary union. It is not at all surprising that this question proved difficult for both the Werner and the Delors committees to answer. The strategy of parallelism was adopted as a compromise for the intermediate stages leading to a complete EMU, while both committees opted for a close economic union as the necessary counterpart to the centralization of monetary policies, the unhindered mobility of capital, and the irrevocable fixity of exchange rates. But this is exactly the point which provoked the strongest reactions from member states. It is interesting that, while there had been a whole debate about the alleged deflationary bias of the EMS, the Delors report concentrated on the need to control the size and financing of national budget deficits, thus implying a fear of excessively expansionary policies and hence inflationary tendencies inside the monetary union. The rather restrictive approach adopted by the Delors committee, as regards the national margin of manœuvre in the area of fiscal policy, was also very different from the attitude expressed in the earlier Padoa–Schioppa report, where the authors referred to the experience of other federal systems which usually impose no strict controls on state budgets and where 'the effective restraint . . . is the sanction of the market (Padoa–Schioppa, 1987: 85).

The new treaty did not create any new institution for the coordination of fiscal policies. It did, however, make provisions aiming at the strengthening of the existing mechanisms and more particularly the elimination of excessive public deficits. The monetization of deficits will not permitted within the monetary union, nor will there be any 'bailing out' of national, regional, or local authorities by the EC. The ceilings of 3 per cent for public deficits and 60 per cent for public debt in terms of GDP, adopted as part of the convergence criteria for admission into the third stage, will also be used in general as reference values for the assessment of national budgetary policies in the context of multilateral surveillance.

On the basis of reports prepared by the Commission, which is therefore given the role of a watchdog, the ECOFIN Council (Economics and Finance Ministers) will examine the policies of individual member countries. A whole series of measures will be available to the Council in order to exert pressure on pro-fligate members, ranging from public recommendations to the imposition of fines. In between, the Council may resort to other measures, always adopted on the basis of qualified majority votes, such as requiring the EIB to reconsider its lending policy to the particular member country and asking the latter to make non-interest bearing deposits with the Community. Assuming strict implementation, those provisions will represent a major constraint on the spending and taxation powers of national parlia-ments, which will be added to the constraints imposed by tax harmonization, either explicit or through the market (see also Chapter 5).

The long experience of fiscal laxity in some countries and the evident inability of financial markets (see, for example, the international debt crisis of the early 1980s) to act as effective and efficient constraints on sovereign actors can be used as arguments in favour of some central discipline on national deficit financing. The treaty provisions are aimed essentially against 'free riders' in the future monetary union. But there can be some doubts both about the desirability and the effectiveness of the specific provisions. Furthermore, there is nothing in the treaty to prevent a deflationary bias resulting from excessively contractionary policies (Melitz, 1991). What is at stake is about who and how will determine the macroeconomic priorities for the European Community as a whole. In the words of Lamfalussy, in the annex to the Delors report (Lamfalussy, 1989: 101):

The combination of a small Community budget with large, independently determined national budgets leads to the conclusion that, in the absence of fiscal co-ordination, the global fiscal policy of the EMU would be the accidental outcome of decisions taken by the Member States. There would simply be no Community-wide macroeconomic fiscal policy.

The reconciliation of different national priorities, themselves a function of history, economic development, and demography among other factors, requires a political system which is much closer to a federation than a system of intergovernmental co-operation. And the EC does not appear ready as yet to give birth to such a system.

The decentralized nature of EC fiscal policy will also have implications for the representation of Community interests in international fora. Until now, the Big Four of the EC, plus the Commission, have participated in the Group of Seven (G-7) within which the leading industrialized countries have tried, albeit with limited success, to co-ordinate their economic policies. In a future EMU, the President of the ECB will be able to speak on behalf of the Community with respect to monetary policy; but what about fiscal policy? The problems experienced by US Administrations in the past in terms of being able to commit themselves and eventually deliver the goods in international policy co-ordination will pale into insignificance compared with those to be faced by Community representatives in the future.

Another very important aspect of the economic counterpart of monetary union will be the size of the common budget and the amount of internal transfers through taxes and expenditure. In existing federal systems, interregional transfers, largely of an automatic nature, play a major redistributive role, thus reducing internal disparities (Eichengreen, 1990; Commission, 1990e). The difference between the EC budget and central budgets in those federal systems is simply enormous, and this situation is unlikely to change dramatically in the foreseeable future, despite the efforts made by the economically weaker countries and the Commission itself (see also Chapter 8). This is a very similar argument to the one developed above: monetary union requires a much more developed European political system to be effectively managed. Could countries abandon completely the exchange rate instrument without the compensating effects of an effective transfer system and without even a high degree of labour mobility? The result could be a large increase in regional disparities and the creation of more depressed regions.

The EC countries have committed themselves to the establishment of an EMU before the end of the decade; and, unlike the 1970s, this commitment is now backed by a new treaty which defines in some detail the final stage, including the institutional framework. Yet, the political courage shown in the definition of the final stage of EMU has not been entirely matched by treaty provisions for the intermediate stages nor by the length of the transitional period. The big decisions have been left for later; and five or seven years is a very long time in politics.

On the other hand, the new treaty has left several important questions unanswered; or at least, the answers provided are not entirely convincing. They refer to the costs and benefits of EMU in a Community which is still characterized by a high degree of economic diversity and relatively limited political cohesion; and also in a Community where political institutions fall far short of economic ambitions. This problem could be tackled more effectively in the new intergovernmental conference planned for 1996. EMU constitutes a long-term and, perhaps also, a high-risk strategy in which wider political objectives have always counted more than strict economic ones. But this is not very different from the overall approach towards European integration during the last four decades.

8. Regional Policies and Redistribution

Traditional economic theory concentrates on questions of efficiency and the maximization of global welfare, while considerations about equity and the distribution of the economic pie are usually left to more 'normative' disciplines; alternatively, they are simply assumed away as problems. The theory of customs unions is a good example of this eclectic approach. But everyday politics is largely about the distribution of gains and losses among participants in any system. Depending on the nature of the latter, the relevant participants can be countries, regions, different social groups and classes, or even individuals.

A relatively equitable distribution of the gains and losses, or at least the perception of such an equitable distribution, can be a determining factor for the continuation of the integration process. Regional integration schemes in other parts of the world have often foundered precisely because of the failure to deal effectively with this problem. It would have been surprising if distributional politics had not entered the European scene. Indeed, its absence could have been interpreted as an unmistakable sign of the irrelevance of the EC as an economic and political system. But there should be no cause for alarm among 'Euro-enthusiasts'. The distributional impact of integration has been paramount in the minds of national politicians and representatives of various pressure groups; and it has strongly influenced negotiations within the common institutions from a very early stage. This was evident in the first package deals on which the Paris and Rome treaties were based. Yet those package deals made few provisions for explicitly redistributive instruments. This is a more recent development brought about through the international economic recession of the 1970s, the increased internal divergence of the EC, caused mainly by successive rounds of enlargement, and, last but not least, the progressive deepening of the process of integration.

Redistribution is one of the central elements of the European mixed economy at the national level; and this has become increasingly true of the EC as well, although still to a limited extent. Redistribution can also be considered as an index of the political and social cohesion of a new system; large transfers of funds across national frontiers not being a normal feature of international organizations. The objective of the EC extends, in fact, beyond a balanced distribution of gains and losses associated with integration. The explicit objective enshrined in the treaty is the reduction of existing disparities between regions, not only countries. This chapter will examine the nature and size of the regional problem inside the Community, the link between integration and regional disparities, and the development of redistributive instruments at the European level. The role of the EC budget will be discussed mainly with respect to its redistributive function. The relaunching of integration in the second half of the 1980s was partly based on the agreement reached on the broad outlines of the budget for the period 1988–92. The same remains true for the future. The proposed establishment of EMU will be closely linked with public finance issues at the Community level.

Regional Policy: In Search of Effectiveness

'Regional problems are difficult to define but easy to recognise' (Robson, 1987: 168). Perhaps the main difficulty does not lie so much with the identification and definition of those problems as with their explanation and even more so with the ways of solving them. Regional problems refer to the persistence of large disparities among different regions of the same country in terms of income, productivity, and levels of employment, to mention only some of the most representative economic indicators. To understand the nature of these problems, one usually needs to go beyond neoclassical economic theory and the host of simplifying assumptions on which it is founded, including perfect competition, full employment, constant returns to scale, and perfect mobility of factors of production. The literature on regional economics concentrates, precisely, on various forms of market failure which constitute a radical departure from the strong assumptions of neoclassical models. It stresses the existence of economies of scale and learning curves for individual firms. It points to external economies such as location advantages associated with easy access to large markets, centres of administration and finance, and sources of skilled labour and technological knowledge. It argues that the imperfect nature of labour markets can lead to situations in which money wages in different regions do not necessarily reflect differences in productivity rates. This is referred to in the literature as differences in 'efficiency wages', defined as money wages over productivity. Furthermore, inter-regional mobility of labour, which is certainly far from perfect, can have perverse effects in terms of regional disparities to the extent that migration to fast developing areas is usually led by the most dynamic and highly skilled members of the labour force in the lagging regions (Myrdal, 1957; Robson, 1987; Begg, 1989).

Under these conditions, initial differences in productivity and economic development or simply an autonomous shift in demand for the goods produced by a particular region can lead to 'circular and cumulative causation' and thus growing polarization between different regions; hence the creation and perpetuation of regional problems. This is what Myrdal calls the 'backwash' effects. On the other hand, the growth of dynamic regions will also have 'spread' effects arising from an increased demand for imports and the diffusion of technology from those regions, and eventually also from diseconomies of location associated with over-congestion in the rapidly growing centres. The relative importance of 'backwash' and 'spread' effects will determine the development of regional disparities within a country.

The main message from regional economic theories is that there are no strong reasons to expect the elimination of regional problems through the free interplay of market forces. On the contrary, such problems could be aggravated without the counter-vailing influence of government intervention. Interestingly enough, there is a close similarity between the literature on regional economics and the new theories of international trade which also place the emphasis on the role of economies of scale, imperfect competition, differentiated products, and innovation. Comparative advantage is no longer seen as the result of different factor endowments. Instead, the reasons for the large intra-industry trade which characterizes relations among industrialized countries, including members of the EC, seem to lie 'in the advantages of large-scale production, which lead to an essentially random division of labour among countries, in the cumulative advantages of experience which sometimes perpetuate accidental initial advantages, in the temporary advantages conveyed by innovation' (Krugman, 1986: 8). Although the new theories do not reject completely the old Ricardian premise regarding the welfare improving effects of free trade, associated in those theories mainly with economies of scale and increased competition, this conclusion is now hedged with many 'ifs' and 'buts'. Tariffs, subsidies and strategic trade policies can make perfect economic sense from the point of view of an individual country, although they can be disastrous when pursued by all countries concerned. On the other hand, gains and losses from trade liberalization are unlikely to be distributed evenly among different countries and regions.

The dividing line between regional and international economics becomes blurred as economic interdependence among different countries increases. After all, what basically distinguishes an intra-country from an inter-country problem in terms of economic disparities is the higher degree of labour mobility within a country and the automatic transfer of resources through the central budget, factors which are meant to compensate for the lack of independent trade, monetary, and exchange rate policies for a region. It is those factors which differentiate the problem of the Mezzogiorno inside Italy from that of Portugal or Greece in the EC.

However, with the progressive deepening of European integration, and especially with the establishment of the internal market and later EMU, this distinction will become increasingly less obvious, particularly with respect to policy instruments aimed at influencing relative prices and, ultimately, the inter-country allocation of resources. Independent trade policies have long since been merged in the context of the common commercial policy of the EC. Monetary policy and the exchange rate will also need to be sacrificed at the altar of monetary union in which case intra-EC trade will be determined by absolute and not comparative advantage; thus effectively turning inter-country disparities in the EC into regional problems. On the other hand, linguistic and cultural frontiers are likely to remain for long a major barrier to labour mobility inside the EC. In fact, it would not even be politically desirable for labour mobility to act as an important adjustment mechanism among member countries. Thus the avoidance of serious regional problems inside an increasingly integrated Community would have to depend essentially on two factors, namely the flexibility of product and factor markets and compensating measures, most notably in the form of regional policy.

The Community of Twelve is characterized by large economic disparities among countries and regions, which greatly exceed those inside the United States (Boltho, 1989). There are various reasons why international comparisons of income levels should be treated with considerable caution. They include the distortions created by exchange rates which do not adequately reflect relative purchasing power over goods and services, a problem which is only partially dealt with by attempts to establish purchasing power standards (PPSs), and the large differences in the size of the unrecorded sector of the economy. Cross-country regional data are even less reliable given the non-comparability of some administrative regions in EC member countries (the *Land* of Hamburg is, for example, hardly comparable to the autonomous region of Andalucia). Yet available data can be treated as a rough indicator of income disparities inside the EC. On the basis of PPSs, per capita income levels in Greece and Portugal are less than half those enjoyed in Germany and Luxembourg (Table 8.1). Interregional disparities are much more pronounced. Thus the difference between Hamburg and Groningen on the one hand, and the poorest regions of Greece and Portugal on the other is more than 4:1 (Commission of the EC data). It should come as no surprise to anybody at all familiar with the European economic scene that income disparities have a strong centre–periphery dimension, the poorest regions of the EC of Twelve being concentrated in the southern and western periphery (Map 8.1; the incorporation of the new German *Länder* has now added an eastern dimension to the periphery of less developed regions of the EC).

With respect to regional policy, three main phases can be distinguished since the establishment of the EEC in 1958. The first phase, which lasted until 1975, was characterized by the lack of any common regional policy worth the name. The second was marked by the creation of new instruments, the strengthening of

Table 8.1. Divergence of GDP per capita, 1960–1990

	1960	1970	1975	1980	1985	1990
Belgium	95.4	98.9	103.1	104.1	101.6	102.6
Denmark	118.3	115.2	110.5	107.8	115.8	108.2
Germany	117.9	113.2	109.9	113.6	114.2	112.8
Greece	38.6	51.6	57.3	58.1	56.7	52.6
Spain	60.3	74.7	81.9	74.2	72.5	77.8
France	105.8	110.4	111.8	111.6	110.6	108.6
Ireland	60.8	59.5	62.7	64.0	65.2	69.0
Italy	86.5	95.4	94.6	102.5	103.1	103.1
Luxembourg	158.5	141.4	126.7	118.5	122.4	125.6
Netherlands	118.6	115.8	115.5	110.9	107.0	103.1
Portugal	38.7	48.9	52.2	55.0	52.0	55.7
United Kingdom	128.6	108.5	105.9	101.1	104.2	105.1
EC-12	100.0	100.0	100.0	100.0	100.0	100.0

Note: Per capita GDP is given at current market prices and purchasing power parities.
Source: Commission of the EC (1991*a*).

the regional policy dimension of others already available, and the steady increase in the amounts of money spent. The third phase is connected with the reform of the so-called Structural Funds in 1988. It constitutes a turning-point in the search for greater effectiveness of common instruments, coupled with a substantial further increase in EC expenditure with a regional bias.

The original six members of the EC constituted a relatively homogeneous economic group, with the exception of the south of Italy; a problem which was, in fact, recognized in the protocol for the Mezzogiorno, attached to the Treaty of Rome. Article 2 of the treaty referred to the objective of a 'harmonious development of economic activities, a continuous and balanced expansion', while in the preamble the contracting parties went even further by calling for a reduction of 'the differences existing between the various regions and the backwardness of the less favoured regions'.

There were only a few provisions made in the treaty for the creation of instruments which could contribute towards this 'harmonious development' and the reduction of regional disparities. The European Investment Bank (EIB) was intended as a source of relatively cheap interest loans and guarantees for the less developed regions of the Community. Provisions for the free

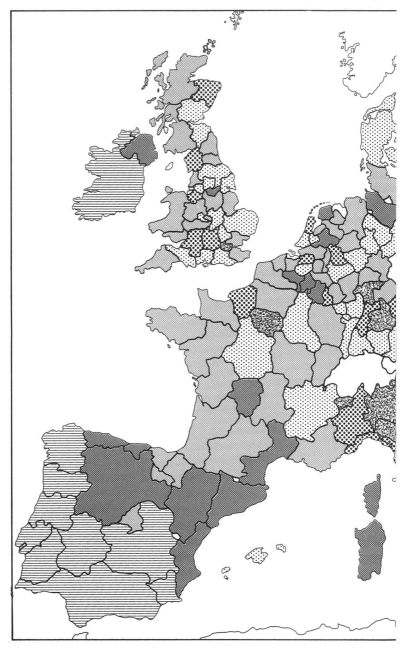

Map 8.1. Regional GDP per Inhabitant, 1988
Source: Commission of the EC, 1991*c*.

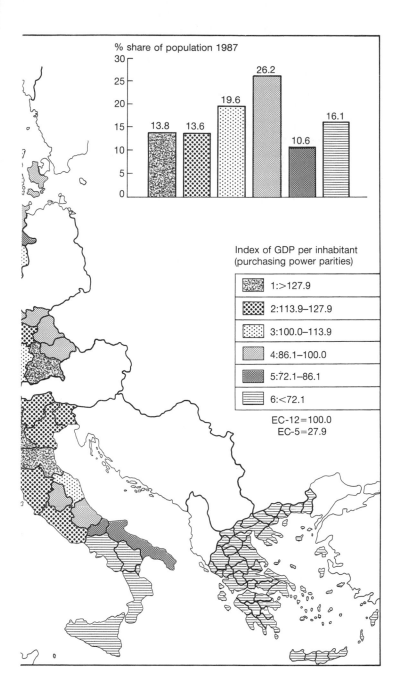

% share of population 1987

Index of GDP per inhabitant
(purchasing power parities)

	1:>127.9
	2:113.9–127.9
	3:100.0–113.9
	4:86.1–100.0
	5:72.1–86.1
	6:<72.1

EC-12=100.0
EC-5=27.9

movement of labour also had an indirect regional dimension in the sense that labour mobility would help to deal with the problem of high unemployment in a less developed region such as the Mezzogiorno. Last but not least, the setting up of the CAP should be expected to contribute towards the reduction of disparities, since farm incomes were generally much below the national or EC-6 average, while economic backwardness was most often identified with a heavy regional concentration on agriculture.

There was no explicit reference to regional policy, albeit in the form of derogations from general provisions in the different policy areas dealt with by the treaty. This is true of social, transport, and agricultural policies. The best known derogation which bears upon the regional dimension can be found in Article 92 of the treaty. It indirectly accepts state aids for intra-country regional development purposes, thus making a big exception to the application of the common competition policy. The various derogations, together with the lack of any separate chapter on a common regional policy, suggest that the authors of the Treaty of Rome, while recognizing the regional problem and the need to employ special instruments to deal with it, had decided to leave the responsibility basically to the hands of national authorities. The role of the Community would remain marginal in this respect, while the relevant institutions were asked to show some flexibility in the development of other common policies in order to accommodate the regional policy objectives of national authorities.

Regional disparities were not as yet generally recognized as a major policy concern at the time of the signing of the treaty. On the other hand, large transfers of money across frontiers were considered as politically impossible. Since the redistributive mechanisms could only be very modest, the six signatories tried to 'control and distribute the gains and losses which might arise in the particular sectors involved in such a way as to determine beforehand the extent to which the national interest of each party to the agreement would be satisfied' (Milward, 1984: 498).[1] This explains the complicated and perhaps economically 'irrational' nature of some of the treaty provisions. On the other hand, equitable distribution of gains and losses basically referred to the distribution among countries.

---

[1] This comment by Milward was in fact made with reference to the Paris Treaty of 1951. It can equally well apply to the EEC treaty signed 6 years later.

Although regional policy had its heyday in most Western European countries during the 1960s and the early 1970s, with large sums of money spent in this direction (Nicol and Yuill, 1982), very little happened at the EC level. The EIB did, as originally envisaged, orient its lending activities mainly towards the less developed regions of the Six, and the south of Italy in particular. However, the sums of money involved were relatively small and the attraction of EIB loans consisted entirely of the preferential rates charged on loans. The EIB could not, according to its statutes, offer any capital grants or subsidies to inputs. The ESF and, to a lesser extent, the EAGGF (European Agricultural Guidance and Guarantee Fund) also played a modest redistributive role during this period. On the other hand, the first serious attempt towards the co-ordination of national regional policies was made as late as 1971, with the aim of preventing an 'overbidding' between regions, which would normally be at the expense of the poorer ones.

Distributional issues in general did not become a serious political problem in the early years of European economic integration. But this was the golden age of the Western European economies, characterized by rapid economic growth, high employment rates, and relative monetary stability. The overall size of the cake grew constantly bigger and European integration continued to be perceived as a positive-sum game in which there were gains to be made by all the countries involved. The reduction of inter-country income disparities among the Six during the same period helped to allay earlier fears about the effects of trade liberalization on the weaker economies (Perroux, 1959; Vanhove and Klaassen, 1987; see also below for further discussion of intra-EC disparities).

On the other hand, national governments pursued active regional and redistributive policies inside their borders aiming at a reduction of income disparities. At the time of prosperity and the rise of social democracy, equality became a widely respected political objective, while the rise of autonomist movements in several European countries surely strengthened the political will for an effective reduction of regional inequalities.

Interest at the Community level grew as a result of the first enlargement of the EC and the rapid deterioration in the international economic environment, both coinciding in the early 1970s. The accession of three new members brought countries with serious regional problems inside the EC. On the other hand, clearly dissatisfied with the overall economic package which had taken shape prior to the UK's accession to the Community, the Government in London searched for other mechanisms which would partly compensate for the budgetary loss arising from the operation of the CAP. The creation of the European Regional Development Fund (ERDF) soon became the spearhead of this effort (H. Wallace, 1983). Regional policy and redistribution have been used almost interchangeably ever since.

The ERDF was set up in 1975. Its birth signalled the growing concern with intra-EC disparities. Strangely enough, this growth of interest in EC regional policy has almost coincided with the noticeable decline in the popularity of regional policies at the national level. Is this contradiction simply another of the peculiarities of the contemporary European scene? What is beyond doubt is that different developments at the two levels do not indicate any conscious transfer of powers and responsibilities to the emerging European centre.

The ERDF started with small sums of money which were initially distributed among member countries on the basis of quota allocations determined by the Council of Ministers. This meant a dispersal of funds among countries where the regional problem of one was the dream of economic development of another. A 5 per cent, non–quota element, allocated at the discretion of the Commission, was first introduced in 1979, while, with the 1984 reform of the ERDF, quotas were replaced with indicative ranges for each country's allocation of funds. Intermediate changes in terms of country allocations were introduced in 1981 and 1986 as a result of the second and third enlargement of the EC.

Funds available through the ERDF grew steadily over the years. Disbursements were in the form of matching grants for the financing of investment projects, with almost exclusive emphasis on infrastructural investment (approximately 85 per cent of total expenditure for projects for the period 1975–88; see Table 8.2). There was also a clear redistributive bias in favour of countries with more severe regional problems and an increasing concentration of resources on the least developed regions. However, the total sums of money remained small, when compared with expenditures in terms of regional policy at the national level. In 1988, the year before the implementation of the latest reform,

Table 8.2. ERDF Commitments, 1975–1988

(Mio ECU)

	Programmes	Projects					Studies	Total Commitments
		Industry and Services	Infrastructure	Other	Total			
Belgium	26.70	41.67	131.97	1.35	174.99		2.00	203.69
Denmark	10.26	25.35	140.42	0.57	166.34		6.14	182.74
Germany	77.01	502.96	347.39	–	850.34		0.19	927.55
Greece	263.38	34.51	2149.65	–	2184.16		0.24	2447.78
Spain	75.59	11.20	1945.23	2.29	1958.72		0.09	2034.40
France	261.90	370.42	2156.66	13.11	2540.19		12.04	2814.13
Ireland	144.01	311.27	841.94	3.67	1156.88		0.99	1301.88
Italy	136.19	975.72	6777.60	0.58	7753.90		21.95	7912.04
Luxembourg	1.94	–	14.69	–	14.69		–	16.63
Netherlands	14.18	32.42	200.85	0.01	233.28		0.19	247.65
Portugal	117.20	–	1080.56	1.48	1082.04		0.58	1199.82
United Kingdom	465.17	1000.02	3620.97	6.43	4627.42		10.66	5103.25
EC-12	1593.53	3305.54	19407.93	29.49	22742.42		55.07	24391.56

Notes: ERDF funds have been available to Greece, Portugal, and Spain for a shorter period than for the other member countries because of more recent dates of accession. Country figures are therefore not strictly comparable.

Sources: Commission of the EC. 1990*f.*

ERDF assistance amounted to only 0.09 per cent of EC GDP and 0.46 per cent of gross fixed capital formation (GFCF) (Commission of the EC, 1990*f*).

On the other hand, while EC regional expenditure increased, there was little evidence until the 1988 reform to suggest that the money spent by the ERDF was in addition to regional aid which would have been given by national governments in its absence. This is again the so-called problem of additionality to which reference has been made earlier with respect to ESF expenditure. On the contrary, many governments seemed to consider EC money as a means of replacing national expenditure for regional development. The strict quota system in terms of national allocations in the beginning and the limited powers of discretion enjoyed by the EC Commission in the selection of projects did not help to solve this problem. It is highly indicative that for years the overwhelming majority of projects receiving ERDF assistance had begun before application for funds was made by the national governments. In the attempt to achieve high absorption rates of funds allocated by the Commission, national and regional administrations were often tempted to sacrifice economic efficiency. Regional aid by the EC was generally viewed by national governments as a means of redistribution across national boundaries, with little effect on actual regional policies or the total sums of money involved. This led a member of the European Parliament from Northern Ireland to argue that 'there never was a regional policy at all, it was a Regional Fund operated on a Red Cross basis, with handouts here and there' (quoted in Shackleton, 1990: 44). Nor was there any serious co-ordination with other instruments of EC policy, and this further reduced the effectiveness of ERDF assistance as a means of tackling regional problems in the Community.

While the ERDF operated on the basis of grants, the EIB continued its lending activities, drawing from national contributions and its high credit rating in international capital markets, which in turn enabled it to raise money at attractive rates of interest. With the setting up of the EMS in 1979, the so-called New Community Instrument (NCI) was created in order to provide through the EIB subsidized loans to the less prosperous members, namely Ireland and Italy. Although, traditionally, the bank has favoured large loans, attention has also been paid to small and medium enterprises (SMEs) which have received

Table 8.3. EIB Financing within the EC, 1986–1990

(Mio ECU)

	Sector			Total
	Industry, services, agriculture	Energy	Infrastructure	
Belgium	218.7	—	16.4	235.1
Denmark	191.3	949.1	984.4	2124.9
Germany	1124.9	729.2	851.7	2705.7
Greece	366.9	283.4	372.4	1022.7
Spain	1562.7	460.0	3036.7	5059.4
France	1985.8	252.0	2758.5	4996.3
Ireland	49.4	220.0	706.5	975.9
Italy	7042.6	4118.6	6134.4	17295.6
Luxembourg	11.8	—	19.8	31.6
Netherlands	497.9	3.2	367.3	868.4
Portugal	911.3	603.9	988.5	2503.7
United Kingdom	1401.3	1885.8	3610.4	6897.6
Other (Article 18)[a]	—	198.5	660.8	859.3
Total	15364.5	9703.7	20507.8	45576.0

[a] Article 18 of the Statute of the EIB allows for the provision of loans for investment projects in the non-European territories of member countries.
Sources: EIB (1991).

assistance from the bank through the system of global loans. These are large lump sums lent to a financial institution in a member country, which in turn lends out the money to SMEs locally (Pinder, 1986). This indirect form of lending has been considered necessary in view of the limited administrative resources of the EIB.

Table 8.3 gives a breakdown of EIB lending between 1986 and 1990 by country and sector. There is a clearly discernible bias in favour of the poorer countries, with Italy far ahead of the rest. After all, it was with Italy, and the Mezzogiorno in particular, in mind that the EIB had been originally set up. As with ERDF expenditure, the figures show a strong bias in favour of energy and infrastructural investment. In 1989 EIB lending accounted for 6.17 per cent of GFCF for Portugal; the corresponding figures for Ireland and Greece were 3.61 per cent and 2.76 per cent respectively.

A new approach to regional policy was introduced in 1985 with the Integrated Mediterranean Programmes (IMPs), intended for the Mediterranean regions of France, Italy, and the whole of Greece. The creation of IMPs was in recognition of the special development problems of these regions and the relative bias of the CAP against southern agricultural products. The Iberian enlargement, which was generally expected to accentuate these problems, acted as the catalyst, strengthening the need for policy adjustment measures (Yannopoulos, 1989). Each IMP would last for three to seven years and a combined total of 6.6 billion ECUs was to be allocated to them, coming partly from the various EC Funds and the EIB and partly from the creation of additional resources through the budget. The main innovation of IMPs was that finance would be based on medium-term development pro-grammes, instead of individual project submission, and on a close co-ordination of different EC instruments. They were the pre-cursor of more general reforms to be introduced later with respect to the various Community Funds. The new Iberian entrants have not benefited from IMPs, although special pro-grammes were designed for Portugal in order to assist the adjust-ment of the Portuguese economy to the requirements of EC membership.

The ground was gradually prepared for a major qualitative change in the use of policy instruments coupled with a major shift in the scale of EC intervention. The decision to establish the internal market and the dramatic improvement in the economic and political environment in Western Europe provided the catalyst. This eventually led to the reform of Structural Funds in 1988. But the legal foundations had been laid earlier with the SEA. The latter introduced Title V to the Treaty of Rome under the heading 'economic and social cohesion'. It was a formal recognition of the greater political importance of the redistri-butive function, while also constituting an integral part of the overall package deal behind the SEA and the relaunching of European integration. The new Articles 130a to 130e were the first attempt to link the objective of 'harmonious development' and the reduction of regional disparities, previously mentioned only in a very general way in the preamble and Article 2 of the original text of the treaty, with specific EC instruments, namely the ERDF, the ESF, the EAGGF-Guidance Section (all three referred to now as Structural Funds), and the EIB. The ERDF was entrusted with the principal task of redressing intra-EC regional imbalances. The new articles called for the effective co-ordination and rationalization of the activities of Structural Funds and the Commission was invited to submit proposals in this direction. This eventually ushered the Community into a new phase in terms of regional policy.

The reform of the Structural Funds started with the agreement reached at the European Council meeting in Brussels in February 1988, when a decision was also taken for the doubling, in real terms, of the resources of the three Funds between 1987 and 1993. The doubling of the resources of the Structural Funds, undoubtedly the most important decision taken until then by the Community in terms of internal redistribution, was part of a package of measures, including the reform of the CAP and the EC budget. The so-called Delors package was presented as a necessary precondition for the successful implementation of the internal market programme. Interestingly enough, the link established between the internal market and the doubling of resources through the Structural Funds also meant an implicit recognition of the danger that the weaker regions of the Community could end up as net losers from further market integration. The creation of new resources was the necessary side-payment for the political acceptance of the internal market programme and a means of preparing weaker regions for the new cold winds of competition.

The doubling of resources went hand in hand with an effort to improve the effectiveness of EC action through the adoption of clearer objectives, an improved co-ordination of different financial instruments, and a close monitoring of jointly financed programmes. The main outlines were agreed upon by the Council of Ministers through the adoption of several regulations in June and December 1988. Five priority objectives were assigned to the Funds, and the EIB was also expected to contribute to those objectives. The latter related to:

1. The less developed regions
2. Areas of industrial decline
3. The long-term unemployed
4. Employment of young people
5a. Adjustment of agricultural structures
5b. Development of rural areas.

The first objective is intended to deal with regions where income falls below 75 per cent of the EC average, and which are characterized by a relatively high percentage of the labour force engaged in agriculture. The list approved by the Council of Ministers includes the whole of Greece, the Republic of Ireland, Northern Ireland, Portugal, the greater part of Spain, the Mezzogiorno, and the overseas departments of France and Corsica. A few regions which are slightly above the 75 per cent rule have been included; after 1993, the list will also be extended formally to the new German *Länder*. The areas under the first objective represent approximately 21 per cent of EC population and they are expected to receive 80 per cent of ERDF expenditure and in excess of 60 per cent of the total expenditure of the Structural Funds by 1993. The areas under the second objective represent approximately 16 per cent of the population of the Community. They are situated mainly in central and north-western Europe, with a high concentration in the UK. The first two objectives require co-ordinated action by two or more Funds. The third and fourth objectives are the exclusive domain of the ESF, following on the steps of the policy established earlier for the Social Fund. As regards objectives 5*a* and 5*b*, the funds allocated are significantly smaller (Van Ginderachter, 1989).

Another feature of the 1988 reform was the emphasis on multi-annual programming as a means of identifying and quantifying economic priorities over a period of five years. It may be interesting that at a time when liberal economic ideas seemed to prevail, the notion of economic planning, even in a mild form, was still very much alive in Brussels and in other places where policy-makers were trying to tackle problems of regional development. For those countries and regions which came under Objective One, regional development programmes were submitted to Brussels, and these in turn provided the basis for the adoption of Community Support Frameworks. The latter set the main guidelines for expenditure through the Structural Funds in each region for the period 1989–93. The switch from the financing of individual projects, which had been the main characteristic of EC action in the past, to medium-term 'operational' programmes and global grants has now been generalized. Programming had been introduced, in a modest fashion, with the 1984 reform of the ERDF and the Integrated Mediterranean Programmes. Strong emphasis has also been placed on the close monitoring of different programmes in an attempt to deal with earlier criticism regarding the wastage of resources and widespread malpractices. The other important feature of the 1988 reform was the concept of partnership which was intended to lead to close co-operation between EC, national and regional authorities both at the planning and implementation stage, thus also establishing direct lines of communication between regional authorities and the EC Commission.

The 1988 reform has led to a remarkable increase in overall expenditure. For the five-year period 1989–93, a total of 60.3 billion ECUs (in 1989 prices) has been committed to be spent through the three Structural Funds; and another three billion ECUs were added at the end of 1990 for the new German *Länder*. By 1992, expenditure through the Structural Funds had risen to 27 per cent of overall EC expenditure, as compared with 17 per cent in 1987. The redistributive impact has been further strengthened through the greater concentration of resources. Thus, by 1992 annual transfers through the Structural Funds represented 3.5, 2.9, and 2.3 per cent of GDP for Portugal, Greece, and Ireland respectively (EC Commission figures). The main emphasis remained, as before, on infrastructural investment which accounted for 29 per cent of overall expenditure for Objective One regions. Naturally, the corresponding figure in terms of GFCF was much higher (exceeding 10 per cent in the case of Greece). Since the outlays through the Structural Funds are in the form of matching grants, ranging from 75 per cent to 25 per cent of total expenditure, the implication is that a very substantial part of public investment in the main beneficiary countries and regions has been tied to EC-approved programmes.

The experience until 1992 points to a high absorption rate of the financial resources committed through the Structural Funds, although high absorption is not always directly related to the most efficient use of resources. The operation of the Structural Funds has, in fact, raised some general questions which can be summarized under the following four headings: subsidiarity, transparency, efficiency, and additionality. Several trade-offs have appeared between those objectives. For example, the division of competences between Brussels and national or regional authorities has always been a moot point. Why should Brussels know better than national capitals and regional authorities about their own development needs and priorities? This is not only a question of principle; it is also linked to the very serious administrative limitations of the EC executive. The small staff and the internal organization of the Commission do not match the ever growing sums of money which it has been asked to administer; and this has become increasingly clear in recent years. The problem of efficiency in the use of financial resources is added to that of limited transparency in the Community system of decision-making, not to mention the problem of accountability.

There are, however, powerful counter-arguments which can be used. First of all, the objective of additionality, namely that EC expenditure should represent a net increase in the amounts spent on development and not on consumption, militates in favour of some central control. This point has been very clearly expressed in the Delors report on EMU (Committee for the Study of Economic and Monetary Union, 1989: 22–3):

The principal objective of regional policies should not be to subsidize incomes and simply offset inequalities in standards of living, but to help to equalize production conditions through investment programmes in such areas as physical infrastructure, communications, transportation and education so that large scale movements of labour do not become the major adjustment factor.

The experience of the ERDF and the other Structural Funds suggests that the efficiency of some regional and even national administrations leaves much to be desired, thus making the EC Commission look like a model of efficiency. After all, low levels of development are not only manifested in income statistics. Administrative inefficiency is often compounded by short-term political considerations which are not always consistent with long-term development needs. Thus, the Commission has often been forced to navigate dangerously between the Scylla of national sovereignty and the Charybdis of administrative inefficiency and political short-termism (or corruption). How much of the responsibility for the planning and the implementation stages, including the monitoring of programmes, could be left to outside experts? This will remain an awkward question for the future.

The 1988 reform has acted in some cases as a catalyst for decentralization within member countries and the strengthening of regional administrations in order to be able to cope with the demands made in Brussels. On the other hand, it represents an important step in the development of genuinely common policies and a shift of power towards the centre. In this respect, national capitals seem to be increasingly squeezed between the other two competing levels of authority.

The discussion about the role of Structural Funds in the post-1993 period has already started; and it is inevitably linked to EMU. The Maastricht treaty has made two new additions to the institutional set-up of the EC, namely the Committee of the Regions and, more importantly, the Cohesion Fund which should come into operation before the end of 1993. This new Fund will make financial contributions to projects in the fields of environment and trans-European networks and only to those member countries with a per capita GNP below 90 per cent of the EC average. Thus, the list of beneficiaries will be limited to Greece, Ireland, Portugal, and Spain; with Italy being, for the first time, left out of an EC regional development scheme. There is also a new element of conditionality introduced with respect to the Cohesion Fund: the countries benefiting from it will need to have a programme of economic convergence approved by ECOFIN in the context of multilateral surveillance. Thus, the link with EMU has been directly established. The Cohesion Fund, which will be added to the already existing Structural Funds, has been mainly a concession to Spain, as part of the overall package deal of Maastricht. It has a distinct Spanish flavour both in terms of the priorities adopted and the GNP ceiling for the beneficiary countries, and this is very likely to be translated into a large Spanish share of total expenditure.

Following on the steps of the Delors package of 1988 which had contributed so much to the budgetary peace and successful implementation of the internal market programme during the intervening years, the EC Commission submitted the so-called second Delors package in February 1992. The aim was to agree on the main budgetary guidelines for the next five years (1993–7). As regards the operation of the Structural Funds, the Commission did not propose any major changes. It did, however, call for more flexibility in the planning and implementation stages; more decentralization towards the regional level; more money to be spent on the so-called Community initiatives (programmes operated by the Commission and not specifically linked to any country or region); higher EC participation rates, especially in cases of budgetary constraints linked to the convergence effort undertaken by member countries; and the extension of EC action into new areas, such as health and education. Even more importantly, the Commission called for a further substantial increase in the overall resources of the Structural Funds, which should raise expenditure to approximately 33.5 per cent of the EC budget in 1997 (compared with 27 per cent in 1992). For the less developed countries and regions, which would also benefit from the new Cohesion Fund, the aim would be for a further doubling of resources in real terms. At the time of writing, the negotiation on the second Delors package is still in its early stages; but it promises to be a tough negotiation taking place in times of economic recession and financial stringency.

Until now, the private sector has benefited directly from only a relatively small share of the expenditure undertaken by the Structural Funds. This may be changed in the future, although such a change will require new and more flexible forms of EC intervention. An important aspect of regional policy is the co-ordination and control of regional aids at the national level. Commission policy has been in the direction of greater trans-parency of aid systems, the imposition of ceilings on aid intensity, and the shift from general to regionally and sectorally specific aids. Commission control in this area has become more effective over the years in the context of the progressive strengthening of its competition policy; but there are still too many holes and *ad hoc* exceptions. Supervision over national aids is very important for the poorer regions of the Community which stand little chance of winning in a free-for-all subsidy game for the attraction of investment, especially if there is further progress towards EMU. The Community may in the future consider the use of regionally differentiated aids and incentives, with more EC money being directed in the form of matching grants towards private investment in the less developed regions.

Country Profiles: Ireland, Greece and Spain

Intra-EC income disparities are very large, and they have become wider with each successive enlargement. The figures in Table 8.1, which are based on estimates of purchasing power parities, point to a considerable stability in the rank order of EC countries in the GDP league over a period of thirty years. But they also highlight the remarkable performance of the three southern European NICs (Greece, Portugal, and Spain) between 1960 and

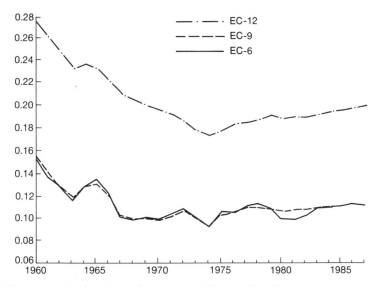

Fig. 8.1. Evolution of Income Disparities between Member Countries, 1960–1987
(Standard deviation of real GDP per capita as a ratio of the Community average in percentage terms)

Source: Padoa-Schioppa *et al.*, 1987: 163.

1975, when they also enjoyed the fruits of limited reciprocity in terms of trade liberalization with the two main European trade blocs, the steady rise of Italy, and the decline, with some ups and downs, of Denmark, Luxembourg, the Netherlands, and the UK.

On the other hand, there is little evidence to suggest that economic integration has itself contributed to the widening of disparities, although it is virtually impossible to isolate the effect statistically. Broadly speaking, it appears that inter-country disparities in the EC were reduced during the 1960s and early 1970s. The trend was reversed during the long recession and this continued to be true until the mid-1980s when growth rates in most of the less developed members of the Community picked up again, leading to another reduction of disparities which coincided with the economic boom of recent years (Table 8.1 and Fig. 8.1; see also Vanhove and Klaasen, 1987; Molle, 1990; Commission of the EC, 1991*a*, 1991*c*). As far as interregional disparities are concerned, the evidence is less clear and also, certainly, less reliable; but it generally points in the same direction as indices for inter-country disparities.

The conclusion one may be tempted to draw from the above data is that there is a positive correlation between economic growth and the reduction of income disparities, especially at the inter-country level. This has also been true of the latest phase of integration. Using the terms introduced by Myrdal (1957), it would appear that the 'spread' effects of economic growth are stronger than the 'backwash' effects. However, there is only insufficient evidence to support a general hypothesis along these lines. Such generalizations are of limited use, if they are not sometimes positively misleading. They leave little room for the special economic characteristics of individual countries and regions and the political choices made by each one of them. Historical experience suggests that although structural factors play a significant role, there is always a certain margin of manœuvre in terms of economic policy. This proposition will be illustrated below with a brief summary of the experience of three among the poorest members of the Community.

The Republic of Ireland joined the Community in 1973, together with Britain and Denmark. It brought with it a large agricultural sector, heavy dependence on the British economy, and a model of industrialization which heavily relied on the attraction of foreign investment through generous grants and tax incentives. The Special Protocol, signed together with the Treaty of Accession, enabled Ireland to continue acting as a gateway to Europe for foreign, and especially US, multinationals.

Between 1973 and 1980, the Irish economy fared better than the EC average in terms of growth and employment creation. Exactly the opposite happened in subsequent years and until 1987 when economic activity started picking up again. During the period of EC membership, Ireland's position in the income league has improved considerably, and especially in more recent years when the country has experienced exceptionally high rates of growth (see also Table 8.1). The large and continuous inflow of funds from the Community (Ireland being the biggest net beneficiary of EC transfers per capita) has, of course, contributed to this improvement in living standards. Participation in the CAP has had positive budgetary and trade effects for Irish farmers, which has meant large transfers of funds through EC intervention and higher prices for agricultural exports. This cornucopia did not, however, last for very long as the gradual alignment of Irish farm prices to the higher EC prices more or less coincided with more determined efforts to bring CAP expenditure under control.

In the manufacturing sector, the increased exposure of the domestic economy to international competition has strengthened its dualistic character, with modern, capital-intensive and high technology-oriented sectors dominated by subsidiaries of multi-national firms, while traditional, more labour-intensive sectors have remained the privileged domain for Irish entrepreneurs. The National Economic and Social Council of Ireland (1989: 208) refers to a 'continuous output and employment decline in a long list of exposed industries, and their replacement by foreign firms in a narrow range of manufacturing activities'. This dualism is also manifested in Ireland's foreign trade: the internationalization and diversification of market outlets has relied mainly on foreign firms, while indigenous entrepreneurs have remained more domestically and UK-oriented (McAleese and Matthews, 1987).

In 1989, exports of foreign-owned firms accounted for approximately 30 per cent of GDP and for a very large part of output and employment in the manufacturing sector (OECD, 1991). Cross-linkages with the domestic economy have remained limited, and this explains the large dependence on imported inputs which reduces the contribution of those firms in terms of the balance of payments. Furthermore, capital inflows for productive investment have been followed by a large repatriation of profits. In 1989 repatriated profits reached almost 14 per cent of GDP. On the other hand, the capital-intensive nature of foreign investment in Ireland has not helped much to alleviate the very serious problem of unemployment; and industrial policy has been partly responsible for distorting the relative price between capital and labour.

Membership of the EC has contributed to the further opening of the Irish economy. In fact, the change has been quite dramatic: exports of goods and services rose from 38 per cent to 67 per cent of GDP between 1973 and 1989 (Eurostat data). The internationalization of the economy has gone hand in hand with an increasing concentration on European markets and a corresponding reduction of the old, virtually exclusive, dependence on the UK. The constraints of a small and very open economy were not, however, immediately realized by Irish policy-makers. Expansionary fiscal policies and accommodating monetary policies sustained, during the second half of the 1970s, rates of economic growth and employment creation which were substantially higher than the EC average. This survived the second oil shock and the decision to join the ERM in 1979. The result was an explosion of the public sector deficit coupled with a large increase in inflation and the current account deficit.

Thus, the change of gear became inevitable. The stabilization policies pursued in the 1980s have produced impressive results, especially since 1987. They have been based on a broad consensus of the main political parties and the social partners, which took a concrete form through the Programme for National Recovery. Participation in the ERM has provided the anchor and the external discipline for anti-inflationary monetary policy. Fiscal consolidation has led to a reduction in the public debt/ GDP ratio, although the latter still remains at much higher levels than those permitted by the convergence criteria established at Maastricht (Fig. 7.7). On the inflation front, the progress has been absolutely remarkable: between 1981 and 1987, Irish inflation had dropped by more than fifteen percentage points, reaching close to the average of the narrow band countries of the ERM (Fig. 7.1). With such progress on the inflation front, interest rate differentials were bound to follow, as domestic stabilization policies became more credible.

The combination of macroeconomic stability, declining unit labour costs and the favourable external environment of the late 1980s enabled Ireland to achieve a sustained recovery in recent years, thus reversing the miserable performance of the first half of the previous decade. High growth went hand in hand with a considerable increase in investment and also a big improvement in the current account. The same is not, however, true of unemployment. Despite the creation of new jobs during the boom and the continued exodus of Irish people in search of employment abroad, the rate of unemployment has remained around 16 per cent, with the prospect of a further worsening due to the more recent downturn in economic activity. High unemployment has been the price for stabilization, and it has been a very heavy price indeed (see also Dornbusch, 1989).

Ireland's economic experience as a member of the Community points to some of the unequal effects of integration between countries which start from different levels of development, technology, and scale of production. The opening of frontiers has led to the elimination or contraction of many indigenous firms in the manufacturing sector, while the 'gateway' policy towards foreign multinationals has been at best a mixed blessing. Large transfers from EC Funds had been until recently directed mainly towards consumption and an internal redistribution of income in favour of the farming community. Serious errors in macroeconomic policy and the attempt to postpone economic adjustment have increased the cost of the latter. Although there has been considerable improvement of the macroeconomic scene since 1987, Ireland is still faced with major structural problems and difficult policy dilemmas. The accumulated public debt can act as a serious constraint on economic growth (not to mention the continuation of generous grants and tax concessions to foreign firms), and this constraint will become tighter as Ireland tries to meet the conditions for entering the final stage of EMU. Rigidities in the labour market and the large excess in the supply of labour, compounded by the highest birth rates in the Community, present a formidable economic and social problem. Large emigration could once again in Irish history act as the safety valve, although this would further weaken the demographic structure by increasing an already large percentage of non-active population (both young and old), while also depriving the economy of much local talent. The search for jobs abroad is nowadays led by graduates and skilled workers who have received an expensive and high quality education at the expense of Irish taxpayers.

Greece's accession to the Community in 1981 was in some respects the next logical step after a long period of association with the EEC (the Treaty of Athens being the first association agreement signed by the EEC in 1961), which had contributed to the gradual opening of the Greek economy, despite the limited reciprocity as regards trade liberalization between the two sides (Tsoukalis, 1981). Yet one of the most successful European NICs of the 1960s ended up as the biggest problem case of the EC during the 1980s. Membership of the Community coincided with a steady deterioration of Greece's position in the European income league (Table 8.1). In the diplomatic words of an international organization, 'the performance of the Greek economy has been one of the least good in the OECD area' (OECD, 1990: 88). It has been, in fact, disastrous. Although it may be tempting to link Greece's membership of the EC with negative developments in the domestic economy in a cause and effect relationship, relatively few economists have succumbed to this temptation.

The problems of the Greek economy started much earlier. The oil shock of 1973 was almost immediately followed by the fall of the military dictatorship; and democratic consolidation took, perhaps naturally, precedence over economic adjustment. There was, in fact, a very strong similarity to political reactions in both Spain and Portugal to the rapid deterioration in the international economic environment (Diamantouros, 1986). While in other Western European countries measures of economic adjustment were taken, in southern Europe government expenditure was growing rapidly, and so too were wages and salaries, leading to a wage–price spiral and a deterioration of international competitiveness.

The resistance to economic adjustment proved to be particularly stubborn in Greece. The second oil shock coincided with the intensification of the domestic political struggle on the eve of the 1981 elections which brought the Socialists to power. The new Government put the emphasis on rapid structural adjustment and internal redistribution, with the State expected to act as the main driving force of economic development. But in the pursuit of its economic policies, the new Government took little notice of the external constraints imposed on a small and open economy and the gross inefficiency of the Greek public sector (see also the relevant chapters in Tsoukalis, 1992).

While the rest of Western Europe quickly shifted into a deflationary gear, which brought unemployment to unprecedented levels, Greece continued strolling happily down the road of expansionary policies; and the latter succeeded only in financing a consumption boom. Rapidly widening public sector deficits, partly financed abroad, and high rates of inflation were combined with extremely low rates of growth and investment. The balance of payments constraint forced the Government to adopt a stabilization programme at the end of 1985, aided by an EC balance of payments loan and following a devaluation of the drachma. But political considerations soon put an end to economic stabilization: the electoral cycle came back in full swing, and the story was repeated once again in 1989–90 when public sector deficits and inflation rates reached new peaks in a period of political instability. Thus the divergence of Greek macroeconomic policies from those pursued in the rest of the EC became even more pronounced.

Although statistics are not very reliable, unemployment in Greece appears to have remained consistently below the EC average, thus offering the only bright spot on an otherwise dark economic horizon. But this has been achieved through the continuous expansion of the public sector, characterized by excessive overmanning and the operation of political patronage, and the artificial survival of many debt-ridden firms which have been kept alive through state subsidies. This policy was hardly sustainable, as public sector deficits reached record levels (exceeding 20 per cent of GDP in 1990).

Import penetration of the Greek market by EC suppliers has grown rapidly, especially in more traditional sectors which had enjoyed in the past high rates of protection, mainly of the non-tariff kind. There has been both a trade creation and a trade diversion effect. On the other hand, Greek exporters have proved unable to secure higher export shares of European markets. The inevitable result has been the steady worsening of the trade deficit and a depressing effect on domestic industrial production. The opening of the economy also seems to have contributed to the gradual shift of Greek entrepreneurs back to traditional sectors such as food, beverages, textiles, and clothing, thus further strengthening the inter-industry division of labour between Greece and the rest of the Community (Giannitsis, 1988; see also Neven, 1990). Greece's main competitors in the manufacturing sector are Third World NICs which operate on very low wage costs; a factor which will continue to act as a major constraint on the further expansion of Greek exports.

Similarly to Ireland, Greece has benefited from large transfers from EC Funds which have contributed to the financing of the large trade deficit. The bulk has gone to the agricultural sector in the form of price intervention, and it has been translated into higher consumption levels. This has, however, changed following the 1988 reform of the Structural Funds. Unlike Ireland, the positive budgetary effects of Greece's participation in the CAP have been partly compensated by negative trade effects due to a different product distribution and the fact that Greece has become a net importer of food products. Unlike also the experience of other new members of the EC, foreign investment inflows into Greece during this period were hardly influenced by accession. Presumably, this needs to be interpreted as a vote of no confidence by foreign investors.

The structural weaknesses of the Greek economy have been, undoubtedly, an important factor behind the difficulties in adjusting to strong European competition. Low levels of development and technology, small size, poor infrastructure, and a long history of external protection have created serious handicaps for Greek industrialists. This also explains the continued reference to infant industry arguments and the appeals for greater flexibility in the application of EC rules made by some Greek economists (Giannitsis, 1988). Such arguments were repeatedly used in the past by the Socialist Government as a justification for the lengthening of the transitional period, following Greece's accession, and the granting of derogations from EC rules which, while causing serious friction with the Commission, were used, almost invariably, as a means of postponing adjustment (Mitsos, 1989).

Structural weaknesses are, however, only one part of the explanation for Greece's difficulties in adjusting to the new European and international environment. The struggle for international competitiveness has been strongly undermined by persistent resistance to modernization, especially of the public sector, structural rigidities and macroeconomic mismanagement. The latter has also led to the overvaluation of the currency since the exchange rate has been used as an anti-inflationary instrument (see also the experience of some of the more inflation prone countries of the ERM), while large public deficits continued to work in the opposite direction. Economics is intimately linked to politics: a polarized society and a predominantly inward-looking political class acted as important constraints on economic adjustment.

Under a conservative government since April 1990, based on a small parliamentary majority, Greece has started a major stabilization effort aiming at the correction of both domestic and external financial imbalances. This stabilization effort will require a substantial shift of resources away from consumption and towards exports and investment; and it will need to continue well into the 1990s, if Greece wants to join the other EC countries into the final stage of EMU. In February 1991 a new loan was given by the Community in order to assist the stabilization programme, although with tighter conditions this time. The required drastic reduction in budget deficits, which started in 1991, promises to be a very painful exercise, because of the large cost of servicing the accumulated debt. Ireland's earlier experience is quite indicative. Macroeconomic stabilization has been accompanied by policies aiming at deregulation and privatization, although those policies have so far proceeded at a relatively slow pace. Thus, Greece has started, with considerable delay and much hesitation, to catch up with economic changes in the rest of Europe.

Because of major structural and macroeconomic problems, Greece has not been able to benefit much from the '1992 effect'. The opening of Eastern Europe, and the Balkan countries in particular, should provide Greece with a valuable economic 'hinterland' and improved prospects for trade and co-operation. But this is more of a medium- and long-term prospect. The serious political instability which has followed the collapse of communist regimes in Greece's neighbouring countries has produced an additional economic burden. This is particularly true in the case of Yugoslavia: reduced bilateral trade has been combined with higher transport costs for a good part of Greece's exports and imports from its EC partners as well as the loss of receipts from tourism. The problem of large numbers of economic refugees from Albania has been added to the misery list, thus making the early 1990s a very difficult period for Greece.

On the basis of the early years of EC membership, which is still too short a period to draw any firm conclusions, Spain's experience contrasts sharply with that of Greece. Spain became a member of the Community in 1986, together with Portugal. The signing of the treaty of accession constituted a major turning-point in a long and sometimes tumultuous phase of relations with Brussels, which had started back in 1962 with an unsuccessful application for association with the EEC.

Spain is, undoubtedly, one of Europe's most successful NICs. The stabilization plan of 1959 marked the beginning of a more outward-looking phase of industrialization, characterized by high rates of growth and rapid structural adjustment, which gradually produced a relatively diversified industrial base and the progressive internationalization of the economy through external trade, labour migration, and foreign investment. Yet high external protection remained for a long time one of the distinguishing features. The two successive oil shocks of the 1970s and the world economic recession fundamentally changed the external economic parameters, while the domestic political scene was also radically transformed after the death of the Francoist dictatorship. For approximately ten years (1975–85) the Spanish economy stagnated. Growth rates lagged behind even the depressingly low EC average. Unemployment reached record levels for Europe (21.8 per cent in 1985), while inflation remained above the EC average. Public expenditure and deficits grew rapidly, although Spain never approached the levels of indebtedness reached by countries such as Belgium, Greece, Ireland, and Italy (Fig. 7.7). Furthermore, unlike some of the weaker European economies during this period, Spain did not experience a severe balance of payments crisis.

The first half of the 1980s was characterized by a major stabilization effort and a substantial restructuring of the domestic economy. Stabilization relied mainly on monetary policy instruments and wage restraint. An extensive programme for the reconversion of the industrial sector, including large capacity cuts, financial restructuring, and technological modernization, was launched in 1984; public firms under the aegis of the Instituto Nacional de Industria strongly felt the pinch (OECD, 1989b). Measures aiming at a greater flexibility of the labour-market were introduced, and the first steps were taken towards the liberalization and deregulation of the financial sector. The impending accession to the EC acted both as a convenient excuse and a catalyst for a basically unpopular package of measures. Some of the results were quickly evident: inflation fell rapidly (it picked up again in 1989), profits rose steadily, and this eventually had a positive effect on investment. However, the price paid in terms of job losses was very heavy indeed. Restrictive macroeconomic policies and industrial restructuring, combined with rigidities in the labour market, the continuous exodus from land, and the termination of traditional patterns of emigration led to levels of unemployment which had been literally unheard of in Europe for some decades; even when some allowance is made for jobs in the black market.

The reward came soon in the form of a big economic boom coinciding with the first years of EC membership. Between 1986 and 1990, Spain registered the highest rate of economic growth among the Twelve. Investment grew at spectacular rates and unemployment began to fall (close to 16 per cent in 1990). With domestic demand growing faster than the demand for exports and the investment boom drawing ever increasing imports of capital equipment, the trade deficit widened rapidly. The progressive overvaluation of the peseta and the EC effect also worked in the same direction. Trade creation and trade diversion, as a result of EC membership, mainly benefited EC exporters to Spain. There was a significant shift in the geographical distribution of Spain's external trade, especially imports, in a very short period of time. Manufacturing imports from the EC more than doubled in real terms between 1985 and 1988. On the other hand, there are strong elements of both intra-industry and inter-industry trade between Spain and the rest of the Community, itself a sign of the more developed state of the Spanish economy as compared with that of Greece and Portugal (Hine, 1989; Neven, 1990; Commission of the EC, 1990c).

Trade and current account deficits were more than compensated by large inflows of capital which, in turn, produced a substantial increase in official reserves and the appreciation of the peseta in foreign exchange markets. Foreign capital was attracted to Spain for two main reasons; that is, in addition to the attraction of a sunbelt country which has been traditionally translated into purchases of real estate. On the one hand, high nominal and real interest rates, the product of a restrictive monetary policy, attracted short-term capital; and this was reinforced by the entry of the peseta into the ERM. On the other, the favourable business climate and the improved export prospects to the rest of the EC (accession plus the '1992 effect') acted as a big incentive for foreign direct investment and the buying of participation in Spanish firms. Political stability, surplus skilled labour, low wages by European standards, attractive living conditions for foreign managers, a booming economy, and the prospect of free access to the large European market helped to turn Spain into one of Europe's last frontiers in the eyes of foreign investors. There was an explosion of foreign direct investment in Spain (and also Portugal) during the last years of the 1980s; and, unlike Ireland, this investment came mostly from European countries. Foreign direct investment accounted for approximately 35 per cent of total investment in the manufacturing sector in the period betwen 1986 and 1988. For the so-called strong demand sectors, the corresponding figure was as high as 88 per cent (Viñals, 1990).

Foreign investment also seemed to reinforce internal regional disparities, with the bulk going to Madrid and Catalonia.

Net budgetary transfers from the EC have played a minor role, although rising in more recent years because of the gradual incorporation of Spanish agriculture into the CAP and the doubling of the resources of Structural Funds, of which Spain is a net beneficiary. Yet, in relative terms, Spain cannot look forward to the size of net transfers directed to Ireland, Greece, and, increasingly, Portugal. In the case of Spain, the market is likely to continue playing a much more important role than inter-country transfers of funds.

Why has Spain's experience from the early years of EC membership been so different from that of Greece and, to a lesser extent, also Ireland? One explanation must surely be that stabilization and restructuring in Spain had preceded EC accession and thus had created the appropriate environment for the positive effects of integration to materialize. This may in turn point to the wisdom of the Socialist Government which introduced the necessary and unpopular measures in time; a wisdom coupled with the strong ambition of Spain's young political leadership to turn the country into one of the big industrial (and political) powers of Europe.[2] Wisdom being rarely a sufficient condition for political success, the Government of Mr Gonzalez also took advantage of its virtually unchallenged position in the domestic political system. A heavy price for stabilization and industrial restructuring has been paid in terms of unemployment, and the subsequent benefits of economic recovery have been unequally distributed. Thus the continuation of the new 'economic miracle' in the future should largely depend on political stability and the development of industrial relations.

There are also other factors which have played a significant part in the positive experience of the early years of EC membership. The large size of the economy, its geographical location, a reasonable infrastructure, and a relatively diversified industrial base clearly distinguish Spain (and, particularly, the most prosperous northeastern part, together with Madrid) from the other less developed and more peripheral countries and regions of the

---

[2] In an interview to the *Financial Times* (20 Nov. 1991), the Spanish Minister of Economics and Finance, Mr Solchaga, declared: 'We are a province of Europe, and that is the best thing that has ever happened to us.' Although, perhaps, uncharacteristically Spanish for its modesty, this statement reflects the importance attached to Europe, and the EC in particular, in post-Franco Spain.

Community. Last but not least, accession to the EC coincided with economic recovery all over Europe and the fall in oil prices. The time was, therefore, highly propitious. On the other hand, a large part of the economic adjustment to EC membership and the creation of the internal market still remains to be done, and this is likely to take place under a less favourable macroeconomic environment. Domestic economic policies will also have to be geared increasingly towards meeting the convergence criteria agreed at Maastricht. But this seems to be well understood by the Spanish political class.

The EC Budget: A Reluctant Acceptance of Redistribution

An easy extrapolation of 'invisible hand' ideas to the real world of regional economics in the presence of market-opening measures would be unwarranted in the light of economic history and theory (Padoa-Schioppa, 1987: 93).

Historical experience suggests, however, that in the absence of counter vailing policies, the overall impact [of EMU] on peripheral regions could be negative (Committee for the Study of Economic and Monetary Union, 1989: 22).

Redistribution constitutes an important function of national budgets in all advanced industrialized countries. Through taxation and government expenditure, resources are transferred automatically to the poorer regions and lower income groups, thus bringing about a reduction of disparities within a country. Federal grants in existing federations within the OECD group of countries constitute between 15 per cent and 30 per cent of regional revenue (Commission of the EC, 1990e). According to an earlier study (MacDougall report: Commission of the EC, 1977), primary income differentials between regions in those federations are reduced by 30 per cent to 40 per cent through the workings of central public finance. A strong redistributive role of the federal budget is a function of political cohesion and a developed sense of *Gemeinschaft*. In this respect, the European Community is still in its early stages. The gradual acceptance of the redistributive function of the EC budget and hence the link between economic integration and solidarity across national frontiers makes the

Community very different from traditional international organizations, and can be interpreted as a sign of the EC slowly acquiring the traits of a proper political system.

The distributional impact of integration does not, of course, operate only through the budget. Trade effects through the establishment of the customs union and the CAP have been much more important than any explicit redistribution through the budgetary mechanism of the EC. Yet the latter has attracted a great deal of public attention which is disproportionate to its real economic significance; and this explains the prominence given to budgetary disputes in the more recent history of the EC. Gains and losses through the budget are much more easily identifiable than, for example, the welfare effects of intervention prices in the agricultural sector, thus leading to the unavoidable trench warfare in Council meetings.

In the early years of European integration, the common budget remained completely in the background. Unlike national budgets, which represent the main economic policy statement of the government, EC budgets were the end-product of decisions taken almost independently and in a non-co-ordinated fashion by different Councils of Ministers in Brussels in the context of EC action in separate policy areas. The overall expenditure, arising from these decisions, was then financed through national contributions, and the keys for those contributions had been established by the treaty. As explained earlier, the common budget was not seen as an instrument of redistributing resources across national frontiers, since equity in terms of the distribution of benefits from integration was to be achieved essentially through the package of different policies which had been agreed upon in advance.

The creation of own resources for the Community and the extension of the powers of the European Parliament in this area were major items in the Commission's proposals of 1965 which led to the first serious constitutional crisis inside the Community. But the real dispute, provoked by General de Gaulle, was about questions of national sovereignty and the division of power between Brussels and the national institutions. The budgetary questions were only of secondary importance. In fact, both proposals were to be adopted some years later, with the Luxembourg agreement of April 1970.

The financial independence of the Community has had in fact a much longer history. The Paris Treaty of 1951 gave the High Authority of the ECSC the right to raise levies on coal and steel production. The creation of own resources was envisaged in Article 201 of the Treaty of Rome. This has always been a very important subject for the EC Commission and European federalists more generally for both practical and symbolic reasons. The financial independence of EC policies and institutions has been considered as a source of political power and an unmistakable sign of the qualitative difference of the Community from other international organizations (Strasser, 1990).

The 1973 enlargement, the economic recession, and the plans for the creation of an EMU combined to usher budgetary issues to the forefront of Community politics. The efforts of the UK to reduce its own net budgetary contribution opened the Pandora's box by turning net national contributions to the budget for the first time into an important, and also explosive, political issue, and thus raising alarm in Brussels about the negative consequences on further integration of a possible entrenchment of the *juste retour* principle (meaning that each country would strive for at least a zero balance in its net budgetary contribution and thus risk the transformation of the integration process into a zero-sum game). With the gradual swing to the right in European politics in the early 1980s and the strong emphasis on budgetary consolidation, an ideological factor was later added to the intra-EC debates about the budget.

The bargaining position of the UK was progressively strengthened as the EC budget approached the limits imposed by the 1970 treaty revision in terms of own resources, because of the rapid increase in agricultural expenditure. A new revision required a unanimous decision by the Council and the subsequent ratification by all national parliaments. It thus provided an excellent opportunity to strive for a more wide-ranging restructuring of the EC budget as part of the general reform exercise of common policies in which the Community became gradually engaged. The new political climate was also more propitious to such an attempt.

The first important step was taken at the Fontainebleau summit of June 1984. The package agreed then included new measures to control the growth of agricultural surpluses, an increase of the VAT rate (as part of the Community's own resources), and the introduction of a permanent mechanism for the partial compensation of the UK based on the difference between its VAT contribution and its overall receipts from the budget. Special provision was also made for the reduction of the net contribution of the Federal Republic. Thus the objective of a certain degree of equity (not the same as redistribution) in terms of net national contributions to the EC budget was in the end officially recognized (Denton, 1984). It had taken a great deal of pressure on other members by the then British Prime Minister, Mrs Thatcher, for this principle to become finally, and reluctantly, accepted. But it was a very important turning point.

The Fontainebleau agreement did not, however, succeed in imposing an effective control over agricultural expenditure. Thus the new own resources were exhausted even before the increase in the VAT rate could be implemented; and this led to the opening of new negotiations on a subject which had deeply divided EC countries for years. But this time, the negotiations took place in a much more favourable political environment in the aftermath of the signing of the SEA and the adoption of the internal market goal. Furthermore, the new budget negotiations could build on earlier agreements, especially as regards the UK compensation.

Then came the agreement reached by the European Council in Brussels in February 1988. It covered the creation of new budgetary resources, much tougher measures with respect to CAP expenditure, and a considerable strengthening of redistributive policies. The agreement also incorporated provisions for the UK compensation basically along the lines established earlier at Fontainebleau. The European Council decision was followed in June of the same year by an inter-institutional agreement on the budget. This meant that member countries and EC institutions had reached an agreement both on the overall size of the budget as well as on the main items of expenditure for the period extending to 1992. It was unprecedented, and the ensuing budgetary peace has contributed significantly to the new momentum of integration.

Two basic characteristics of the EC budget are its very small size, compared with national budgets, and the legal requirement for zero balance between revenue and expenditure. Thus its role in terms of the traditional functions of allocation, stabilization, and redistribution can only be extremely limited. In 1992, EC expenditure was expected to reach 1.15 per cent of Community GNP. Since 1973, total expenditure had more than doubled as a percentage of GNP. However, this rapid upward trend, admittedly starting from a low base, has slowed down in recent years in view of the ceiling imposed by the agreement reached at the Brussels summit of 1988 (1.2 per cent of EC GNP for payments until 1992). On the other hand, important financial operations are outside the EC budget. They include the borrowing and lending activities of EC institutions, such as those of the EIB, and aid to developing countries through the European Development Fund (EDF). There is also a separate budget for the operations of the ECSC.

As shown in Fig. 8.2, EC revenue consists of customs duties on imports from outside the Community, agricultural levies (variable taxes on imports plus sugar and isoglucose levies on production), VAT contributions calculated on a harmonized base, and GNP-related national contributions since 1988. The item 'miscellaneous' includes national contributions of new members during the transitional period and intergovernmental advances for the years 1984–6 when *ad hoc* measures were needed in order to balance the books. Due to the progressive lowering of the common external tariff, as a result of GATT negotiations (the latest being the Tokyo Round), the relative importance of customs duties was bound to decline over time, and this also is evident in Fig. 8.2. The rapid increase in EC self-sufficiency in agricultural products largely explains the declining trend in agricultural levies as a source of revenue; declining but also variable, since those levies depend on the difference between world and EC prices for agricultural goods, and thus also on the exchange rate of the dollar in which most farm products are still quoted. Only the revenue from sugar levies has remained fairly steady over time.

Customs duties and agricultural levies constitute the traditional own resources of the Community. Their declining importance will be accelerated by a future agreement for further trade liberalization in the context of the Uruguay Round and CAP reform which should reduce the difference between EC and world prices.

The structural weakness of the above sources of revenue forced the Community to rely on a constantly growing share of VAT contributions in order to finance expenditure. The upper ceiling of VAT contributions had been set in 1970 at 1 per cent of a theoretical harmonized base, while waiting for the harmonization of indirect taxation, which may prove to be like waiting for Godot. As growing expenditure hit against the VAT ceiling, the

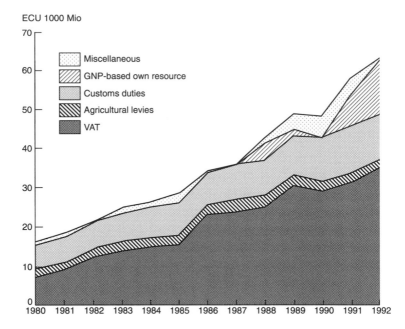

ECU 1000 Mio

Fig. 8.2. Structure of EC Budget Revenue, 1980–1992

Source: Commission of the EC, 1991*a*; for 1992: 'final budget as adopted by the European Parliament', O.J. L 26 of 3 Feb. 1992.

latter was raised to 1.4 per cent at Fontainebleau, with effect from 1986. In fact, the VAT rates actually payable differ between countries because of the operation of the compensation mechanism for the UK and the various complicated provisions which have accompanied it. The so-called fourth resource was introduced in 1988. It consists of national financial contributions, based on a GNP key, which cover the difference between total expenditure, within the limits defined by the 1988 agreement extending to 1992, and the revenue raised by traditional own resources and the 1.4 per cent VAT contribution (Fig. 8.2).

Before the introduction of the fourth resource, any redistributive impact of the EC system of revenue had been totally haphazard: the distribution of tax revenue among countries depending on their propensity to import from third countries (customs duties and agricultural levies), their propensity to consume (there is no VAT on investment), and the efficiency of their tax collection systems (southern countries in particular suffer from widespread tax evasion). An important step towards the application of the ability to pay principle, which still does not contain any form of progressivity in terms of the tax burden, was made with the 1988 decisions. The fourth resource is more closely related to the ability to pay than any other source of EC revenue. It also has the advantage of flexibility in order to meet further increases in expenditure. Furthermore, the European Council in Brussels also decided to limit the VAT base, in terms of which VAT contributions are calculated, to a maximum of 55 per cent of GNP in order to avoid an excessive penalization of countries with a high propensity to consume.

On the expenditure side, the predominance of the EAGGF Guarantee Section has continued all along, although steadily declining in relative terms after 1988 (Table 8.4). The CAP has not met earlier expectations regarding its redistributive role. Not only have income differentials in the agricultural sector remained large, but the CAP has, in fact, contributed through its operation to the worsening of regional and income inequalities inside the Community (Buckwell *et al.*, 1982; Rosenblatt *et al.*, 1988). This was the unavoidable result of a policy which relied for years almost exclusively on price support for unlimited quantities, thus favouring the large and efficient producers. Structural measures through the EAGGF Guidance Section have always accounted for only a small percentage of total agricultural expenditure.

Expenditure per farmer has been a function of farm size, productivity, and product specialization. Thus the subsidy received by the average Dutch farmer is a multiple of the amount paid to his less fortunate colleagues in southern Europe. As a consequence, the overall receipts of the Netherlands from the EAGGF have always exceeded those of Greece, despite the much larger number of farmers in the latter country (Table 8.5). The figures for Spain and Portugal are not comparable, because of transitional arrangements still being in operation. Despite the relative bias of the CAP, among the poorer countries both Ireland and Greece are important net beneficiaries of expenditures through the EAGGF Guarantee Section, precisely because of the

Table 8.4. Structure of EC Budget Expenditure, 1980–1990
(in Mio ECU and as % of total budgetary expenditure)

Sector	1980 Mio ECU	%	1983 Mio ECU	%	1985 Mio ECU	%	1988 Mio ECU	%	1990 Mio ECU	%
EAGGF Guarantee	11485.5	71.0	15811.0	63.1	19955.0	70.2	27500.0	62.8	26431.0	56.3
Agricultural structures	328.7	2.0	653.4	2.6	687.7	2.4	1222.0	2.8	2073.5	4.4
Fisheries	64.1	0.4	84.4	0.3	111.7	0.4	281.0	0.6	358.6	0.8
Regional policy	722.7	4.5	2383.0	9.5	1697.8	6.0	3201.4	7.3	5209.7	11.1
Social policy	768.8	4.8	1495.1	6.0	1626.2	5.7	2845.3	6.5	3667.0	7.8
Research, energy	379.5	2.3	1386.5	5.5	706.8	2.5	1153.6	2.6	1733.4	3.7
Development co-operation	641.6	4.0	992.2	4.0	1043.7	3.7	870.5	2.0	1489.6	3.2
Administration	938.8	5.8	1161.6	4.6	1332.6	4.7	1967.2	4.5	2381.3	5.1
Miscellaneous	852.8	5.3	1093.8	4.4	1271.8	4.5	4779.4	10.9	3584.3	7.6
Total budget	16182.5	100.0	25061.1	100.0	28433.2	100.0	43820.4	100.0	46928.4	100.0

Sources: Commission of the EC, 'The Community Budget: The Facts in Figures', various issues. For 1990 budget: Court of Auditors Annual Report, O.J. C 324, 13.12. 1991.

Table 8.5. EAGGF Payments to Member Countries, 1980–1990

(in Mio ECU)

	1980	1983	1985	1988	1990
Belgium	596.3	630.0	928.7	732.8	877.9
Denmark	639.0	701.2	842.3	1183.8	1110.6
Germany	2593.5	3183.5	3706.6	4638.6	4100.9
Greece	–	1029.3	1276.3	1452.0	2174.4
Spain	–	–	–	1870.7	2292.0
France	2960.6	3748.6	4755.6	6294.1	5388.4
Ireland	603.4	703.5	1239.9	1072.7	1676.9
Italy	1921.0	2923.5	3586.2	4314.1	4170.8
Luxembourg	12.6	4.8	6.6	5.1	10.7
Netherlands	1565.2	1740.0	2065.5	3774.7	2648.8
Portugal	–	–	–	256.9	459.4
United Kingdom	984.4	1840.6	2004.1	1929.5	1886.8
EC-12	11876.0	16505.0	20411.8	27525.0	26804.8

Notes: EAGGF payments consist of guarantee and guidance expenditures. Outlays on a number of specific structural measures have not been included. The figures for Spain, Portugal, and Greece should be interpreted with caution. Transitional arrangements after accession mean that receipts from EAGGF during this period are significantly lower than under conditions of full participation.

Sources: Court of Auditors annual reports.

sheer size of their farming sector. But so also are Denmark and the Netherlands, which are at the top of the EC income league.

According to the 1988 decision, the rate of growth of EC expenditure on agriculture under the Guarantee Section of the Fund should not exceed 74 per cent of the annual rate of growth of EC GDP. Special provisions were made for exchange rate fluctuations and the running down of accumulated stocks. The aim was to bring the figure down to approximately 55 per cent of total expenditure by 1992, which is very likely to be met. The attempt to control agricultural expenditure was helped in the beginning by the strengthening of the dollar and the more favourable conditions applying in world agricultural markets. Thus collective political will in Europe was given a welcome boost by US monetary policy and climatic conditions in North America (a possible sign that God was in favour of European integration); but, alas, not for very long. As supply started growing again in international markets, so did EC surpluses and the overall cost of the CAP.

With still more than half of total EC expenditure going to agriculture, there is a limit on how much can be spent on other policies. Clearly, there is an enormous difference from national budgets where a large bulk of total expenditure goes to defence and social security. Expenditure through what is now called the Structural Funds covers a major part of the rest; and this is very likely to continue increasing further beyond 1992. This is, in fact, the only item of the EC budget with a clear redistributive bias. According to Table 8.4, other policies, including industry, energy and research account for only a small percentage of total agricultural expenditure.

The EC Court of Auditors undertakes a painstaking work every year in an attempt to trace receipts and payments through the budget. This in turn makes possible the calculation of net national contributions based on approximately 90 per cent of total expenditure, that is excluding administrative costs and development aid. Those calculations are subject to the usual qualifications about accuracy, including the so-called Rotterdam–Antwerp effect.[3] Table 8.6 shows that in absolute figures, the Federal Republic has remained all along the biggest net contributor to the budget, thus justifying the popular image of Germany as the paymaster of Europe. But the overall sums are clearly very small. It is followed, at some distance, by the UK, despite the operation of the compensation mechanism. In relative terms, Belgium and Luxembourg are also important net contributors, although those two countries receive other pecuniary and non-pecuniary benefits, not incorporated in those figures, by having the seat of EC institutions. The net contribution of France has risen significantly in recent years, and Italy has also become a net contributor. Those developments are not unrelated to the entry of poorer countries into the EC, while, in the case of Italy, it is also connected with the upward revision of income statistics in recent years to take account of the large underground economy in that country. Thus, the prestige of a higher position in the European income league has been paid for by the Italians in hard cash.

---

[3] This refers to the general problem created by large transit trade in calculating net national contributions to the EC budget. Should the revenue raised from imports through a Dutch port, with Germany as the final destination, be attributed to the Netherlands or the Federal Republic? A substantial amount of trade in northern Europe goes through the ports of Antwerp and Rotterdam.

Table 8.6. Net Transfers from EC Budget, 1980–1990

(Receipts minus contributions expressed in Mio ECU and as % of national GDP)

	1980		1982		1984		1986		1988		1990	
	Mio ECU	%	Mio ECU	%	Mio ECU	%	Mio ECU	%	Mio ECU	%	Mio ECU	%
Belgium	−273.4	−0.32	−499.0	−0.57	−398.2	−0.41	−283.9	−0.25	−995.0	−0.79	−773.9	−0.54
Denmark	333.9	0.70	228.3	0.40	487.2	0.70	421.1	0.50	350.9	0.38	422.5	0.42
Germany	−1670.0	−0.28	3171.7	−0.47	−3033.1	−0.39	−3741.8	−0.41	−6107.1	−0.60	−5550.4	−0.49
Greece	—	—	604.3	1.53	1008.2	2.34	1272.7	3.10	1491.6	3.30	2470.2	4.90
Spain	—	—	—	—	—	—	94.9	0.04	1334.3	0.46	1711.3	0.47
France	380.4	0.08	−827.3	−0.15	−459.8	−0.07	−561.5	−0.07	−1780.9	−0.22	−1804.9	−0.20
Ireland	687.2	5.00	671.5	3.50	924.1	4.01	1230.3	4.90	1159.3	4.30	1892.5	6.19
Italy	681.2	0.21	911.4	0.22	1519.0	0.29	−130.3	−0.02	124.2	0.02	−416.7	−0.05
Luxembourg	−5.1	−0.15	−24.3	−0.67	−40.1	−0.93	−59.3	−1.20	−67.4	−1.20	−60.0	−0.97
Netherlands	394.5	0.32	86.9	0.06	434.8	0.06	167.5	0.10	1150.0	0.60	368.4	0.17
Portugal	—	—	—	—	—	—	219.0	0.73	514.9	1.45	600.8	1.44
United Kingdom	−1364.6	−0.35	−1193.6	−0.24	−1337.0	−0.24	−1438.4	−0.26	−2070.0	−0.26	−3386.9	−0.45

Notes: Approximately 10% of all expenditure is absorbed by administrative costs or is used for development aid. These outlays cannot be apportioned among the member states and consequently the aggregate of member states contributions to the EC budget exceeds the total for receipts.

Sources: Court of Auditors annual reports.

Ireland and Greece have been the only important net beneficiaries, with net inflows from the EC, excluding loans, accounting now for 5–6 per cent of GDP. Ireland's relatively high net inflow, which in absolute per capita terms is the highest in the Community, is evidence of the advantage of being a small member of the Community. Those figures will be even higher in 1992 and 1993, thus making the EC contribution to economic welfare in these two countries far from insignificant. On the other hand, the overall effect of these transfers in terms of economic development is reduced by the very fact that a large bulk is still devoted to agricultural price support and not to structural measures.

Portugal is also expected to join gradually the group of important net beneficiaries, as it approaches the end of its transitional period following accession to the EC. Thus the three poorest members among the Twelve will end up as beneficiaries of sizeable transfers of resources through the budget. The corresponding figures for Spain, the fourth poorest country of the EC, are not expected to reach similar levels as a percentage of GDP, partly because of the considerably higher level of economic development of that country and also partly because of its relatively large size.

In this respect, the redistributive function of the budget is already a reality; certainly limited, when compared with the size of interregional redistribution of resources inside member countries and federal systems outside the EC, but no longer negligible. Important anomalies, however, still remain, being basically the product of financial arrangements in the agricultural sector which still play a dominant role in the EC budget. Given the size of the budget, the net contribution of Germany and the UK is disproportionately large, both countries being big net importers of agricultural products. Exactly the opposite is true of prosperous countries such as Denmark and the Netherlands which continue as net beneficiaries from the EC budget, because of their large agricultural production which is concentrated mainly in products characterized by heavy intervention.

The battles waged by the UK for many years helped to bring budgetary issues to the forefront of public attention. UK representatives waved the flag of equity which for some Europeans was almost indistinguishable from the banner of *juste retour*, associated with infidels. In more recent years, the battleline has shifted considerably: the poorer countries of the Community have been fighting for ever bigger budgets, waving the flag of redistribution. The European Parliament has been a valuable ally in this respect in its effort to make the best possible use of its limited budgetary powers set by the treaty revision of 1975. Through its sustained attempts to increase the so-called non-compulsory expenditure, defined as expenditure other than that necessarily resulting from the treaty, on which it has greater powers *vis-à-vis* the Council of Ministers, the European Parliament has been in effect pushing consistently for more money for regional and social policy, thus acting as an important lobby in favour of redistribution and bigger budgets (Shackleton, 1990).

Back in 1977, the MacDougall group (Commission of the EC, 1977) had stressed the redistributive role of the central budget in existing federations such as the United States and West Germany, which accounted for a substantial reduction of inter-state income disparities. This was due to the differential impact of taxation and expenditure policies and the specific and general purpose grants from the federation to the states. The German system of fiscal equalization (*Finanzausgleich*) is a good example of how relatively small amounts of transfers can have a large effect in terms of the tax revenue of each individual *Land* and the reduction of disparities. This is due to the importance of tax sharing arrangements and horizontal transfers between the *Länder*, which bring the 'fiscal capacity' of the poorer *Länder* to a minimum of 95 per cent of the federal average, the general aim being to achieve a homogeneous supply of public services across the country (see also Biehl, 1988c).

Aware of the political constraints in what it called the pre-federal state of European integration, the MacDougall study group had proposed a relatively small and high-powered budget. An EC budget accounting for 2–2.5 per cent of GDP, with heavy concentration on structural, cyclical, unemployment, and regional policies as well as external aid coupled with progressive taxation, could result in a 10 per cent reduction of interregional disparities inside the Community, while also providing an effective insurance policy against short-term economic fluctuations. True enough, the MacDougall group had written about the EC of Nine; the task would be much bigger for the EC of Twelve. In any case, those proposals stood little chance of being adopted in the political climate of the late 1970s and early 1980s.

This issue was later taken up by the Padoa-Schioppa group which argued in favour of a long-term social contract between the Community and its member states based on competitive markets, monetary stability, redistribution, and growth. According to this group, redistribution through the budget should become virtually automatic. Thus net national contributions should be progressively (and of course inversely) related to national income per capita. The need for stronger redistribution in the context of EMU was again reiterated in the Delors report and, in a somewhat milder form, in the Commission's study on the costs and benefits of EMU. Meanwhile, Brussels was getting increasingly worried that the intergovernmental conference on EMU could become tied up with demands from the weaker countries for large transfers (Commission of the EC, 1990e).

As with the earlier package deal, which led to the adoption of the SEA and the internal market programme, redistribution is an integral part of the new package of which the other most important part is EMU. Following again the example of the SEA, the Maastricht treaty does not attach any specific price tags. It does, however, include Title XIV on 'economic and social cohesion' and, more importantly, a separate protocol on the subject, which refers among other things to the correction of regressive elements in the own resources system, the setting up of the Cohesion Fund and to several changes in the operation of the existing Structural Funds aiming at greater flexibility in financing and higher levels of Community participation.

The second Delors package, submitted to the Council of Ministers in February 1992, constitutes a first attempt by the Commission to translate those general objectives into concrete proposals with figures attached to them. The Commission has proposed a further increase in the budget which would raise the ceiling to 1.37 per cent of Community GNP by 1997, implying an increase of 5 per cent *per annum* in real terms. Although this would certainly not alter substantially the distribution of financial power between Community and national institutions, it would, nonetheless, lead to a significant further strengthening of EC intervention in certain areas.

Although their distributional impact is largely haphazard, customs duties and agricultural levies (including sugar levies) naturally belong to the EC system of own resources; hence being called natural or traditional own resources. They are the end product of common policies in the fields of agriculture and external trade. On the contrary, there is little economic logic in the Community's reliance on a small percentage of total receipts from VAT, calculated on a notional common base, because of the lack of harmonization of national tax systems; unless it is meant as a reminder of the role of the EC in the introduction of VAT itself. In the previous budgetary reform and in search for a more equitable system, the EC had partly gone back to national contributions based on a GNP key. The EC Commission is now ready to move further in this direction, by reducing the VAT element in the budget and thus relying even more on national contributions. This may be regarded with some apprehension by traditional federalists, although perhaps misguidedly, since European integration has gone much beyond the stage where reliance on national contributions may endanger its financial independence and the development of common policies. A stronger redistributive bias on the revenue side could be, of course, achieved through progressive taxation related to GNP. This would simply mean transposing to the EC level principles which have long guided taxation policies within the European nation-state. But the Community does not seem ready as yet to take such a step; nor does it appear ready to introduce a proper EC tax.

On the expenditure side, the Commission has proposed substantial increases in three main areas, namely structural policy, industrial policy, and external aid. More than half of the total increase would go to the Structural Funds, including the new Cohesion Fund. This would, therefore, further strengthen the redistributive impact of the budget. For the least developed countries and regions of the Community (Objective One), the Commission wants to go as far as doubling again in real terms the resources available. Since the Commission has not proposed any radical changes in the priorities of the Structural Funds, the emphasis should remain on investment rather than the compensation of losers, although some element of the latter can be found in existing policies and especially in some of the actions undertaken by the Social Fund. However, the choice between the two may become more pressing and, indeed, difficult in the future, if unemployment rises again. Should the compensation of losers be left entirely to national governments, especially in view of the limitations to be imposed on national budgets by the convergence criteria?

There is another difficult choice to be made in the next few years, with the prospect of a limited increase in overall EC resources, namely between intra-EC redistribution and economic aid in favour of third countries, and most notably countries in Eastern Europe and the southern part of the Mediterranean (see also Chapter 9). Implicit in the Commission's proposals is a strong choice in favour of intra-EC redistribution — a sign, perhaps, of political realism — even though a considerable increase in external aid is also being proposed. The third area singled out for a major increase in EC expenditure for the next few years is that related to the competitiveness of EC industry. But the EC Commission does not have any ambitions for an active industrial policy. Instead, the emphasis has been on R & D and the development of infrastructure networks.

Agricultural policy will remain, for some time at least, the biggest item of EC expenditure, although continuing to decline in relative importance. In the second Delors package, CAP expenditure would fall below the 50 per cent line by 1997. At the time of writing, the chances for a radical reform of the CAP, leading to a substantial shift from price support to income subsidies, appear to be very high indeed. But since there will be an attempt to preserve existing income levels for all farmers, this reform is unlikely to change much the redistributive dimension of the CAP, at least in the short- and medium-term.

An agreement on the main budgetary guidelines for the next few years should contribute a great deal to the successful implementation of the Maastricht package and the smooth functioning of the Community in general. It will not be, however, an easy agreement to reach in times of economic recession and also given the budgetary difficulties of the biggest net contributor, namely Germany. The budgetary constraints imposed by the convergence criteria will further temper the enthusiasm of other net contributors to accept a new substantial increase in expenditure; and any re-opening of the question of the UK rebate would make matters worse. Furthermore, as EC policies acquire a stronger redistributive dimension and the number of net beneficiaries (in terms of countries) becomes more limited, agreements for further increases in EC expenditure may also prove more difficult to reach. For all these reasons, the Commission's proposals in the second Delors package may be very close to the maximum in terms of political feasibility. On the other hand, one of the main arguments against a rapid transition to irrevocably fixed exchange rates has been based on the large adjustment costs for the weaker economies, with higher propensities to inflation and also greater reliance on the inflation tax as a partial substitute for a weak tax base. Would the Community budget be able to help much those countries faced with a huge task of fiscal consolidation? And where are the automatic stabilizers to be found in order to deal with asymmetric external shocks in an EMU?

On the other hand, the strengthening of the EC budget and its redistributive function raise more general questions regarding the efficiency in the use of scarce resources, transparency and democratic accountability. On all these questions, the Community's record until now has not been impressive. The deficiencies of the system will become more glaring as more power and money are transferred to the centre; and this will automatically lead to further discussion about the powers of the Commission and the European Parliament. These are the inevitable questions faced by a fledgeling political system, which have not been answered adequately by the last intergovernmental conference. Not surprisingly, they are centred on public finance.

9. The External Dimension

European economic integration does not take place in an international vacuum. The countries of the EC are open economies, with close interaction with the rest of the world. Therefore, intra-EC developments are bound to be seriously affected by economic events and political decisions taken outside the Community boundaries. On the other hand, the combined size of EC countries makes them a major economic actor on the world scene, an actor whose actions have a significant impact on the international economic system. The small country assumption frequently employed in economics textbooks hardly applies to the large majority of EC member countries taken separately; it makes absolutely no sense for the regional bloc as a whole.

European regional integration has coincided with the post-war process of international liberalization which has in turn contributed, together with autonomous economic forces, to the remarkable growth of interdependence. As national economic frontiers are progressively eliminated in Western Europe, the question immediately arises as to whether this liberalization should be extended to the rest of Europe and the world. What, if any, is the specificity of Europe, who should define it, what and where should be the common frontier against third countries? These are old questions which acquire new significance during the current phase of deepening of regional integration. A group of people usually defines itself collectively with reference to outsiders, and the same applies to political and economic systems. There is little reason to assume that the new regional economic system in Western Europe will behave differently.

The EC started basically as an incomplete customs union, which meant that the common external tariff constituted the building-block of its fledgeling international role. With the gradual deepening of integration, new common instruments have been created, while there has also been a shift in the division of external policy competences between EC and national institutions. There is a wide range of policy instruments which can be used by political authorities in order to influence cross-border economic exchange. Some of these instruments have long since been transferred to the EC level, while others have remained in the preserved domain of the nation-state. In between, there has always been a wide grey area where the division of power between different levels of authority has been ambiguous but also changing. This is the counterpart of the grey area which exists in terms of internal policies, the boundaries of which are in the process of being redefined because of the internal market programme; and the same will be true in the future because of EMU.

The picture is further complicated by the wide discrepancy between political objectives and economic instruments at the Community level, which has often made it difficult to understand and even more difficult to evaluate EC policies in terms of purely economic criteria. Duchêne (1973) coined the term 'civilian power' to describe the reality of a powerful economic bloc which was heavily constrained in the traditional area of foreign and defence policy, in other words in what is generally referred to in the academic literature as high politics. The Community has frequently resorted in the past to economic instruments for the pursuit of wider political objectives. The concept of the 'civilian power' may therefore be considered, at least partly, as an *ex post* rationalization of the constraints imposed by an underdeveloped political system and the international balance of power; and sometimes as an attempt to make virtue out of necessity, by offering this as a new model of international political behaviour. Such ideas have been largely forgotten since the EC has moved more and more, albeit with much trepidation, into the area of high politics. But the discrepancy between objectives and instruments has remained as true as ever. The treaty of Maastricht is supposed to open wide the gates leading to a common foreign and security policy, although the journey beyond this point still remains unknown.

This chapter will examine the development of EC policies *vis-à-vis* the rest of the world. This covers an extremely wide area in terms of both objectives and instruments. We shall concentrate in the first two sections of this chapter on trade policies, where the competences of EC institutions are most clearly defined and also effectively exercised, and the use of common instruments as means of influencing the Community's position in the international division of labour. Special attention will be paid to the link between the internal market programme and international trade negotiations in the context of GATT. Preferential relations with certain countries will be examined in the third section. This will inevitably lead us to the discussion of wider economic and political aspects. Membership of the club is the ultimate preference that can be accorded to outsiders. The spectacular changes of recent years on the European scene have opened once again the prospect of further enlargement of the Community. This subject, which is likely to figure prominently on the European agenda for several years, will be discussed in the last section. Enlargement is linked to internal policy and institutional changes. It is also inextricably linked to the search for a new European architecture and the place of the Community in it.

Common Commercial Policy: A European Mosaic

With successive enlargements, the EC has become by far the biggest trading bloc in the world. In 1989, extra-EC exports accounted for approximately 20 per cent of world exports; the corresponding figure for the United States was 16 per cent and for Japan 12 per cent. Thus the three biggest trading blocs, the triad as it is increasingly referred to in the literature, represented almost one-half of total world trade. Interestingly enough, the openness of EC economies towards the rest of the world in terms of goods has hardly changed in the course of the last three decades, with the exception of a period of approximately ten years (starting around 1973–4) during which there was an upward kink in the curve (Fig. 9.1). The reasons for this relative increase in extra-EC trade have been discussed earlier. The openness of the EC as a whole (that is, excluding intra-EC trade) is comparable to that of Japan and the United States; the latter country having registered a substantial and steady increase in its trade openness since the early 1970s. But averages of imports and exports can conceal wide differences between the two, and this has long been true, especially of Japan which in terms of imports,

% of GDP

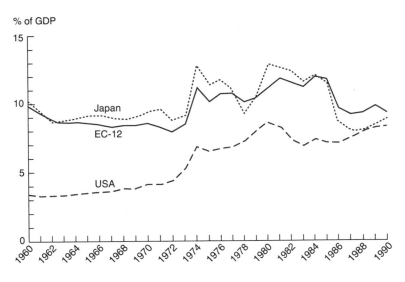

Fig. 9.1. Exports and Imports of Goods of EC-12, USA, and Japan, 1960–1990
(Sum of exports and imports divided by 2, as % of GDP)
Source: Eurostat, (GDP figures for the USA and Japan: OECD).

especially in manufactured goods, still appears as a relatively closed economy (the import penetration rate in the manufacturing sector being around 3 per cent, as compared with almost 7 per cent in the case of the EC (Commission of the EC, 1989*a*).

If the openness of the EC as a group has hardly changed over a long period, the openness of individual member countries has been growing steadily due to the increase in intra-EC trade as a percentage of GDP (see also Table 7.7). This is true of both goods and services. There was a partial reversal of this rising trend during most of the 1970s which was, however, largely compensated by the faster growth of extra-EC trade during the same period. Accession effects, following the enlargements of 1973, 1981, and 1986, have contributed to the growth of intra-EC trade, which now represents approximately 60 per cent of total trade for the average EC country. It is true that hardly anybody talks about the trade openness of Arizona or Nebraska, and the relevant statistics would be hard to find. But the EC is in an intermediate stage. Trade statistics for individual member countries are relevant, since many of the macroeconomic policy instruments still remain in the hands of national authorities. This also explains a certain degree of confusion in the literature, where total trade and extra-EC trade statistics are often used interchangeably.

Table 9.1 shows the geographical breakdown of extra-EC trade in goods. More than half of the Community's external trade is conducted with the rest of the OECD. This is mainly intra-industry trade. The small and highly industrialized countries of EFTA constitute by far the biggest trading partner of the EC, followed by the United States. In other words, for an average EC country almost three-quarters of its total trade takes place within the wider area of Western Europe; and the same applies to EFTA countries. Geographical contiguity, similar economic structures, and long historical links largely account for the high degree of economic integration of Western European countries. On the other hand, geographical contiguity cannot explain the relatively low share of Eastern Europe (including the former Soviet Union) in EC external trade. The cold war and major differences in the economic systems had, until recently, played a more important and negative role. Developing countries account for a large and widely fluctuating share of trade, the bulk of which is represented by OPEC countries which are also largely responsible for this fluctuation. As for the dynamic economies of South-East Asia, they still represent a small, although steadily rising, share of EC external trade.

One first conclusion which can be drawn from Table 9.1 points to the strong regional character of the Community's external trade, which is even more pronounced when member countries are treated separately. The regional concentration of trade will increase further with the opening of Eastern European economics. Trade with neighbours (EFTA and Mediterranean countries) has also usually generated the surpluses which have helped to finance at least in part the large trade imbalances incurred with the countries of OPEC and East Asia. The trade deficit with Japan is very large and it has been steadily rising over the years. The geographical distribution of the Community's external trade has been strongly influenced by two highly unstable factors, namely oil prices and the exchange rate between the US dollar and European currencies which has an impact not only on transatlantic trade but also on trade in most commodities.

Table 9.1. Breakdown of EC Trade by Trading Partners

(Imports and exports as % of extra-EC trade)

Trading partner	1970			1980			1984			1988			1991		
	% of imports	% of exports	Exports /imports	% of imports	% of exports	Exports /imports	% of imports	% of exports	Exports /imports	% of imports	% of exports	Exports /imports	% of imports	% of exports	Exports /imports
Western industrialized countries (non-EC)	54.6	59.3	0.95	46.1	49.6	0.83	51.0	55.0	0.97	61.6	61.2	0.93	59.4	57.2	0.83
EFTA countries	17.4	25.1	1.26	17.0	25.5	1.15	19.4	21.8	1.01	23.4	26.6	1.06	22.4	25.7	0.98
United States	21.7	18.0	0.73	16.9	12.8	0.58	17.2	21.0	1.10	17.6	19.8	1.05	18.6	16.8	0.77
Japan	3.4	2.6	0.68	4.9	2.2	0.34	6.6	2.7	0.36	10.7	4.7	0.41	10.5	5.2	0.43
Eastern Europe[a]	6.4	7.3	1.00	7.3	8.0	0.84	9.2	6.3	0.62	6.4	5.7	0.83	7.0	7.5	0.92
Developing countries	38.0	31.0	0.71	45.7	41.2	0.69	38.9	37.4	0.86	30.1	31.3	0.97	30.4	33.7	0.95
ACP countries[b]	8.9	7.6	0.75	7.3	7.9	0.82	7.2	5.2	0.66	4.5	4.3	0.89	3.9	3.8	0.83
Mediterranean basin[c]	9.4	10.3	0.96	8.3	13.4	1.23	10.2	12.3	1.09	7.8	9.8	1.17	8.8	10.8	1.05
OPEC countries	16.3	7.5	0.40	27.2	18.1	0.51	18.5	15.6	0.76	8.2	8.6	0.98	9.5	9.3	0.84
Asian NICs[d]	1.5	2.1	1.19	3.5	2.7	0.58	3.6	3.3	0.82	6.3	5.4	0.80	6.2	6.1	0.84

[a] Soviet Union, German Democratic Republic (until 1990), Poland, Czechoslovakia, Hungary, Romania, Bulgaria, and Albania.
[b] This group includes 68 African, Caribbean, and Pacific countries.
[c] Malta, Yugoslavia, Turkey, Albania, Morocco, Algeria, Tunisia, Libya, Egypt, Cyprus, Lebanon, Syria, Israel, and Jordan. Gibraltar, Ceuta, and Melilla are also included.
[d] Hong Kong, South Korea, Singapore, and Taiwan.

Source: Eurostat.

Some of the figures in Table 9.2 are quite revealing of the changing position of the EC in the international division of labour. Machinery and transport equipment which has always accounted for the big bulk of EC exports shows a steadily increasing import penetration. This is certainly true of motor vehicles which represent a significant share of this category; but it is also true of several dynamic sectors, with a high technology content, such as telecommunications equipment, office machines, and data-processing. With respect to those sectors, the decline of the EC export/import ratio, as shown in Table 9.2, has been very rapid indeed. The Community has remained a big net exporter of chemicals and steel; and it has also retained a strong presence in the up-market end of relatively weak-demand sectors such as textiles and clothing, although the export/import ratio has been steadily declining.

Several studies have pointed to the unfavourable pattern of export specialization for the EC, the poor geographical spread because of the heavy reliance on slow-growing economies in the developing world, and the increasing import penetration (Buigues and Goybet, 1989; Lafay *et al.*, 1989). The conclusion reached by the EC Commission on the basis of external trade trends in the 1980s is also sufficiently alarming for the Europeans: 'for sensitive high-tech sectors such as computers, telecommunications and consumer electronics, the Community, firstly, seems to be concentrating on its own market, but, secondly, even in its own market is losing part of its market shares. In the markets of third countries, the loss of volume shares seems even more pronounced' (Commission, 1989a: 215).

Although the above observations are generally true of the EC as a whole, including the most advanced of its member countries, generalizations of this nature conceal enormous differences among the Twelve. Obviously, the commodity structure of external trade for a country such as Portugal or Greece is closer to that of some developing countries than the commodity structure of Germany's external trade. The same is true of the geographical orientation of trade. EFTA countries are much more important for Denmark and the Federal Republic, UK exports to the United States represent a much higher than average share of total exports, while the presence of Spain, the UK, and France is relatively more pronounced in the developing world. This explains why member countries have tended to squint in different directions.

The EC's position in the international division of labour, as summarized in the above tables, has been the joint product of autonomous economic forces and government policies. Gilpin (1987: 223) has argued that '[in] a world where who produces what is a crucial concern of states and powerful groups, few are willing to leave the determination of trading patterns solely up to the market'. European countries have certainly not been innocent of such intentions and practices, although the policy instruments used have not always been mutually consistent. Although hardly anybody could accuse national governments of internal consistency in their policies, the performance of the Community has been much worse because of the highly decentralized nature of political power.

The common external tariff (CET) forms an integral part of the customs union, and it has therefore provided the basis for the Community's common commercial policy (CCP) and its role as an international actor. This was, indeed, enshrined in the Treaty of Rome where Articles 110–16 referred to the progressive establishment of the CCP during the twelve-year transitional period; unlike its predecessor, the Paris Treaty, which had made no provisions for a common external policy in the sectors of coal and steel. Article 113 provides the legal basis for the common policy which is conducted by the Commission on the basis of mandates agreed upon by the Council of Ministers. But a commercial policy does not only rely on the use of tariffs, and this is explicitly recognized in Article 113 which also refers to other instruments such as liberalization measures (mainly quotas), export policy (credits), anti-dumping, and countervailing measures against subsidies. On the other hand, there is little to negotiate with third parties when there are no common internal policies. Thus the lack of provisions in the original treaty for a common macroeconomic policy also meant, unavoidably, the lack of an EC international presence in this area. The EC started as a trading bloc, and it has continued acting as such for a long time.

Successive GATT negotiations have contributed to the liberalization of international trade mainly through the reduction of tariff levels and the elimination of quantitative restrictions. This is mainly true of goods, since services have only recently started figuring on the agenda of GATT negotiations. The EC has played an active part in all those negotiations on the basis of common positions which have frequently been the result of long and

Table 9.2. Structure of EC Trade with the Rest of the World by Commodity Type[a]

(As % of total extra-EC imports or exports in each year)

Commodity type	1970			1980			1984			1988			1991		
	% of EC imports	% of EC exports	Exports/ Imports	% of EC imports	% of EC exports	Exports/ Imports	% of EC imports	% of EC exports	Exports/ Imports	% of EC imports	% of EC exports	Exports/ Imports	% of EC imports	% of EC exports	Exports/ Imports
Food products, beverages, and oils (SITC 0 + 1 + 4)	18.7	8.2	0.39	9.9	8.7	0.67	10.2	8.5	0.75	9.3	7.4	0.75	8.1	7.7	0.82
Crude materials (SITC 2)	18.8	2.4	0.11	10.6	1.8	0.13	10.0	1.9	0.17	9.2	2.1	0.21	6.7	1.9	0.24
Fuel products (SITC 3)	17.1	3.0	0.15	34.4	4.5	0.10	30.6	5.0	0.15	12.3	2.3	0.17	14.4	2.5	0.15
Chemicals (SITC 5)	5.0	10.6	1.84	4.2	10.3	1.89	5.2	10.8	1.89	6.5	12.3	1.77	6.6	12.0	1.57
Manufactured goods (SITC 6)	18.3	23.0	1.10	12.7	21.8	1.32	11.8	19.5	1.49	15.1	18.1	1.13	14.2	16.9	1.02
Textile and Fabrics (SITC 65)	2.0	5.1	2.27	2.1	3.2	1.20	2.0	3.1	1.40	2.5	3.0	1.12	2.4	3.1	1.10
Iron and steel (SITC 67)	3.3	6.5	1.70	1.8	5.7	2.51	1.5	5.2	3.18	2.0	4.1	1.96	1.7	3.3	1.68

Machinery and transport equipment (SITC 7)	14.2	40.4	2.50	13.6	37.1	2.10	17.6	35.8	1.83	27.5	38.9	1.32	30.0	40.8	1.16
Office machines and data processors (SITC 75)[b]	1.7	1.5	0.75	1.9	1.4	0.59	3.6	2.0	0.50	5.3	2.5	0.44	5.2	2.5	0.42
Telecommunications equipment (SITC 76)[c]	0.7	2.0	2.45	1.7	1.9	0.87	2.0	1.6	0.74	3.5	1.7	0.44	3.6	1.8	0.44
Miscellaneous manufactured articles (SITC 8)	5.0	10.3	1.79	7.4	9.9	1.02	8.5	11.7	1.24	12.9	13.4	0.97	14.9	13.4	0.78
Apparel and clothing accessories (SITC 84)	1.2	1.9	1.36	2.4	1.7	0.53	2.6	2.0	0.67	4.1	2.4	0.55	5.2	2.5	0.41
Goods not classified elsewhere (SITC-9)	2.8	2.2	0.68	7.2	5.9	0.63	6.2	6.8	0.99	7.3	5.5	0.71	5.1	4.8	0.81

[a] All figures are for EC-12. Columns 'Exports/Imports' give the value of EC exports divided by the value of EC imports.
[b] For 1970, sub-category 714 of the SITC system as by that time.
[c] For 1970, sub-category 724 of the SITC system as by that time.

Source: Eurostat, External Trade Statistics. Figures for subcategories 65, 67, 75, 76, and 84 for years 1970 and 1980 have been calculated using the OECD Statistics of Foreign Trade, Series C (Trade by Commodities).

painful intra-EC bargaining (Hine, 1985; Murphy, 1990). The CET has been progressively reduced, reaching an average of about 5 per cent after the implementation of the Tokyo Round agreements. Of course, wide differences in tariff rates lie hidden behind the low average; higher tariff rates have survived in some sensitive sectors such as textiles and clothing as well as consumer electronics. Furthermore, effective rates of protection calculated on the basis of value added are almost invariably higher than nominal rates because of low or even zero rates for the import of raw materials.

However, the internal experience of the Community suggests that tariff barriers are only one part of the story, while the rest is usually much more complicated. Different forms of government intervention, either inside or at the border, create NTBs which play an important role in terms of international economic exchange. Over the years, the attention of GATT has increasingly shifted towards NTBs both because of the progressive reduction in tariff levels and because of the growth of 'new protectionism' which coincided with the deceleration of economic growth in the early 1970s. As explained earlier, this 'new protectionism' has relied mainly on non-transparent forms of intervention (Kelly *et al.*, 1988). The Tokyo Round which lasted between 1975 and 1979 was mainly about NTBs, and the result was the adoption of various codes on technical regulations and standards, customs valuation, import licensing procedures, anti-dumping duties, subsidies and countervailing duties, and public procurement among others. These codes were not signed by all Contracting Parties of GATT. Moreover, subsequent experience suggests that they have only had a very limited effect on national practices; the same issues have figured again prominently on the agenda of the Uruguay Round. Thus the mountain of Tokyo seems to have given birth to a mouse, perhaps not surprisingly, in view of the nature of the problems tackled.

Many international trade practices have developed on the margin or even completely outside GATT legality. The formal emphasis on transparency is difficult to reconcile with the multitude of protective barriers and the *de facto* exclusion of whole sectors from the GATT framework. On the other hand, the principle of multilateralism, based on the Most Favoured Nation (MFN) clause, has been steadily undermined by the proliferation of bilateral agreements. The result has been the general weakening of the GATT order which has not, however, prevented the continued growth of international trade, albeit at a slower pace. The EC has not been an innocent bystander at these developments, but neither have most of the other Contracting Parties of GATT.

Agriculture has been outside the GATT framework from the very beginning, largely because of the insistence of the United States which was granted a waiver in 1955 allowing it to keep its domestic agricultural policies outside international control. The subsequent development of the Community's CAP has made the Americans regret bitterly the early reluctance to submit to international rules in this area. Western European protectionism in agriculture has a very long history: it dates back to the end of the previous century and is associated with the loss of comparative advantage in extensive farming to the New World and the political pressure from large landowners. It was further strengthened during the Great Depression (Tracy, 1989). The main principles of the CAP were laid down between 1958 (Stresa conference) and 1962. Decisions came at the end of extremely long and difficult negotiations which set some of the more notorious precedents in intra-EC decision-making, such as the 'marathon sessions' and the 'stopping of the clock'. During this period, and in fact for much longer, regional integration in general often seemed to depend on the outcome of negotiations on the CAP, a clear indication of the absolutely crucial importance of the latter in the overall European package.

Community preference, as one of the main principles of the CAP, works through a system of variable import levies and export restitutions. The former are calculated as the difference between threshold prices and the price offered by low cost third country suppliers.[1] They ensure that imports cannot sell below domestically produced goods. This is clearly the most effective form of protection. Tariffs and quantitative restrictions also form part of the EC armoury. Export restitutions are the other side of

---

[1] Although market organizations vary from one product to the other (and they have become increasingly complex over the years), there are some basic principles which are common to most of them. The unity of the market relies on a system of price support which usually includes a target price, as the upper end of the range within which producer prices are left to fluctuate, and an intervention price which operates like a floor (a minimum guaranteed price at which producers can sell their products to intervention). Threshold prices are set between the target and the intervention price.

the coin of Community preference. They are meant to cover the difference between EC and world prices.

In view of the size of the EC and the strong emphasis on Community preference, it was only natural that the CAP would become the object of much criticism from third countries and a major issue in international trade negotiations. Agiculture proved to be the big non event of both the Kennedy and the Tokyo rounds, mainly because of the EC's successful resistance to outside, and mainly American, attempts to make inroads into the CAP. The codes adopted in the Tokyo Round have not had much tangible effect on world agricultural trade; and this should not come as a surprise in view of the general vagueness of their contents. According to the relevant code, export subsidies (read CAP) should not be used as a means of obtaining 'more than an equitable share of world export trade'. But what constitutes an 'equitable share' is in itself a multi-million ECU question which cannot be easily settled by lawyers. And it has not been.

At the early stages of the CAP, the attention of large agricultural exporters, and especially the United States, was drawn by the trade diversion effects of the common policy and the main effort was therefore concentrated on weakening the Community's external protection. However, with rapidly growing self-sufficiency ratios and the emergence of the Community as a major exporter of agricultural products (see also Table 9.2), although still the largest world importer, the emphasis has gradually shifted to the negative effects of the CAP on international markets for cereals, sugar, and dairy products. While insulating European farmers from world price instability, the CAP has a depressing effect on world prices, as a result of subsidized exports, while also adding to the instability in world markets. Export restitutions have enabled EC exporters to increase their share of world markets at the expense of other producers. While the effect of the CAP has been clearly negative on exporting countries, such as the United States, Canada, Australia, and New Zealand, most developing countries seem to have benefited, at least in the short term, from subsidized imports of food (Rosenblatt *et al.*, 1988; Tarditi *et al.*, 1989).

Of course, the EC has not been the world's only culprit in terms of income subsidization of its farmers. According to an OECD study, the total amount of agricultural producer support represented in 1986 35 per cent of output in the United States, 46 per cent in Canada, 15 per cent in Australia, and 75 per cent in Japan. The corresponding figure for the EC was 49 per cent (Tracy, 1989: 353).[2] However, the extensive use of export subsidies, combined with the size of EC surpluses, have turned the CAP into the main target of international criticism. The 1988 reform has gone only a small way to meeting the demands of other agricultural producers who have used the Uruguay Round as a means of exerting pressure on the Europeans for a more radical reform of the CAP.

The combination of external pressure, linking the successful conclusion of the Uruguay Round to the substantial reduction of EC prices and export subsidies, and the internal budgetary constraint seem to have finally produced the political agreement for the reform of the CAP which should lead over the next few years to a considerable narrowing of the difference between EC and world prices, especially for the most tradable of agricultural commodities such as cereals. An agreement to that effect was reached at the Council of Ministers in May 1992. This agreement should also lead to a progressive shift of EC intervention from price support to income aids which will have a more neutral effect in terms of production.

Textiles and clothing are another sector where protectionism in industrialized countries has a long history. This is the sector which is closely associated with the first stage of industrialization, hence the pressures for adjustment in advanced economies resulting from growing import competition from developing countries. Textiles and clothing represent by far the biggest source of export earnings for developing countries in terms of manufactured goods. The large majority of Western European countries have experienced heavy job losses in this sector over the years. However, generalizations about textiles and clothing as a whole can also be highly misleading. The European experience suggests that there are important and highly profitable segments of the market where differentiation and the high quality of products can more than compensate for the relatively high cost of labour. The large amount of EC exports in this sector bears testimony to it; in textile and fabrics, the EC has remained an

---

[2] These figures need, however, to be treated with some caution since the OECD calculations of Producer Subsidy Equivalents (PSEs) use world prices in dollars as the standard of reference (OECD, 1987). In view of large-scale dumping, world prices are not exactly representative of free market prices.

important net exporter, while in clothing EC exports are far from insignificant (Table 9.2).

The so-called Long Term Arrangement (LTA), dating since 1962, was replaced in 1974 by the first multifibre arrangement (MFA) which provided a framework for bilateral agreements between importing and exporting countries for quantitative restrictions on low-cost traded products originating from the developing world. As with GATT itself, temporary arrangements have often shown a tendency to become permanent. The fourth MFA is more restrictive than its predecessors. Some 70 per cent of total EC imports are subject to quotas. Overall EC quotas for different products are usually further subdivided into national quotas for member countries. This treatment of the EC as twelve separate national markets automatically leads to the application of rules of origin and the imposition of restrictions in intra-EC trade. The unity of the EC market has been further undermined by the extensive use of national subsidies. On the other hand, the protectionism of industrialized countries (the EC is certainly not alone in this area) has not prevented the steady increase of import penetration ratios from developing countries; but it has, undoubtedly, slowed down the process (Kelly *et al.*, 1988; Murphy, 1990).

Another example of a sector where trade flows have been heavily regulated is steel, especially after 1974 when the growing challenge of Third World countries (Korea, Brazil, etc.), following on the heels of Japan which had earlier shown the way, coincided with a major decline in demand; the improvement of market conditions in the late 1980s has led to some liberalization of trade in steel. Growing EC intervention domestically, marked by a rapid escalation of measures leading to the imposition of compulsory production quotas on EC producers, was coupled with the signing of bilateral VERs with the main suppliers of steel to the Community. The obligation of countries signing these agreements was to respect both a certain price discipline and a ceiling in terms of the total amount of exports. These VERs were renewed annually for several years. On the other hand, the EC has remained a major net exporter of steel products (Table 9.2). Submitting to divine justice and the new rules of the game, the EC was forced in turn to sign similar agreements in order to restrain its own exports of steel to the United States which had threatened to impose heavy countervailing duties on European producers accused of subsidized or dumped imports into the US market (Crandall, 1986).

The policy on steel has been the only example of a developed industrial policy at the Community level. It has been criticized, along similar lines as the CAP, as being both inefficient and ineffective, and also as highly protectionist. It is, however, difficult to prove whether EC intervention in steel has been anything more than a damage-limitation exercise in a sector where pure market solutions stood little chance of being accepted by governments (Tsoukalis and Strauss, 1985; Messerlin, 1987). Presumably, the main question is whether EC intervention has facilitated or delayed the internal process of adjustment. With the improvement of market conditions, the policy has been progressively liberalized in recent years.

From the point of view of importing countries trying to slow down the growth of import penetration into their markets, one advantage of VERs in general is that they offer a certain predictability in terms of trade outcomes, including import ceilings and sometimes also some form of 'price discipline'. Being 'voluntary' and informal agreements, they also provide importing countries with a cloak of legality. Having failed to convince their partners in the Tokyo Round of the merits of selective safeguards, EC countries have often resorted to VERs as the only convenient substitute. From an economic point of view, tariffs would have been much preferable, since these 'voluntary' agreements usually imply some form of cartelization on the side of exporters who also benefit from the higher prices. But tariffs are 'bound' in GATT, and any attempt to change them would be politically awkward; hence the recourse to economically second- or third-best solutions.

VERs have been used predominantly in the traditional sectors where industrialized countries have been losing their comparative advantage *vis-à-vis* parts of the developing world, although not exclusively. Two notable exceptions are cars and electronics where VERs have been mainly directed against Japan and more recently also some of the Asian NICs. The European car industry is very important in terms of its share of industrial output and employment. It has been steadily losing world export shares to Japan, thus being forced to concentrate more and more on its large domestic market. The European market itself has been heavily fragmented through different technical standards, taxes, state aids, and the exclusive dealership systems used by car manufacturers. This fragmentation is manifested in the large price differentials, even after having allowed for tax differences. On the other hand, it has been further strengthened by national restrictions, mainly in the form of VERs, which have been imposed on Japanese exports of cars to France, Italy, the UK, Spain, and Portugal. It is quite indicative that the Japanese share of European national markets varies from less than 1 per cent in the case of Italy, which has the most restrictive policy, to more than 30 per cent in some of the small countries with no local production and thus no interest to discriminate against Japanese exports (Smith and Venables, 1990).

The division among the Europeans, faced with the Japanese challenge in the car sector, extends beyond the field of external trade. Thus, while some members of the EC have tried to restrict Japanese exports, while also trying to encourage their national champions though various financial incentives to accelerate the process of restructuring and rationalization, others have been offering incentives to attract Japanese investment at home (Ishikawa, 1990). The failure of the British champions as mass producers of cars has thrown their government into the arms of the Japanese who have been eager to circumvent the import barriers raised by several European countries, including the UK. Thus the latter has tried to act as a gateway to the large European market for cars, as it has also done with respect to other sectors of the economy. And Britain has certainly not been alone in pursuing such a policy. The setting up of 'transplants' by Nissan and, later, Toyota in the UK has been seen as a Japanese Trojan horse by French and Italian car manufacturers; and this has led to attempts by their respective governments to include the intra-EC exports of these 'transplants' into the overall Japanese quota, thus disputing the 'European' origin of these exports. Long intra-EC disputes about rules of origin and minimum local content have created a picture of complete disarray with respect to one of the most important industrial sectors where the customs union has remained largely a myth.

Despite the setting up of the CCP, a number of national restrictions have survived in other sectors, especially with respect to Japan and Eastern Europe. National restrictions on intra-EC trade for imports originating from third countries are also permitted under Article 115 of the treaty. This article has been used largely as a complement to the MFAs, and to a lesser extent with respect to other sensitive sectors such as agriculture and steel (Sapir, 1989). In all these cases, the EC has behaved more as a free trade area than a customs union.

During the 1980s, the EC made frequent use of another policy instrument for the protection of domestic producers, namely the anti-dumping instrument based on legislation first introduced at the EC level in 1968. European legislation was amended as a result of the relevant code adopted in Tokyo, and then again in 1987 when anti-dumping duties were extended to imported components as a means of dealing with the so-called 'screwdriver plants', that is, assembly plants with low value added set up by third country exporters inside the EC allegedly in order to circumvent the imposition of anti-dumping duties on the imported final products. Japan was once again the main target of this new measure. This has in turn led to the imposition of rules on minimum local content in order to determine a genuine EC-made product, thus further adding to friction with foreign producers. In the end, this legislation was found incompatible with GATT rules.

Both the EC and the United States have made active use of the anti-dumping instrument in order to protect themselves from 'unfair trade practices' in third countries. Views on what constitutes legitimate protection in this respect vary widely among countries, depending almost entirely on which side of the fence they find themselves on each particular occasion. Neither economic theory nor international legislation offer clear guidance on this issue, and the resulting gap is filled with an explosive mixture of unilateralism and arbitrary action. Such luxury items can usually be afforded by large countries which are less concerned with the threat of retaliation.

Anti-dumping action by the EC needs to be based on evidence of discriminatory pricing by exporters and injury caused to domestic producers. In many cases, however, especially in the context of oligopolistic competition and economies of scale, differential pricing can constitute a perfectly rational and legitimate form of economic behaviour which has little to do with predatory pricing. The line of distinction between the two is virtually impossible to define. On the other hand, the definition of injury is again largely arbitrary. Messerlin (1989) has found that EC anti-dumping duties, frequently a multiple of the relevant CET, have had a large effect on imported quantities. Although the number of cases in which the EC Commission has imposed anti-dumping duties does not indicate an increasing recourse to this protective instrument during the 1980s, it could be argued that the potential threat posed to exporters, which became increasingly credible through the cumulative effect of earlier action, has sometimes led to voluntary restraint both in terms of prices and quantities exported to the European market. In addition to acting as a deterrent, anti-dumping investigations have sometimes also led to the conclusion of 'voluntary export restraint' agreements. Steel and chemical products figure prominently in European anti-dumping action, while increasing recourse has also been made to this instrument with respect to high technology products in more recent years; hence the criticism of anti-dumping as an instrument of European industrial policy. According to Davenport (1989), anti-dumping action has been taken mainly in highly concentrated and slow-growing sectors, and its principal victims have been Eastern Europe, Japan, and the NICs.

Thus the trade liberalism manifested through the low level of the CET needs to be qualified in several respects because of the extensive use of other instruments of external protection which varies considerably from one sector to another. Jacquemin and Sapir (1990) distinguish between two categories of products subject to relatively high protective barriers in Europe. On the one hand, there are labour-intensive products, with low R & D intensity, such as textiles and clothing, footwear and shipbuilding, characterized by growing import penetration by the developing countries. In these cases, Jacquemin and Sapir argue that international trade is close to the Heckscher-Ohlin paradigm of comparative advantage based on different factor endowments. On the other, there are products with high R & D, large economies of scale, and learning curves, such as telecommunications, consumer electronics, and office equipment, where the steady loss of market shares by European producers is due to competition from the United States and Japan. Modern strategic trade theories are more relevant to those products. Thus European countries seem to be squeezed from both sides, and their immediate response has been to raise the protective walls. One important sector does not, however, fit into the above classification, and this is agriculture which apparently forms a category of its own.

Discrimination and internal disunity have been the two main

characteristics of EC policy towards Japan, while the European trade deficits have continued to grow and so, at an even faster pace, has the import penetration in specific sectors, because of the high degree of specialization of Japanese exports (the so-called laser beam policy). Serious friction has marked Europe's economic relations with Japan, although this has also been true of US–Japanese relations. We shall not enter here into the long debate as to whether the discriminatory policies adopted by both Americans and Europeans have been a legitimate response to the unfair trade practices of the Japanese or simply a defensive reaction against a superior economic system; perhaps the truth lies somewhere in between. There is, however, a certain contrast between the basically defensive attitude of the Europeans and the more aggressive stance of the Americans who have made an active use of the notorious Section 301 of the 1974 Trade Act (and its 1988 successor) in order to prise open the market of a difficult trading partner. The Structural Impediments Initiative signed between the United States and Japan in 1988 is a good example in this respect. A similar instrument at the EC level, namely the New Community Instrument adopted in 1984 and intended as a weapon against 'unfair trading practices' of third countries, has never been put to similar use. This difference between the EC and the United States has much to do with the nature (to be more precise, the weakness) of the European political system which has contributed a great deal to the Community's essentially defensive and reactive role in international trade. It would have been difficult for the EC to throw its weight around, even if it had felt the inclination to do so.

Friction in Euro-American relations has generally proved much more manageable. Agriculture has remained all along the sore point in trade relations between the two sides. Close political and defence ties, the lack of substantial and persistent trade imbalances, and large amounts of cross-investment have always helped to cool down the temper of obstreperous negotiators, be they European or American. Some statistics found in Hufbauer (1990) provide a very good illustration of the nature of economic interdependence between the United States and Western Europe. In 1988, US exports of goods to the EC amounted to 75 billion US dollars, while the production of affiliates of US-owned multinationals in the EC reached the figure of 620 billion dollars. There were comparable figures from the European side. In fact, the stock of European FDI in the United States is nowadays much bigger than the stock of American FDI in Western Europe. The contrast with EC–Japanese relations, and even US–Japanese relations, is very sharp. Cross investment between Western Europe and Japan has been extremely limited, and despite the rapid growth of Japanese foreign investment during the 1980s, the European share has remained relatively small (15 per cent of the stock of the Japanese world total in 1988; Commission of the EC data). To make matters worse, the Europeans did not have a clear and united view as to whether such investment should or should not be welcome.

1992 and the Uruguay Round

The Community's aspirations as an international actor and the expectations of third countries have been disproportionate to the instruments available. Several holes have remained in the CCP armoury, and many of them have been the result of the incomplete nature of internal policies and the customs union. Some instruments of foreign economic policy have been shared between the EC and national governments, foreign aid being one important example, while other instruments had remained until recently at the national level. This applies to the control of labour and capital movements, including foreign direct investment. It still largely applies to macroeconomic policy despite the existence of the EMS. In the macroeconomic field, the international presence of the EC as a whole has been virtually non-existent, although this is bound to change in the future with the establishment of EMU. On the other hand, the Community has long since entered the area of high politics through EPC, but co-operation within the latter has remained intergovernmental and *à la carte* (Pijpers *et al.*, 1988). In international relations there has never been a neat separation between economic and political instruments and objectives; the attempt to impose such an artificial separation in the context of the EC long acted as a major constraint on the Community's international role, and this has become increasingly recognized by member countries which have therefore started to bring down the barriers. The implementation of the Maastricht treaty, together with external pressures, should help to accelerate this process.

The external policy of the EC, and more particularly its external trade policy, will be influenced both by the establishment of the internal market and the outcome of international negotiations, most notably the Uruguay Round. The former implies the adoption of a new approach to old NTBs, leading to their elimination, and the extension of the *acquis communautaire* to new areas, services being the most important among them. This in turn requires an internal agreement regarding the terms of access to the European internal market for third countries. On the other hand, important aspects of the *acquis communautaire*, comprising both old and new elements, are subject to negotiation either bilaterally or in international fora.

The White Paper of 1985 launched the latest stage of internal liberalization in the Community. Yet one remarkable feature of the White Paper is its apparent introspection manifested through the absence of any serious consideration of the external dimension of the internal market. The White Paper referred to the various intra-EC restrictions in the free movement of goods resulting from the remaining gaps in the CCP and the recourse of national governments to Article 115, restrictions which would have to be eliminated as part of the abolition of internal frontiers. But, otherwise, no reference was made to the external policy implications of the internal liberalization programme, and this remarkable abstention continued for some time to characterize Commission pronouncements. Perhaps wisely so, to the extent that this abstention represented a deliberate attempt to sidestep a potentially controversial issue at a delicate stage of the intra-EC negotiations. One step at a time, following popular wisdom. The first important policy statement on the external dimension of the internal market was issued in October 1988, and this was followed by the Rhodes declaration adopted by the European Council in December of the same year. Both were meant to reassure the Community's trading partners that there was no protectionist undercurrent in the internal market programme.

The absence of explicit references to the external dimension in the White Paper should not necessarily be interpreted as lack of awareness or interest. Far from it. From the early years of the EEC, member governments had been faced with the question as to whether regional integration and international liberalization were complementary or competitive objectives. The absence of a broad consensus on this issue has meant that Community participation in international trade negotiations has usually been based on difficult and fragile compromises. Prior to the publication of the White Paper, the long period of economic stagnation and steadily rising unemployment had been accompanied by the strengthening of protectionist pressures which threatened to undermine both the incomplete customs union and the common commercial policy. The US initiative for a new round of multilateral trade negotiations launched at the GATT ministerial conference in November 1982 had met with remarkably little enthusiasm from the Europeans. True, the US initiative was at least partly a response to domestic problems closely associated with the overvaluation of the dollar at the time. It was perceived as a means of deflecting mounting protectionist pressures at home, as it had also happened with the Kennedy and the Tokyo rounds. But the European response was entirely defensive. In the depths of the recession and 'Euro-pessimism', the EC was hardly in the mood to contemplate measures of international trade liberalization, especially since agriculture and high technology were among the priority items on the American agenda. Fully conscious of their fragile unity, the EC countries also preferred not to expose it to the test of difficult and long-drawn negotiations.

The link between regional integration and international liberalization was stressed in the French memorandum of 1983 which, starting from an acknowledgement of Europe's lagging behind in several high technology areas and the progressive loss of world market shares, called for closer intra-EC co-operation in the industrial field and the elimination of internal barriers (Pearce and Sutton, 1986). For the French Socialist Government, a European industrial policy implied a considerable degree of intervention, including not only close collaboration in R & D but also the active use of public procurement and the simultaneous relaxation of competition rules in an attempt to promote European champions. External protection would also be an integral part of this new industrial policy. The elimination of internal barriers should be accompanied by the raising of the external wall, as European producers set about 'la reconquête du marché intérieur'.

Traditional infant industry arguments were employed to support the calls for more effective external protection. Such arguments were given more credence at the time and also much wider scope through the increasingly fashionable strategic trade theories which concentrated mainly on high technology sectors, thus touching a raw nerve in European sensitivities. More effective external protection was presented as an essential part of a medium-term strategy aiming at greater competitiveness of the European industry, with emphasis on the so-called strategic sectors. The call for external protection was partly a reflection of the depressed economic climate and the old Colbertian tradition in France. It was also consistent with the general predisposition of the European Left which saw external protection as a means of preserving a certain model of the economy and society. Hager (1982), one of the most eloquent defenders of Europe's mixed economy and the social democratic order, drew a distinction between the economies of Western Europe (and to some extent also the United States) on the one hand, characterized by relatively free capital markets and unfree (regulated) labour markets, and Japan, together with the smaller Asian tigers and the other NICs on the other, where Hager saw the opposite combination, namely a strong guidance of the 'invisible hand' in capital markets and free labour markets (in the case of some NICs, economic freedom being guaranteed by throwing trade union leaders in jail). According to Hager, no free trade equilibrium could exist between such different social and economic systems. It was the defence of a certain model of society that was at stake.

Such ideas were quickly swept aside by the large deregulation waves originating from the Atlantic and continuing through the Channel. They were swept aside but not completely destroyed, remaining in the background as the debate on Europe's internal market slowly gathered momentum. The loss of international competitiveness, with the main emphasis being placed on the high technology sectors, was one of the main driving forces behind the new consensus for the further deepening of European integration. As the emergence of national champions had been Europe's response to the American challenge of the 1960s, closer collaboration at the production level and the eventual development of European champions were seen as the right answer to the new challenge of the 1980s and the 1990s, dressed usually in kimonos rather than blue-jeans. Although the main emphasis has been on closer intra-EC collaboration, stronger competition, and supply-side measures as the main elements of the European response to the external challenge, the policy towards third countries has only become gradually clarified, and it is highly unlikely to be the same in all sectors. On the other hand, the attraction of external protection has never completely disappeared.

The adoption of the internal market programme and its implementation have coincided with the Uruguay Round launched at Punta del Este in September 1986 and still continuing at the time of writing. European integration has been once again linked to international trade liberalization; and this link has operated in more than one way (Murphy, 1990; Bourguinat, 1989). The agenda of the Uruguay Round is long and ambitious. It contains much of the unfinished business of previous GATT negotiations and also new items which have emerged as a result of more recent developments in the field of international trade. The inclusion, not only in formal terms, of sectors such as agriculture, textiles, and clothing in GATT is part of the old, unfinished business; so is the further lowering of tariffs for manufactured goods and the more effective tackling of the multitude of NTBs. On the other hand, the Uruguay Round is intended to extend the jurisdiction of GATT to new areas, notably services, intellectual property rights, and trade-related investment. The general aim is to strengthen the legal and institutional framework of GATT by giving, for example, new muscle to existing dispute–settlement procedures; and all this while the membership of GATT is steadily growing.

The Uruguay Round will be at best only a first step in the above direction. Most of the items on the agenda are very familiar to any observer of the European scene who has witnessed the long effort to turn the incomplete customs union into a real internal market. The tortuous path of intra-EC liberalization can serve as an indicator of the difficulties in reconciling the push and pull factors resulting from international economic forces on the one hand, and national political realities on the other; if only magnified several times. GATT is hardly comparable to the EC in terms of the powers of central institutions, the solidity of the legal order, the shared political commitment of their respective members, not to mention the degree of internal economic and political diversity.

As with the Kennedy Round more than twenty years earlier, the intra-European liberalization process has acted as a catalyst for multilateral trade negotiations, as the Community's trading partners tried to minimize the trade diversion effects arising from

internal EC decisions. The sheer weight of the Twelve in international trade helps to explain this reaction. On the other hand,
the virtuous circle created by internal political decisions and the
autonomous change in the economic environment have gradually
brought about a shift in European attitudes and consequently a
more positive response to the extension of international liberalization to new areas where the EC has come to play the role
of a pioneer. Had it not been for the emphasis placed by the
Americans and several other countries brought together in the
Cairns group on the liberalization of agricultural trade, where the
Europeans inevitably have found themselves on the defensive
(and largely giving in at the end), the EC could have appeared as
one of the most eager participants of the Uruguay Round.

Reciprocity has been one of the foundation stones of GATT
and the post-war international trading system. It has encouraged
mutual tariff concessions which were then extended to all Contracting Parties through the MFN clause. But with the progressive weakening of multilateralism, the term has acquired a
different meaning (Cline, 1983c). Reciprocity has been applied
increasingly to specific sectors and countries. Furthermore, the
emphasis has been placed more on market outcomes and less on
processes and policies. Thus governments have tended to refer to
market shares and import penetration rates instead of simply
concentrating on more or less transparent forms of protection.
The United States has acted as a pioneer in this area, while
Japan has been the most frequent target of this new approach to
reciprocity, since it has been identified by its trading partners as
the main culprit in the use of non-transparent forms of protection.

Reciprocity has therefore been used as a weapon to prise open
the markets of specifically targeted partners, which although contrary to the spirit of GATT may also be seen as another sign of
the latter's weakness to provide a working framework for international trade. The term reciprocity is very relevant in the case of
the internal market programme because the latter extends into
policy areas where international legislation is very thin. The
question therefore arises whether and to what extent intra-EC
liberalization should be extended to third countries. This applies
to the whole area of services, and also to public procurement and
technical standards among others.

The reciprocity provisions in the early Commission draft of
the Second Banking Directive provoked a strong reaction and

intensive political lobbying on behalf of governments and banks both inside and outside the Community. Those provisions were also largely responsible for the term 'Fortress Europe' which was coined on the other side of the Atlantic in order to focus international attention on the allegedly protectionist character of the internal market. Reciprocity provisions had existed earlier in several national banking legislations. Thus the reaction to the Commission proposals focused on two aspects: the definition of reciprocity and the transfer of powers from the national to the EC level. Foreign banks and governments were concerned about ensuring an unrestricted access to the European market, while member governments with a large presence of international banks on their territory, most notably the UK and Luxembourg, showed little keenness on transferring the ultimate power of decision regarding the establishment of foreign banks to the Commission (Bisignano, 1990; Golembe and Holland, 1990). The possibility that the exclusion of, say, a Greek bank from Tokyo might force the British authorities to take similar measures against a Japanese bank wanting to open a subsidiary in the City was not equally appreciated by everybody.

The definition of reciprocity was the other issue of contention. The proposed EC legislation was one of the most liberal to be found anywhere in the world, both in terms of the activities which European banks would be allowed to undertake and in terms of their ability to open up branches anywhere on EC territory. Should the Community require similar treatment for EC banks wishing to operate in foreign countries? This would have necessitated, for example, the abolition of both the Glass-Steagall and the McFadden Acts in the United States which impose restrictions on the activities and branching of US and foreign banks alike. Although, as usual, the Americans did most of the talking, it is clear that they were not meant to be the main target of the reciprocity provisions; the latter were directed more against Japan and several developing countries which had kept their domestic markets shut to foreign financial institutions.

In the end, the reciprocity provisions of the Second Banking Directive were clarified and also heavily diluted. In the final text, reciprocity was defined in terms of 'national treatment' of EC banks (meaning that there should be no discrimination between national and European banks in third countries) and 'effective market access' which can be more substantial in those cases where discrimination has taken less transparent forms (read Japan). Only when the criterion of 'national treatment' is not fulfilled may the Commission initiate direct negotiations with the government of the country concerned; otherwise, it will have to wait for a Council mandate. Since the new directive should apply from 1 January 1993 and the reciprocity provisions have no retro-active character, any foreign banks can precipitate their entry into the EC market through what they consider as the most convenient national door.

Taken in conjunction with the decision to liberalize all capital movements and to extend this liberalization unconditionally to third countries, the new EC legislation in the financial sector constitutes an important step towards liberalization which is effectively extended to foreign establishments without much of a *quid pro quo*. Thus any foreign bank with a subsidiary in an EC country can take advantage of the single banking licence and the liberalization effects of the Second Banking Directive. This may be consistent with the global nature of the industry. Moreover, in view of the decentralized nature of political power inside the EC and the divergence of economic interests, the Community is unlikely for some time to adopt an aggressive stance in the pursuit of reciprocity with respect to the financial sector. And this may apply to other policy areas as well.

If nothing else, new EC legislation with respect to public pro-curement is unlikely to make access more difficult to third country producers. Until now, discrimination by public authorities has applied equally to EC and non-EC producers. The new legislation covers the four 'excluded' sectors (water, energy, transport, and telecommunications) which have always remained outside the GATT framework. There is a clause in it which allows the authorities of member countries to ignore offers from companies with less than 50 per cent EC local content, although this pro-vision is only optional. The legislation also gives a 3 per cent margin of preference in favour of European companies against foreign bidders, in analogy with the 6 per cent rule in the United States. However, both provisions are viewed as bargaining weapons for the Uruguay Round. Strange (1988) sees little possi-bility of the Europeans, without one strong political authority at the centre, using government procurement as part of an industrial strategy.

Some concern has been expressed, especially from the American side, about technical standards being used as a protectionist device by the Europeans in the technological race. To avert such a development, they have tried to obtain a seat in the European standard-setting organizations, although without success. However, the interests of some foreign companies have been indirectly represented in organizations such as CEN, CENELEC, and ETSI through the European affiliates of multinationals. Technical standards and regulations have been used by most industrialized countries as an instrument of promoting the interests of domestic producers, thus acting as an important NTB in international trade. To the extent that this NTB in intra-EC trade will be tackled through mutual recognition of national standards, foreign producers can only gain since, from now on, they will need to go through one instead of twelve approval procedures. In those cases where a European standard will replace the old national ones, there is little reason to expect that the twelve countries collectively will be more protectionist in the use of technical standards than they have been in the past when acting separately. A related matter which may have to be negotiated bilaterally with third countries concerns the mutual recognition of tests and certification.

A more general issue which has already arisen in the past and which is likely to be given more prominence with the progressive strengthening of the European regional system is the definition of 'Europeanness' of economic agents. Article 58 of the Treaty of Rome states that companies and firms with registered offices in a member state should be treated in the same way as natural persons who are nationals of those states. This article has enabled the European affiliates of foreign companies to obtain free access to the European market. According to Sandholz and Zysman (1989), this provision formed part of the original bargain between the six founding members and the United States. The Community has traditionally adopted a liberal interpretation of this article, but problems have arisen in recent years with several attempts to apply rules of minimum local content to the subsidiaries of foreign firms, one prominent example being the treatment of Japanese car 'transplants' in the Community. Local content provisions have also appeared with respect to anti-dumping duties and the new directives on public procurement.

These developments may suggest that the EC will gradually adopt a more restrictive definition of what constitutes a 'European' company. On the other hand, difficult political decisions need to be taken regarding the participation of companies in EC-financed research programmes. If EC industrial policy, even in the low-key form in which it has developed in recent years, is intended to strengthen European companies in the struggle for world market shares, a struggle in which both private firms and governments play an active role, what should be the treatment of foreign subsidiaries? For example, is IBM-France more or less French (and European) than Bull? And this may be in fact a relatively clear case, in view of the scale of the French (and European) operations of that particular company; other examples will be much less straightforward. Hufbauer (1990) mentions the admission of IBM to the JESSI (Joint European Submicron Silicon Initiative) programme and wonders whether the United States is likely to follow with a similar opening with respect to its own research programmes, something which had not previously happened. Presumably the Japanese are not faced, at least as yet, with such existential problems. In general, with the increasing globalization of production, the national identity of firms will become more difficult to define. But this may also undermine the somewhat vague strategy for the promotion of European champions.

The issue of extra-territoriality has rapidly emerged out of its legal shell into the political limelight. This issue refers to attempts by nation-states (and emerging federations such as the EC) to apply their laws and implement their policies beyond their own frontiers. The growing integration of the world economy and the inadequacy of international rules have provided the justification for such action almost invariably undertaken by the bigger powers and much resented by the other countries. The US courts have acted as pioneers in this respect as well, especially as regards anti-trust laws and taxation. The ill-fated Vredeling directive had a strong extra-territorial dimension, while Articles 85 and 86 on competition policy have sometimes been used against companies based outside the EC (Demaret, 1986; Rosenthal, 1990). The new powers given to the Commission for the control of big mergers can bring it one day into conflict with foreign authorities. The agreement signed with the United States in 1991 is meant precisely to avoid conflict between the two sides in terms of competition policy. On the other hand, the liberalization of capital movements and the new possibilities created for tax evasion will eventually lead to stronger pressure being exercised on third countries for closer collaboration among national tax authorities, and possibly also lead to unilateral action. These problems do not arise because of European integration; they are simply the product of a major discrepancy between international economic integration and national political realities. Yet the creation of a powerful economic bloc may increase the temptation for unilateral action.

The Community has repeatedly stressed that the 1992 process will not lead to higher levels of external protection, and in fact there is virtually nothing in the internal market programme as such which might suggest the opposite. To the extent that the elimination of intra-EC barriers is not automatically extended to third countries, this may lead to some trade diversion which could, however, be more than compensated by the positive effects through higher economic growth. Thus the fears expressed about 'Fortress Europe' are more closely related to already existing policies of the EC than to any allegedly protectionist aspects of the 1992 programme. They may also be part of a 'crying wolf' strategy pursued by the Community's partners, and especially by the United States. If anything, the internal market programme is likely to have a liberalizing effect both directly through the elimination of intra-EC barriers and indirectly by acting as a catalyst for further international liberalization.

One sector which is closely connected with the establishment of the internal market and where European policies have been generally highly protectionist is that of cars. The simple abolition of national restrictions has not been seriously considered even as a possibility. A study undertaken in 1987 estimated that the removal of all restrictions could lead to an increase of the Japanese share of the European market from approximately 10 per cent to 18 per cent by 1995. 'Arithmetically, this increase of one million units in Japanese imports would equate to the disappearance of one of Europe's six main producers' (quoted in Ishikawa, 1990: 74); and this would have been politically impossible.

In July 1991 an agreement was reached between the EC Commission and the Japanese government, entitled 'Elements of Consensus'. This consensus appeared, however, to be somewhat obscure in places and subject to different interpretations by the two sides; perhaps, a necessary price to pay for the conclusion of an agreement which left some of the difficult problems to be solved later. National restrictions would be phased out after the end of 1992. They would be replaced by a system of monitoring of Japanese exports to the Community as a whole as well as to specific national markets during a transitional period of seven years (1993–9). According to the agreement, Japanese exports to the EC would remain at 1989 levels, but provision was also made for a large increase in the production of Japanese 'transplants' which, according to a Commission internal document, was expected to reach 8 per cent of total EC consumption (plus another 8 per cent for exports; see *Financial Times*, 23 and 26 September 1991). It was not, however, at all clear whether both sides agreed that there should be any limit on Japanese investment in the Community or on the production of Japanese 'transplants'. Difficult negotiations seem to lie ahead for the interpretation of different elements of this rather vague consensus. The new common policy of the Community implies only small and gradual changes from the *status quo* of national restrictions, and the fragile unity of the Europeans will largely depend in the future on Japanese informal undertakings. On the other hand, the difficulties will be further compounded when the US Administration begins to defend actively the interests (and exports) of Japanese 'transplants' on its territory against not only those of Fiat and Renault, but also those of Ford and General Motors (Europe) as prominent(!) European champions.

Other sectoral policies will not be much influenced by the establishment of the internal market, although important changes may come about as a result of the Uruguay Round. The most obvious examples are agriculture, textiles, and clothing. With respect to the former, GATT negotiations have centred on the reduction of overall levels of support, which has been rendered technically and politically easier by the OECD calculation of Producer Subsidy Equivalents (PSEs) for different countries. Generous government subsidies in this sector have been in part a self-defeating exercise. Moyer and Josling argue that '[for] the USA, Canada and the EC, about one-half of the support goes to offset the impact of other countries' policies' (1990: 179). In such a game, it is the smaller players who tend to lose most. As far as the Community is concerned, although the attainment of high self-sufficiency ratios for food, internal redistribution, and the preservation of social structures in the countryside may be legitimate political objectives, it is much more difficult to defend the dumping of surpluses in foreign markets. The successful conclusion of the Uruguay Round largely depends on the reaching of an agreement with respect to agricultural trade, and this seems to be more likely than ever after the EC decision of May 1992 which paves the way for a genuinely radical reform of the CAP.

As for textiles and clothing, the elimination of internal frontiers necessitates the abolition of national quotas. On the other hand, the system and level of external protection will depend on the decision to be reached within GATT regarding the phasing out of MFA. Too rapid a liberalization could lead to heavy job losses in the weaker economies of the EC, with a higher concentration of their industrial production in this sector. When added to the adjustment needs arising from the internal market programme, this could prove politically untenable and hence likely to be resisted strongly.

The Politics and Economics of Preference

Multilateralism is a principle that the Americans fought very hard to establish as one of the foundations of the post-war international economic order. According to Article XXIV of GATT, the Community (and the EEC in particular) is a legitimate exception to this principle. Yet the Treaty of Rome also makes provisions for trade and association agreements with non-member countries and international organizations, which suggests that the founding fathers intended to take this exception further; and those intentions were to be confirmed soon.

Together with transparency, multilateralism has suffered badly from the trading policies of GATT members, and most notably the EC, and this was particularly true in the early years of the latter's existence. Historical legacies, pressures from outside, the shortage of policy instruments available, and, shall we say, a rather half-hearted commitment to such principles imported from the other side to the Atlantic, led the EC to rely on trade preferences as an important instrument of its external policy. The preferential agreements signed by the EC until now can be distinguished as a function of two main variables, namely the level of economic development of the EC partner and whether it is considered as a potential member of the Community or not. In most cases, the classification is the same irrespective of the variable used, which means that the countries with the higher level of economic development are, with few exceptions, also those which are treated as potential members (the ultimate test being the European status of a country). For these countries, preferential agreements are characterized by reciprocal concessions. For the others, which constitute a select group of developing countries, there is limited or no reciprocity at all in the agreements signed with the EC.

EFTA started as a rival organization to the EEC in Western Europe and as an alternative for countries which opted for the intergovernmental model of co-operation. It has ended up partly as an antechamber for countries waiting to join the EC as full members and partly as an organization whose main function is to manage (or try to manage) collectively the relations of its members with the ever-enlarging Community. EFTA consists of small, industrialized economies with standards of living which are higher than the EC average. In terms of growth, the performance of these economies has been close to the Western European average both during the brighter and the darker phases of the post-war period. In terms of unemployment, the performance of most EFTA countries has been much more successful, especially since the early 1970s when unemployment rates in the rest of Western Europe started rising fast. In fact, this relative success used to be often interpreted in the past as concrete evidence of the ability of small Western European countries to survive outside the EC framework.

Important economic characteristics of EFTA countries (Switzerland being the main exception) have been the reliance on Keynesian demand management and the existence of highly advanced welfare systems (see also figures for public expenditure as a percentage of GDP in Table 2.4). In several EFTA countries, restrictions on foreign ownership of the means of production have also survived for many years. The desire to preserve a wide margin of autonomy in economic decisions was for a long time an important factor militating against membership of the EC; and this was further strengthened by the rejection of supranationality in general and the consensus character of domestic politics. However, with the rapid globalization of production, the growth of international trade and the deregulatory wave of recent years, much of the economic orthodoxy in those countries has come under severe challenge; and this has contributed to the increasing attraction of the EC. Neutrality in terms of foreign policy (four of the present members of EFTA are neutral countries) had been the other major constraining factor; but this again has changed radically since the end of the cold war.

For many years, EFTA countries tried to reconcile domestic and foreign policy constraints with the exigencies of international and especially intra-European economic interdependence; and this was not always easy. In 1972, on the eve of the Community's first enlargement, when two former EFTA countries decided to join the other camp, the remaining members (including Iceland and Finland which had joined the EFTA ranks in the meantime) signed individual but similar agreements with the EC. These were free trade area agreements covering most industrial goods, while the agricultural sector was virtually excluded. They were accompanied by the usual safeguard clauses and provisions for the application of rules of origin, the latter being necessary whenever there is a departure from the multilateralist principle in order to avoid trade deflection. Through these agreements, a large free trade area for industrial goods was created in Western Europe within which the EC constituted the hard core. Co-operation with the EC was, however, extended to many other areas. Some examples are the adoption of the Single Administrative Document (SAD) by EFTA countries, their participation in European standard-setting organizations such as CEN, CENELEC, and ETSI as well as the more recent European Organization for Testing and Certification (EOTC), and the close association of some EFTA countries with the snake and later with the ERM. The need to preserve their access to the large European market often forced the EFTA countries to conform unilaterally with EC rules. This explains Nell's phrase the 'EFTA countries have reacted against marginalization, but at the inevitable price of satellization' (1990: 352).

The economies of EFTA are small and very open, with a high degree of trade dependence on the EC which also explains the efforts expended to keep trade flows as unrestricted as possible. The trade interdependence between EC and EFTA countries is very considerable (Table 9.1), but the relationship remains highly unequal. Trade with the EC-12 represents between 55–60 per cent of the total trade of an average EFTA country, which is about the same as for the average EC member. The export specialization of EFTA countries is mainly in low and medium technology products, with the exception of Switzerland which has a stronger presence in the high technology sector and also in services. Both in terms of trade interpenetration and inter-firm co-operation at the production level (including cross-investment), there is very little to distinguish EFTA countries from those of the average member of the EC. The institutional separation between those two groups has not prevented the close economic integration of the countries of EFTA in the wider economic area of Western Europe.

On the other hand, intra-EFTA trade is less than 15 per cent of total trade, and the remaining countries of EFTA are divided in economic terms in two distinct groups, namely the Nordic group, which is characterized by close economic ties and a relatively high degree of trade interpenetration, and the Alpine group, consisting of Austria and Switzerland, countries which form an integral part of the core of the Western European economic area. Despite the long existence of EFTA, history and geography have kept those two groups rather separate.

The further deepening of economic integration inside the EC and its extension into new areas have created a new situation for the EFTA countries. The elimination of many of the remaining intra-EC barriers was seen as likely to lead to trade diversion at the expense of the Community's small and prosperous neighbours. Krugman (1988) compared the effects of 1992 with a reduction of transport costs which does not, however, apply to non-members. Thus, in order to retain their competitiveness and their share of the export markets on which their prosperity largely depends, EFTA producers would need to reduce their prices. *Ceteris paribus*, there should also be some dislocation of investment from EFTA countries to the EC, as the former become economically less attractive; and there is evidence of that already happening in the late 1980s when investment from EFTA-based companies in the EC increased substantially (Leskelä and Parviainen, 1990).

Already in April 1984, the representatives of the two organizations issued the so-called Luxembourg declaration committing themselves to the creation of a European Economic Space which should comprise all eighteen countries of the EC and EFTA (plus the principality of Liechtenstein as part of EFTA). This term was later replaced by the European Economic Area (EEA) in one of those subtle distinctions which only seasoned diplomats can make. The contents remained for long undefined. And then, in the inaugural speech for the second term of his Presidency in January 1989, Mr Delors passed the ball to the EFTA court by calling for a wider Europe and 'a new, more structured partnership with common decision-making and administrative institutions' between the two organizations and the acceptance by EFTA countries of a large part of the *acquis communautaire.* This in turn implied a more unified EFTA which would be able to act as the single representative of the member countries, something which had not happened in the past.

EFTA countries rose to the new challenge, and the negotiations between the two organizations started in June 1990. The agreement for the establishment of the EEA was signed in May 1992, following an unexpected hitch created by a decision of the EC Court of Justice which had found an earlier form of the EEA agreement incompatible with the Treaty of Rome. Assuming that the agreement is ratified in time by all parliaments concerned, it will come into effect in January 1993, thus coinciding with the establishment of the Community's internal market. The EEA agreement implies the acceptance by the EFTA countries of a very large part of the *acquis communautaire*, including the free movement of goods, services, persons, and capital, with some transitional arrangements for sensitive areas. It also includes the application of EC competition rules to the whole area as well as rules applying in the areas of social policy, consumer protection, the environment and R & D. There is also provision for a relatively small EFTA contribution in terms of soft loans and grants destined for the less-developed members of the Community, confirming once again that redistribution is now seen as an integral part of the overall package of European integration.

The EEA agreement will lead, with few exceptions, to the extension of the internal market to EFTA countries; and this should be particularly important with respect to financial services, transport, and competition policy. In the case of Switzerland, and assuming again that the Swiss people ratify the agreement in the referendum, the application of the EEA agreement will eventually put an end to highly restrictive policies on immigration and the acquisition of land by foreigners; not a small change for a deeply conservative country. Two very sensitive items in the negotiations were fish (for Iceland and Norway) and transport (for Austria and

Switzerland as transit countries for much of intra-EC trade). A compromise was finally reached on both those items. On the other hand, the EEA agreement does not include agriculture, where EFTA countries have even more protectionist policies than the EC, and tax harmonization. The common commercial policy is also excluded, the EEA agreement being based on the principle of a free trade area.

Economic participation needs, however, to be distinguished from political participation. Despite the earlier call by President Delors for 'common decision-making and administrative institutions', the political and institutional hybrid created through EEA allows for consultation of EFTA countries but not for their participation in decision-making; unless, this is assumed to operate through some kind of osmosis. And it could not really have happened otherwise, if the legal and institutional autonomy of the EC were to be respected. This is exemplified by the decision of the EC Court of Justice in December 1991 against the creation of a joint court which would deal with cases arising in the EEA area, because this would endanger the very foundation of the EC. According to the final version of the treaty signed in May 1992, the EC Court will remain solely responsible for the application of Community law, including all competition cases in which foreign-based companies generate some of their turnover inside the EC.

In the words of E. Mortimer (*Financial Times*, 23 October 1991): 'the EEA is a kind of second-class citizenship, which allows you to be governed by rules you do not make, but does not exempt you from contributing to the cost of helping other people, poorer than yourself, to conform to them.' Under the new economic and political conditions prevailing in Europe, the EEA agreement cannot provide a satisfactory substitute for EC membership, although it can be seen as an intermediate stage leading to that. The increasing number of applications for EC membership from EFTA countries suggests that this is precisely the assessment made by most of those countries. The subject of further enlargement will be discussed in greater detail below.

Following the dramatic events of 1989, a totally new situation has emerged on the Community's eastern frontier. The rejection of the old system of central planning and administrative prices, together with the collapse of the former communist regimes, has brought about one of the most spectacular ever changes in the economic and political order of a large region over a very short period of time. However, the transition to market economy has not been at all easy or smooth. Domestic stabilization and liberalization measures, coupled with a major disruption of external trade relations, have caused serious economic hardship. Economic adjustment has meant for every single country of Eastern Europe, including the former Soviet Union, a significant decline in production and standards of living and a substantial increase in unemployment. Economic hardship has not, of course, facilitated the political transition; and in some of those countries, the experience with democratic institutions is very limited indeed.

The political and economic changes in Eastern Europe have suddenly opened up the possibility of the EC frontiers being extended eventually to the eastern part of the European continent; but this can only be a gradual and relatively long process. The countries of the EC have a strong interest in the stability of their eastern neighbours and their progressive integration in the international, and more particularly the European, economic system. In the aftermath of the 1989 events, economic and political relations between the two sides were at a low level. Institutional relations had been influenced in the past by the long refusal of the Council of Mutual Economic Assistance (CMEA), otherwise known as Comecon, the now defunct economic organization of Eastern European countries, to recognize the EC (Nello, 1991). A large part of the trade with these countries was subject to quotas, usually national rather than EC-wide, which also contributed to the fragmentation of the customs union. The picture has changed radically since then. In a very short period of time, the countries of Eastern Europe have shifted from negative discrimination to privileged status in their relations with the EC. This privileged status is directly linked to parliamentary democracy, the respect of human rights and economic reform.

The first step was taken with the signing of trade and co-operation agreements, the so-called first generation agreements which aimed at the progressive elimination of quantitative restrictions and the application of MFN treatment (de la Serre, 1991). But in the case of several of those countries, the first generation agreements were very quickly overtaken by events; and this has led to the conclusion of a new form of association agreements, otherwise known as second generation or Europe agreements. At the time of writing, Czechoslovakia (how long will it remain a single country?), Hungary and Poland had signed such agreements with the EC, while negotiations had started with Bulgaria and Romania. Further down the list of candidates, one can find Estonia, Latvia, Lithuania, and Albania, without also excluding the possibility of negotiating in the future similar agreements with some or all of the former constituent republics of Yugoslavia, once the war is over. Thus, a dividing line has been drawn by the EC separating those countries which qualify for an association agreement and the others. It is, however, very probable that this line may shift in the future to include more countries.

With respect to trade, the new Europe agreements are intended to create a free trade area within a period of ten years, with a shorter timetable of liberalization on the EC side. Yet, a distinction is made between 'general industrial goods' and 'sensitive' sectors, such as agricultural products, steel, coal, and textiles, where EC liberalization will be slower or more limited. These are, however, precisely the sectors where the Community's partners have some comparative advantage. The ability to offer in the future improved access to the exports of Eastern associates in sensitive sectors will depend on economic restructuring inside the EC. Another difficult issue of contention between the two sides has been the free movement of workers. Despite strong pressures from the other side, the EC has made virtually no concessions in this area. The persistence of high rates of unemployment in the Community and growing social resistance to immigrants, combined with the strong migratory pressures from the former communist countries, hardly allowed any room for flexibility. The Europe agreements also contain provisions for technical and economic co-operation and the approximation of laws in many areas. Although far-reaching, they do, however, fall short of the provisions made in the EEA agreement. Last but not least, the political dialogue between the two sides is institutionalized and the 'European vocation' of the associated countries explicitly recognized. This means that the associated countries are treated as potential members of the EC, although no specific date is mentioned.

The role of the EC has not been limited to trade agreements. Trying to help the transition of Eastern European countries to democratic, pluralist regimes and market economies, the EC has provided technical assistance through the PHARE programme which, starting with Poland and Hungary, has been gradually extended to all the countries of the region, including the three Baltic republics from the former Soviet Union. Economic loans are also provided through the EIB. On the other hand, the EC Commission has acted as the co-ordinator of bilateral economic aid provided by the OECD countries, the so-called G-24; and the Community has played the role of catalyst for the setting up of the European Bank for Reconstruction and Development (EBRD) which concentrates its lending to the private sector in Eastern European countries. Until now, by far the biggest part of economic aid to Eastern Europe has come from the EC, either in the form of multilateral aid or bilaterally by member countries, and most notably Germany. EC governments have insisted on keeping control of the purse strings and the result has been an uneasy compromise between them and the Brussels executive as regards the formulation and implementation of policy towards Eastern Europe.

For the countries of Eastern Europe, access to EC markets (the share of trade with the EC has been rising very fast in recent years) as well as FDI and financial aid flows from their Western neighbours are of paramount importance for their economic development. Political relations are also intimately connected with economics: institutional links with the EC and the prospect of membership, even in some distant future, can reduce the element of uncertainty in terms of economic decisions in those countries, thus having a positive effect on investment behaviour. Accession to the Community constitutes the main foreign policy goal for virtually every country of Eastern Europe, although, admittedly, this will be a very long term goal, for at least some of them. This subject will be discussed further in the section below.

There is clear hierarchy of treatment accorded to the Community's partners in the developing world. Although different kinds of agreement have been signed with the large majority of developing countries, either individually or with regional groups, a number of countries have been singled out for special treatment by the EC. One such group of countries can be found on the shores of the Mediterranean. Bilateral agreements with Mediterranean countries started soon after the establishment of the EEC in 1958. Most of them were renewed and extended in the context of the so-called global Mediterranean policy during the 1970s. But global does not necessarily mean uniform. Different agreements have survived, which can be divided broadly into two categories: those which involve reciprocity and leave the door open for EC membership and the others. Thus, there is a division between the north and the south of the Mediterranean, since only European countries can aspire to membership of the EC.

Three of the existing members of the Community, namely Greece, Portugal, and Spain, had concluded different kinds of preferential agreements with the EC prior to their accession. Three other Mediterranean countries (Cyprus, Malta, and Turkey) have since joined the queue, with applications for membership. All three have association agreements with the EC, centred on the much-delayed objective of customs union, and accompanied with provisions for financial aid and political dialogue. Very different from one another, they were referred to once by President Delors as the 'orphans of Europe', because they had remained outside the main economic blocs. Since the dissolution of the CMEA, the number of European orphans has increased enormously, and the EC appears as the only credible candidate to adopt them.

The typical co-operation agreement with the countries of North Africa and the Middle East provides for free access to the European market for the industrial products of these countries, without any reciprocity on their part (the only exception is Israel in view of its higher level of economic development), and some special concessions for their agricultural exports as well as technical and financial assistance. The northern members of the EC have been more favourable to trade concessions, while the southern countries, which are more likely to pay the price of those concessions because of competitive products, have expressed a stronger preference for aid.

These agreements are comprehensive in their coverage and at least superficially generous in the treatment accorded to the Mediterranean countries. But free access for industrial exports may not mean a great deal if there is not much to export. The application of rules of origin, together with restrictions imposed on exports of textiles and clothing, significantly reduce the value of EC trade concessions to the Mediterranean non-member countries. Most of these countries are mainly agricultural producers whose entry to the European market is heavily restricted through the application of CAP rules. The accession of Greece,

Spain, and Portugal to the EC has further reduced the export possibilities of other Mediterranean countries to European markets by substantially increasing the EC self-sufficiency ratio in agricultural products.

In fact, the share of Mediterranean countries in the Community's external trade has not risen during the last two decades (Table 9.1); if anything, there has been a small decline which can be largely attributed to the slow pace of economic development of the large majority of those countries. Here again, the relationship between the two sides is very unequal, and this inequality is reflected through the large differences in economic and political weight (even more so since the Mediterranean countries negotiate separately with the EC) as well as the differences in trade and economic interdependence between the two sides. The EC has relied on limited trade and financial instruments for the pursuit of wider economic and political objectives in the area, and, not surprisingly, with limited effects. According to Pomfret (1986), the cost of preferential agreements with the Mediterranean countries has been borne mainly by EC taxpayers (through the financial aid granted, arguably the most important element of these agreements, and the small loss of tariff revenue) and outsiders, due to trade diversion at the expense of third country producers.

The economic stagnation of most countries on the southern shores of the Mediterranean, their heavy debt burden, rigid social and political structures, all coupled with a demographic explosion, have been the cause of growing concern for the countries of the Community, and especially the southern European countries which have felt more directly the brunt of the pressure in the form of growing numbers of illegal immigrants. Will the Mediterranean prove a more effective barrier to population flows from the south than the Rio Grande has been for the United States? The continuous attempts made by Italy to convince its northern partners of the urgency of a special aid programme for the countries of Maghreb and Mashreq is an indication of this concern. The new Mediterranean policy of the EC, which will extend until 1996, puts the emphasis on financial aid and internal structural reforms along the lines also suggested by the IMF and the World Bank.

In 1990, the Community opened negotiations with the six countries of the Gulf Co-operation Council (GCC) with the view of signing a free trade area agreement, after long pressures from those countries and against the strong resistance of Europe's petrochemical industry. The EC needs to reconcile wider political and strategic interests (given much greater prominence by the war in the Gulf), not to mention the need to guarantee its oil supplies, with the fear of growing import penetration in an important and vulnerable sector. The long time it has taken the EC to respond to GCC overtures suggests that the task will not be an easy one, and the free trade agreement risks being negotiated to death.

Some of the observations regarding the limited effectiveness of the Community's Mediterranean policy apply equally well to the other group of privileged partners of the EC, namely the ACP (Africa–Caribbean–Pacific) countries (Lister, 1988). This group, which includes some of the poorest countries in the world, consists almost entirely of former colonies and dependent territories of EC member countries in Africa plus a few island states in the Caribbean and the Pacific; it does not, however, comprise former colonies in Asia essentially because of their higher level of economic development and export possibilities. The numbers have been steadily rising with successive enlargements of the EC, the first enlargement having led to the transition from the Yaoundé to the Lomé agreements. Lomé IV, which was signed in December 1989, is an agreement of ten-year duration between twelve European countries and sixty-eight countries of the Third World. Tariff free access applies to virtually all exports of ACP countries, without any reciprocity obligations from their side. This is coupled with preferential rules of origin, complex arrangements for sugar exports, a scheme for the stabilization of export earnings for several commodities (Stabex) and minerals (Sysmin) as well as financial and technical aid. Lomé IV provides for approximately 12 billion ECUs of grants and loans for a period of ten years, with the bulk of the money provided for in the form of grants.

The Community has met half-way some of the demands made by Third World countries in the context of the New International Economic Order (NIEO), which reached its peak during the 1970s, only to sink into oblivion some years later when it became clear that the demands made by the developing countries had been based on a clear misjudgement of both the internal cohesion of the Group of 77 and the international balance of power. But modest trade concessions and commodity stabilization schemes are unable on their own to bring about a fundamental change in economic structures and hence in international trade. Here again, financial transfers have been by far the most important element of these agreements, although they have not prevented the disastrous economic performance of many ACP countries during the last two decades. This is also reflected in the steadily declining share of ACP countries in the Community's external trade (Table 9.1). Although often criticized as insufficiently generous, these agreements have been judged positively by the ACP countries themselves, and this is also witnessed by the steadily growing number of participants and the interest of other developing countries to join. Outsiders have levelled accusations of neocolonialism and have seen in European policies a deliberate attempt to divide the Third World, although the assistance of the Europeans in this respect would have been hardly essential.

Trade preferences, as part of an active commercial policy, have been subject to two major constraints. The EC has tried to reconcile preferential agreements with Article XXIV of GATT which refers to customs unions and free trade areas. Such agreements need to cover a 'substantial' part of trade, while liberalization should take place within a 'reasonable' length of time. The Community has tended to adopt a rather wide interpretation of both these adjectives. On the other hand, the actual economic significance of preferential concessions has been limited in view of the low level of the CET and the Generalized Scheme of Preferences (GSP) offered to all developing countries. Under the GSP, tariff free access is offered to the industrial exports of all developing countries, although subject to quantitative restrictions beyond which the CET applies as normal. Under those constraints, the granting of preferences to privileged partners has sometimes led to absurd situations. Hine (1985) refers, for example, to seven different tariff rates for canned sprats and frozen prawns, with marginal differences between them, presumably as an example of bureaucratic paranoia. Too many preferences lead to no preferences at all.

Assuming that the EC has indeed tried through the Lomé and the Mediterranean agreements to create its own sphere of influence, it has clearly not chosen the most dynamic, in economic terms, countries in the world. The contrast with Japan and South-East Asia could not have been starker. Perhaps, to be more accurate, history seems to have chosen for the Europeans, and the policies pursued have been subject to several constraints and without much internal coherence; a familiar story, after all, in international politics.

In the earlier years of the Community, the emphasis on trade preferences could be seen as another sign of the large discrepancy between objectives and instruments available. Keen on making its mark as an international actor and also interested in exerting influence in certain areas, the EC frequently resorted to preferential agreements. Since the 1970s, however, the Community has made a greater effort to reconcile its commercial policies with GATT rules. Furthermore, the distinction between European and non-European countries in the group of the Community's privileged partners has become increasingly clear and relevant in terms of policy. This distinction is linked to the prospect of further enlargement of the Community.

In Search of a European Policy

Until now, the EC may have had some kind of a Mediterranean policy and also, arguably, a policy towards the developing world in general and some of its former colonies in particular. But it has never had a European policy, nor has it felt the need to formulate one. For several decades, European countries were divided into different groups, be they EFTA, CMEA, and the Mediterraneans. However, with the end of the cold war and the collapse of the old political order in Europe, the situation has become totally different. Virtually every European country is now an actual or potential candidate for membership of the EC, and this calls for a response from the side of the Community. On the other hand, the early enthusiasm with which the fall of totalitarian regimes was received has been gradually replaced by much concern and anxiety about stability and peace in Eastern Europe.

Are we witnessing the resurgence of many of the problems which followed the Versailles treaty of 1919? The disintegration of Yugoslavia and the bloody war which has accompanied it and which sometimes threatens to spill over to neighbouring countries, the large number of ethnic conflicts in the former Soviet Union, some of which have already produced local wars, the announced separation of Czechoslovakia into two independent units, they all are manifestations of a twin problem: the ethnic mosaic which characterizes most of Eastern Europe and the weakness of political institutions. The economic difficulties associated with the transition to market economy and stabilization policies are added to the litany of political problems. Despite its weakness in the security field, which is unlikely to change radically in the foreseeable future, the EC is a big power in Europe, not purely in economic terms, and the other countries behave towards it accordingly. Its ability to act as a stabilizing force on the old continent will essentially depend on its internal unity and the formulation of a European policy with a minimum of coherence, which also implies a realistic matching of objectives with instruments.

The increasing number of candidates for membership raises the awkward question of boundaries. Only European countries have the right to apply, but is there a consensus about which country is European and which is not? After all, traditional geographical boundaries do not always conveniently coincide with political ones. Should strategic and balance of power factors prevail over cultural or religious ones? The answer to this question would largely determine whether Turkey is considered more European than Russia and the Ukraine or vice versa. Some ambiguity as regards the ultimate frontier of Europe to the east, since all the other frontiers have been clearly demarcated by nature, is likely to persist. Ambiguity is very often a necessary ingredient of a successful policy.

If the boundaries of Europe are still somewhat ambiguous, the criteria for eligibility for EC membership are more clearly defined (see also Michalski and Wallace, 1992). They start with the ability and willingness of candidates to adopt the *acquis communautaire*, after a relatively short transitional period. This should also include the aims set out by the Maastricht treaty in terms of EMU and the two new pillars of the European Union, namely foreign and security policy and co-operation in the fields of justice and home affairs. There is sometimes talk of the need for an explicit acceptance by candidates of the *finalité politique* of the Community, although there is hardly a consensus among existing members as to what precisely this term is supposed to mean. Democratic institutions and the respect of human rights are also a *sine qua non* condition which is now explicitly mentioned in Article F of the Common Provisions of the new treaty.

At the time of writing, seven applications were already waiting at the table of the EC Council of Ministers, and the number could increase further in the future. Turkey applied in 1987; it was followed by Austria in 1989, Cyprus and Malta in 1990; and then came the other EFTA candidates, namely Sweden (1991), Finland and Switzerland (1992). Norway is another potential candidate in the near future. The further enlargement of the Community is expected to take place in different waves, and the EC needs to decide which countries will be included in the first wave, what, if anything, should be done with those who may have to remain in the waiting-room for much longer, and what kind of internal policy and institutional reforms may be required in a Community with more members. The old dilemma between deepening and enlargement has cropped up again. And, having made an exception for the very special case of East Germany, the Twelve have so far insisted on further deepening as a pre-condition for a new enlargement. Thus, the establishment of the internal market and the completion of the ratification process of the Maastricht treaty (January 1993, at the earliest) have been considered as the starting point of negotiations with the first group of candidates.

This first group is expected to include all the candidate countries from EFTA, which means a maximum of five countries, if Norway also decides to apply (Iceland not being a candidate). The EEA agreement already includes a very substantial part of the *acquis communautaire*. Thus, accession negotiations with those countries should start from a relatively advanced level. On the other hand, there is a variety of reasons why those countries should be attractive candidates for EC membership. They include the close affinity of economic, political, and social systems between the EC and EFTA countries; the small size and high level of development of the EFTA economies; their already high degree of integration in the Western European economic system; their macroeconomic performance which brings them closer to fulfilling the criteria for membership in the final stage of EMU than several of the existing EC members; and last but not least, the fact that they are all likely to be net contributors to the Community budget. The main difficulties in terms of the *acquis communautaire* lie in areas such as agriculture and tax harmonization; not surprisingly, matters which have been left out of the EEA agreement. Neutrality may also prove an important stumbling block, even though the radical change of political conditions in Europe seems to have deprived neutrality of much of its meaning.

There is, however, a deeply held suspicion among several EC members about how genuine is the change of heart implied by the applications made by EFTA countries. After all, these countries have for many years willingly (with the possible exception of Finland) stayed out of the political and institutional dimension of the integration process, opting only for intergovernmental forms of co-operation. If for essentially economic reasons, they have now decided to join the EC, they may try to block, once inside, more advanced forms of integration, and by joining in coalition with other members, their accession could bring about a fundamental shift in the intra-EC balance of power. Hence also the emphasis placed by the 'maximalists' in the Community on further deepening as a pre-condition for enlargement and the provision of guarantees (what kind?) by candidates on institutional and foreign policy questions. This is reminiscent of the first enlargement when another group of EFTA countries joined the Community. Should further institutional reforms be considered before 1996, when the next intergovernmental conference is supposed to take place, and should the candidates participate in this negotiation as full members? The answer to these questions will be determined by political developments in the next few years.

The other group of candidates consists of the three Mediterranean countries which have already applied and none of which is expected to be included in the first wave of new entrants. Turkey is by far the most difficult case. In terms of economic structures, it is still closer to many developing countries than the advanced industrialized countries of Western Europe. On the basis of purchasing power parities, the per capita income of Turkey is approximately 60 per cent of that of the poorest member of the EC of Twelve. Economic factors, together with a large and rapidly expanding population, a poor record in terms of human rights, strong cultural and religious differences from the rest of Europe, not to mention bilateral problems with Greece and the link established between EC–Turkish relations on the one hand and the resolution of the Cyprus problem on the other, have all combined to temper any enthusiasm that may have existed about Turkey's early accession to the Community. On the other hand, the strategic importance of the country, its potential role in some of the Central Asia republics of the former USSR, its membership of NATO, and its long institutional relations with the Community make an outright rejection of Turkey's 'European vocation' well nigh impossible, and also undesirable. Hence the enormous difficulties experienced by the EC in providing a clear answer to Turkey.

The case of the two Mediterranean islands, Cyprus and Malta, is very different indeed. Their level of economic development, combined with their very small size, suggests that accession to the EC would present very few problems, at least from an economic point of view. However, the same is not true of the institutional dimension. The prospect of entry of two more countries, the size of Luxembourg, raises difficult questions about representation in the different EC institutions and the rotating six-month Presidency of the Council; and it is doubtful whether the Community is yet ready to tackle them. In the case of Cyprus, EC membership may also be difficult to envisage as long as there is no solution to the internal political problem.

The precise number of potential candidates in Central and Eastern Europe is yet unknown. In fact, even the number of sovereign states in the region is unclear, given the strong tendency towards disintegration which has emerged in the post-communist era. Three groups of countries can be distinguished. The first, which is the most advanced economically and also the most advanced in terms of its institutional and trade relations with the EC, consists of Czechoslovakia, Hungary, and Poland. All three have already signed comprehensive association agreements with the Community and are keen on early membership. However, none of these countries could envisage adopting the *acquis communautaire* in the foreseeable future, unless very long transitional periods were offered which would essentially create a new category of EC membership. An important complicating factor will be the announced divorce between the Czech and the Slovak parts of the existing federation. An independent Slovakia faces a very difficult task of internal economic adjustment.

Trade relations with the Twelve are basically of the inter-industry kind. A large part of the exports from Czechoslovakia, Hungary, and Poland concentrates on the so-called sensitive sectors in the EC, which means that further trade liberalization from the side of the Community will depend on painful internal adjustment. Agriculture is a notable example; and the same applies to textiles and clothing. Internal EC adjustment will not be made easier by the fact that exports from those countries compete more directly with goods produced in some of the less developed countries and regions of the Community. There is also direct competition with southern European countries in terms of FDI and financial aid.

The second group consists of the Balkan countries. With respect to the signing of Europe agreements, Bulgaria and Romania follow closely on the toes of their northern neighbours, while Albania is further behind. Albania's economic structures and level of development bear little resemblance to those of any other European country. For a large part of its population, which has had until now by far the highest rate of increase in Europe, it is basically a question of subsistence. The final form of what used to be Yugoslavia is still uncertain, and so are therefore the prospects of institutional relations of some of the newly in-dependent republics with the Community. The latter has become closely involved in the intra-Yugoslav conflict in a peace-keeping capacity, although with very limited success until now. In fact, it is arguable that the EC may have precipitated the war there by pursuing a policy which has not taken sufficient account of the ethnic complexity of several of the constituent republics and the historical sensitivities of the peoples living there. The EC may be expected to play a stabilization role in the Balkans. Respect of existing international frontiers needs to be linked with guarantees for the rights of minorities in an area characterized by wide ethnic diversity. Thus, bilateral agreements with the countries of the region may need to include strong provisions in the political and foreign policy field, with the Community undertaking the role of external guarantor and stabilizer. This could also involve the creation of a multilateral regional framework of co-operation.

The third group of potential privileged partners of the EC is much less clearly defined. It includes parts of the former Soviet Union, although the line of demarcation separating privileged from less privileged has not yet been definitely traced. The three small Baltic republics are candidates for the signing of Europe agreements, which suggests that they should also be considered as candidates for EC membership in the somewhat distant future. They are certainly not in the same track as countries of the first group. It is also not yet clear whether other newly independent republics in the area should receive a similar treatment in the future. Perhaps, the Ukraine, Georgia, or even Russia? The formulation of some kind of EC policy in the area will at least partly depend on the development of the CIS (Commonwealth of Independent States); and the future is very uncertain indeed. If the Community and its individual members have implicitly or otherwise assumed a leading responsibility in the collective effort to help the countries of Eastern Europe in their transition to market economy and parliamentary democracy, including also the role of key mediator and stabilizer in regional conflicts, this is not also true for the former Soviet Union where the United States is keen on substituting the old superpower rivalry for a special and privileged relationship with Russia and, to a lesser extent, the other independent republics. Added to the weakness of central institutions and the insistence of the bigger members in pursuing separately their interests in the former Soviet empire, this may push the EC to a back-seat position.

The further enlargement of the Community could eventually lead to almost trebling of the existing membership. This, of course, raises, basic questions about institutional and policy reforms and the viability of the old integration model. But this trebling of members is not for tomorrow. Further enlargement will be extended over a relatively long period, which may therefore allow time for internal adjustment and reforms. Under normal circumstances, which however only rarely materialize in real life, we should expect no more than four or five new members to join until the end of the decade. In the case of EFTA candidates, who have been given priority in the queue, their accession to the EC should not create any major economic problems. For those countries, membership of the Community should add a political dimension to an already existing economic reality. True, their accession may accelerate the process of CAP reform and the shift towards income subsidies; and perhaps, it will also increase the pressure towards the renationalization of agricultural policy. But in general, the process of adopting the *acquis communautaire* should be relatively smooth. On the other hand, the effect on the political side of the European Union, in terms of institutions and traditional foreign policy, may be much more significant. This leads to more general questions about the effectiveness of the EC decision-making system in an enlarged Community, the division of powers between different levels of authority (subsidiarity revisited) and the interaction between the State and the market at the European level. Any substantial reforms will need to be discussed in the next intergovernmental conference.

Since the period of waiting in the antechamber for the other European countries promises to be quite long, the EC will need to decide whether the different kinds of association agreements offered to potential members (including the model of Europe agreements adopted for the former communist countries) provide an adequate framework for economic and political relations with its European neighbours. After the next enlargement, those European neighbours will consist almost entirely of countries with much lower standards of living than EC members and where often serious economic difficulties need to be tackled by weak political institutions in a context of wide ethnic diversity. Several ingredients of instability are already there. One concrete manifestation of economic/political problems spilling over the EC frontier will be the pressure of labour migration which has already become a major political issue in the host countries. The same problem exists, if only on a larger scale, in relation to the Community's southern sea frontier.

Access to EC markets will be absolutely crucial for the economic development of other European countries. But further trade liberalization from the EC side will depend on the degree of internal adjustment which is socially and politically acceptable. A trade-off between the exchange of goods and labour migration? This will require difficult political decisions. The same will be true with respect to the distribution of limited financial resources between the Structural Funds for the less developed countries and regions of the Community and external aid for non-member countries on the EC periphery. The claims on financial resources have increased substantially since the events of 1989, although external economic assistance could be seen as a better investment for the future than defence expenditure. Improving the living conditions in the Community's antechamber reserved for associated countries (and potential members) should be much more costly for the EC than allowing some EFTA countries in the Council room.

The Community has powerful instruments at its disposal if it decides to act as a stabilizing force on its eastern frontier. Privileged trade access, capital flows, technical assistance, and an institutionalized political dialogue which is seen as a precursor to EC membership are very powerful instruments indeed. A linkage has already been established in the existing agreements with the internal politics of associated countries, and most notably with the functioning of democratic institutions and the respect of human rights. Questions of security, the inviolability of international frontiers and the respect of minority rights are dealt with by other organizations such as NATO, the Conference on Security and Co-operation in Europe (CSCE) and the Council of Europe. The proliferation of multilateral organizations on the old continent does not necessarily lead to a more effective management of regional problems. The Community may decide, or perhaps it will be forced by external events, to play a more active role on the level of High Politics. The tragic events in Yugoslavia are the obvious example of how a more active role can be thrust upon a rather reluctant Community. On the other hand, the Maastricht treaty opens the road for closer co-operation among member countries in the field of foreign and security policy. Eastern Europe is likely to be given first priority in this respect.

The Europe agreements, and the more traditional association agreements before them, establish a framework for bilateral relations between the EC and the associated countries. In preparation for full membership, which may have to wait for several years, the EC could establish a system of more regular consultations with the associated countries covering a wide range of policy areas. This should make the Community more sensitive to the interests of its associates, while also helping with the socialization process of future candidates. Given the inequality of bilateral relations, when the EC is one of the two partners, and also given the desirability of promoting co-operation, if not only a dialogue, among several of the associates, the creation of one, or perhaps more, multilateral frameworks could be envisaged, in which regular political consultations will take place.

The EC will have a difficult task in trying to reconcile the objective of further internal construction with its role and responsibilities as a global and regional power. This will imply not only difficult choices in economic terms (trade adjustment, financial aid) but also a heavy strain on both the Commission and the Council. Can a weak political system, with limited administrative resources, deliver an effective and highly diversified foreign policy? The discrepancy between objectives on the one hand, instruments and institutions on the other is likely to re-surface once again.

The combination of Maastricht and the multiple challenge presented by Eastern Europe could strengthen the regional character and orientation of the Community. There has already been much talk of regionalism replacing multilateralism as the guiding principle of international trade and economic relations in general. In recent years, the United States as well seems to have succumbed to the charms of bilateralism and regionalism. The Caribbean initiative, the free trade agreement with Israel, and more importantly the North American Free Trade Agreement (NAFTA), with the participation of Canada and Mexico, are major examples of this tendency. Several economists, however, have challenged the alleged incompatibility between regional initiatives and multilateral trade liberalization, arguing that regional integration efforts, including most notably the EC, can complement those multilateral efforts (Lawrence, 1991; Pelkmans, 1993). The rapid process of globalization of produc-tion runs counter to any attempt towards exclusive regionalism. On the other hand, international institutions and arrangements can also play an important role. GATT is a good case in point. It remains to be seen how important a role it will be given after the Uruguay Round, especially by the bigger powers in the system.

10. The Political Economy of New Europe

Political initiatives and autonomous economic forces have brought about over the years a radical transformation of intra-European economic relations as well as economic relations between the region as a whole and the rest of the world. The EC and its different institutions have served until now as the main political instruments for the progressive integration of national economies. The interaction of politics with market forces, which naturally tend to transcend national frontiers, has shaped Western Europe's regional identity in the context of growing international economic interdependence.

The foundations of European economic integration were laid in the aftermath of the Second World War. Since then, three main phases can be distinguished. The first one coincided with the long period of high growth, the golden years of the 1950s and the 1960s, when the creation of common institutions and the establishment of a new legal framework were coupled with the rapid liberalization of intra-European trade, a liberalization which affected mainly trade in goods. In economic terms, the emphasis was on the elimination of border controls, and more specifically the elimination of tariffs and quota restrictions. Active intervention by the common institutions was limited essentially to agriculture, because of the peculiar characteristics of this sector and its large size in the national economies of that period. Provisions for such active intervention also existed in the case of coal and steel, although little use was made of them in those early years.

The second phase was one of stagnation. The long period of the economic boom came to an end as the European economies were buffeted by oil shocks and the rapid disintegration of the international monetary order. It could also be argued that the period of rapid growth and high employment carried the seeds of its own destruction, to use a now unfashionable term of political discourse. At the same time, there was stagnation in terms of integration, and much of the effort of the architects was expended on preventing the collapse of the still very incomplete European edifice. Although those were the years when national economic diversity became pronounced, stagnation is perhaps too strong a term to use since 'l'Europe des petits pas' continued to move along; and some of those small steps, such as the setting up of the EMS, proved in the end to be quite significant. Furthermore, the economic crisis did not prevent more countries from coming into the EC fold.

The beginning of the third phase can be situated around 1985, although some undercurrents of change had been noticeable earlier. The second half of the last decade was marked by the relaunching of the process of regional integration and a remarkable change of the economic and political climate in Western Europe. Integration moved into new areas: economic regulation inside national borders, and also to services and factors of production. The European agenda kept on expanding, and this eventually included the creation of new redistributive instruments. At the same time, the return of high growth and the creation of many new jobs went hand in hand with a major restructuring of industry; and, unlike earlier periods, this restructuring was no longer confined within national boundaries. The improvement in the economic environment was due to several factors, although political initiatives at the European level clearly played an important part in sustaining and strengthening the shift in the market expectations. It was a virtuous circle created by political initiatives and autonomous changes in the market place; and this led to an investment boom and the further expansion and deepening of regional integration.

It may be too early to assess with enough confidence the overall importance of recent developments for European integration. We still live through many of those developments, and there is always the temptation to exaggerate the significance of recent events and to look for major turning points which, with the passage of time, may fade into the grey area of the ordinary. Having said that, we shall, however, venture the prediction that the second half of the 1980s will occupy an important position in the history of European integration, having some similarities with an earlier phase connected with the establishment of the EEC and the first years of the transitional period provided for in the Treaty of Rome.

Steadily, the economic map of Western Europe has been transformed. More than forty years after the setting up of the first regional organizations, Western Europe is characterized by a high intensity of cross-border economic exchange. National economic frontiers have become less and less important, although they are still far from irrelevant. The transcending of economic frontiers applies not only to border controls but also increasingly to the various forms of indirect discrimination between producers and owners of factors of production on the basis of nationality, resulting from different regulatory frameworks in each country. Integration has slowly but steadily penetrated the area of mixed economy, and this has been achieved through a combination of deregulatory measures, the wide application of the principle of mutual recognition and the adoption of common rules at the European level. The emerging new 'regimes' vary considerably from one economic area to the other. Over the years, attention has shifted progressively from customs duties on goods to technical regulations and standards, to supervision rules of financial institutions and the opening of public procurement.

In terms of goods and services, the emphasis has been increasingly on market liberalization and the strengthening of competition. State intervention at the national level has been weakened as a result of the opening of frontiers and the constraints imposed at the European level, and this has not been compensated by similar intervention undertaken by the new central institutions. The latest phase of integration has been characterized by a strong deregulatory element, although the jury is still out as to the extent and the likely effects of this deregulation. The new approach to standards and the liberalization of financial services are two cases in point. Europe's industrial policy consists mainly of competition policy which applies both to private enterprises and also increasingly to state aids and nationalized firms. Otherwise, public intervention at the European level is mainly directed at the promotion of R & D, especially in high technology sectors, and inter-firm collaboration across national borders. This is a very mild European version of the old policy of national champions which it has partially replaced. It remains to be seen whether the relationship struck between the European bureaucracy and several large firms is too cosy for the comfort of European consumers.

Interventionism, in the form of an active industrial policy aimed at influencing the allocation of resources at the sectoral or even the micro level, has been the exception rather than the rule. The most notorious exception is agriculture where a highly elaborate set of rules and a very costly policy date back to the early years of integration. *Laisser-faire* in the farming sector is certainly not on as a political option; there are too many considerations, social and environmental among others, apart from narrow market efficiency, which exclude the complete dismantling of the CAP in the foreseeable future. However, the balance sheet of this sectoral policy, the most advanced form of a common policy at the European level, is not at all encouraging. There has been little correspondence between objectives and instruments, and this has led to much wastage of scarce resources and serious aggravation in relations with third countries. Furthermore, European institutions have shown great inflexibility in adjusting the CAP to changing economic circumstances. At long last, the Community now appears ready to proceed with a radical reform, shifting the burden of intervention to income subsidies.

Some interesting comparisons can be made between land agriculture and fishing, a relatively small sector where a common policy has also existed for years. Here the main problem has been to ration a declining stock among too many fishermen. But unlike agriculture, European policy in this sector has not relied on high support prices as a means of a largely ineffective social policy; and thus high financial costs have been avoided.

In the manufacturing sector, the only important example of a highly interventionist policy has been with respect to steel. It developed as a response to the deep crisis of the sector during the years of the long recession and it has been strongly challenged in terms of its efficiency. However, the only realistic alternative at the time was, arguably, national protectionist policies which could have caused even greater damage. With the improvement of the economic conjuncture, the elaborate system of European controls has been gradually dismantled and it is now difficult to envisage such a system being put together again in the foreseeable future.

Trade liberalization, including the progressive elimination

of a large number of NTBs, has helped to bring about a very high degree of trade interdependence among Western European countries; and this extends beyond the EC. Intra-European trade has grown faster than GDP and also faster than trade with the rest of the world. This interdependence is mainly true of goods, although it also increasingly applies to services which have been at the centre of the latest phase of integration.

The situation is qualitatively different with respect to capital and labour. Capital mobility has grown rapidly over the years, but this is more an international than a European phenomenon. For many years, capital flows between Western Europe and the rest of the world were several times bigger than intra-European flows; this applied both to FDI and hot money. The situation has changed somewhat in more recent years which have witnessed a substantial increase in intra-European capital mobility, aided by significant liberalization measures. The ability of national governments to influence the location of investment and capital flows in general has been curtailed, although it is still far from marginal. Increased capital mobility has been accompanied by a wave of cross-border mergers and acquisitions and the appearance of more and more European and international companies. There has been a progressive weakening of ties between firms and states. Although this trend is not only limited to Europe, it should eventually lead to a further increase in intra-European trade interdependence.

On the contrary, labour mobility across national frontiers has remained low, and most migrants have come from outside Western Europe. Within the region, professionals are more mobile, and their mobility is expected to increase further as a result of the current phase of liberalization. National labour markets are still characterized by wide diversity in terms of legislation and power relations between employers and trade unions; in other words, the European labour market remains highly compartmentalized. The persistence of national social and political realities largely explains the failure to make any serious advances with respect to European social policy, despite increased efforts in recent years. Decentralization and subsidiarity are likely to remain for some time the key principles in this area. Too much and too rapid a harmonization in terms of social policy could undermine the competitiveness of the less-developed economies.

In the macroeconomic field, intra-European co-operation has made significant progress after a slow and difficult start. The emphasis has been on the preservation of stable intra-European exchange rates, and close co-ordination of national monetary policies has been the means to this end. This has been based on a convergence of policy preferences towards more stability-oriented policies. Even greater convergence will be, however, required in order to achieve the objective of a complete EMU formally adopted in the latest treaty revision agreed at Maastricht; and the criteria for being admitted to the final stage are quite strict. This explains the fears of those who see the gates of paradise at the end of the road as perhaps too narrow for them as well as the anxiety of the more virtuous ones who dread too close an interaction with sinners. The creation of a complete monetary union, which should be the key issue of the 1990s, will entail the transfer of important policy instruments from the national to the European level, the political and economic implications of which will be quite considerable. In fact, the effect of a monetary union on the European economic system is likely to be greater than anything else that has happened until now.

In contrast to monetary policy, fiscal policy has remained until now distinctly national. Progress in terms of the harmonization of taxes has been very slow, and short of a sufficient harmonization, the method adopted for the elimination of intra-EC fiscal frontiers, although quite ingenious, carries with it considerable risks. An important achievement has been in the past the harmonization of indirect tax systems, although both the taxable base and the rates still differ considerably among countries. Budgetary policies have also continued to differ widely, as an important remnant of national economic sovereignty. To some extent, making a virtue out of a necesssity, substantial fiscal autonomy has been retained as part of a complete EMU. Weak fiscal power at the centre is the other side of the coin, and this also means that both the stabilization and redistribution functions of the central institutions will remain rather marginal for some time.

High economic interpenetration among Western European countries is combined with considerable openness *vis-à-vis* the rest of the world, although in this respect Western Europe as a whole is not much different from either the United States or Japan. Western Europe is the world's biggest economic bloc and it participates actively in international economic exchange.

Both the large restructuring of the manufacturing sector and the deregulation of financial services which have marked European economic developments in recent years are part of international phenomena; but they have not been, of course, unaffected by political decisions. Defensive policies and relatively high protective barriers exist in some sectors, both at the upper and the lower end of the international division of labour, where the Europeans perceive a loss of comparative advantage. But in general, it would be perhaps unfair to qualify European policies as more protectionist and inward-looking than those pursued by the other two major actors on the world economic scene. European trade and trade policies have traditionally had a strong regional dimension, although trade preferences have sometimes been used as a substitute for the lack of other policy instruments for the pursuit of wider objectives. These are, perhaps, the frustrations of an economic giant who remained for years a political dwarf. The relative weakness of central institutions also explains, at least in part, the defensive and reactive character of European policies.

The emerging European economic system is characterized by a rapidly increasing mobility of goods, services and factors of production; it is also characterized by a high degree of decentralization of political power. Some transfer of power in the economic sphere has, indeed, taken place away from the national level and this has been reinforced by the adoption of common rules. Yet, economic integration has had an undeniable effect on the interaction between the State and the market, bringing about a shift towards the latter. To the extent that there is a general trade-off between efficiency on the one hand and stability and equity on the other, the emphasis has been on the former, especially during the latest phase of integration, which is also consistent with the ideological shift of the 1980s. The new European economic system is closer to the American model: more dynamic perhaps, but also more unequal.

The weakening of state power is as much true of the regulatory and interventionist activities of public authorities as it is of their redistributive role. The examples of standards and financial services relate to the former, although, as mentioned before, the extent of the deregulatory impact of integration cannot as yet be fully assessed and it may be sometimes exaggerated. To the extent that the elimination of intra-European barriers takes place through the mutual recognition of national rules, instead of the adoption of new common ones, integration will also imply intensified competition among national economic systems. Siebert (1989: 6), for example, has argued that 'the arbitrage of consumers and firms will show which national regulatory framework is the best in the eyes of the consumer or the producer: national regulation has to pass a litmus test of private agents voting with their purses and with their feet'. Is this likely to prove a Darwinian process of selection or a search for the lowest common denominator which may mean less regard for economic externalities and the protection of the consumer and the environment?

There is also an external dimension to the reduced role of public authorities in influencing the allocation of resources. Thus, in a world of strategic trade interaction, European institutions are bound to remain for long weak players, while national governments have become less and less able to play this game effectively. However, this weakness of European institutions can only be seen as a true blessing by free traders.

On the other hand, the slow progress in terms of tax harmonization, combined with the growing mobility of goods and factors of production, and most significantly capital, is bound to lead to more arbitrage across national boundaries and hence to lower tax revenues. This will in turn constrain the ability of the national state to provide public goods and reduce income disparities among different regions and social classes through redistributive policies, while the contribution of European institutions in this respect remains limited. A less active role of the State is also true of the monetary sphere where Keynesian ideas have gradually lost their attraction for policy-makers. Thus, monetary instruments have been increasingly geared towards price stability, a tendency which has been reinforced by European co-operation. It will be taken to its logical conclusion by the creation of central monetary institutions and the further diminution of the discretionary power of elected governments over monetary policy.

Increased intra-European interdependence and growing competition inside the region coexist not only with a high degree of decentralization of power but also with wide economic, political and social diversity and large income disparities. In terms of the intensity and nature of economic interaction and also the level of economic development, we may refer to the existence of a core and a periphery in Western Europe (W. Wallace, 1990).

Economic boundaries are much more difficult to trace than political boundaries. Yet, there is broad agreement that Germany (although excluding the former German Democratic Republic), France, the Benelux countries, Denmark, and also large parts of England, Italy, and Spain form part of the core group, while the less developed countries and regions of the Community are situated on the geographical periphery. There is also little doubt that the heart of the European economic system lies in Germany which has acquired over the years some of the traits of a dominant economy. Germany is the biggest trading partner for almost all the other countries in the region, with a long history of surpluses, which has been, however, temporarily(?) halted by the economic effects of unification. Although also weakened by the large cost of integrating the new *Länder* into the German economic system, the Deutschmark remains the undisputed leader in the European exchange rate mechanism and the second largest international reserve asset, after the US dollar. Germany is also the most important 'hub' in terms of intra-European human flows.

Most of EFTA forms part of the core group, which in turn suggests that the underlying patterns of industrial production and financial integration, not to mention the levels of economic development, are not necessarily consistent with institutional arrangements. Although political and institutional arrangements are not, of course, irrelevant, they do not by themselves determine the direction of economic flows. The future accession of several EFTA countries to the EC will further strengthen this economic reality. On the other hand, recent political developments leave the possibility open for a progressive expansion of the European economic system in the future to other countries of central, eastern, and southern Europe.

Decentralization, combined with wide diversity and large disparities, means that private economic agents enter the intra-European competition with very different handicaps. The differences in terms of capital infrastructure between countries and regions are simply enormous; as for the differences in the quality of educational systems and the relative efficiency of institutional structures, they are not much smaller either. Yet, historical experience does not allow us to make any meaningful generalizations about the effect of integration on economic disparities between different countries and regions. If anything, those disparities have shown a tendency to narrow during periods of high growth. Thus, we may have to remain agnostic about the distributional effects of further integration in the future. Serious fears have been, however, expressed in relation to the internal market programme about the possibility of an increased concentration of economic power, the freezing of the intra-European division of labour and the widening of existing disparities, if the invisible hand were left unguided.

The Community attempts to deal with the problem of intercountry economic disparities through a small number of redistributive policy instruments, some differentiation in the application of common rules and policies (longer timetables for the weaker economies and temporary derogations have characterized a good part of the internal market legislation) and the attempt to improve general macroeconomic conditions which may in turn facilitate the task of economic adjustment. Although the redistributive dimension of EC policies has been very much strengthened in recent years, the main responsibility for redistribution still lies with national governments. While European institutions concentrate largely on liberalization measures and the pursuit of economic efficiency, national institutions continue to assume the main burden of income redistribution and welfare provision. For some of the less developed countries of Western Europe, further economic integration, and especially the steps which will need to be taken on the road towards EMU, create a fundamental challenge in terms of the modernization of political and institutional structures as a pre-condition for their successful participation in a new, more competitive environment.

The process of regional integration has largely depended until now on a wide perception of it as a positive sum game in which there are no important losers; and this extends beyond the inter-country distribution of gains and losses. Those who identify themselves as potential losers from integration will always be tempted to rally behind the nationalist flag, and this will become more evident as the economic effects spread to larger sections of the population. Integration, especially in the more recent phase associated with the creation of the European internal market, puts a premium on the flexibility and mobility of factors of production, as the inevitable outcome of the shift towards more competition and the free interplay of market forces. Capital should therefore be expected to gain more than labour, and professionals more than unskilled workers. This is also expressed in terms of relative political support for the integration process. Another consequence could be the further strengthening of economic dualism in some countries and regions, especially the less developed ones, where the more European and/or internationally competitive section of the economy may coexist for several years alongside with the large majority of economic agents operating still under local and national conditions. This economic dualism could be a source of serious political strain.

Historical experience suggests that there is a strong element of spill-over which operates between different areas of economic policy and also between economics and politics. To some extent, the integration process can be viewed as a series of dynamic disequilibria which create the conditions and the pressure for further extension and deepening. Looking for 'level playing fields' in different areas, as a means of protecting the interests of their national economic agents, governments have very often created the need for common rules and the transfer of powers to the centre, even when those same governments were ideologically committed to the much touted principle of subsidiarity. Competition policy is an obvious case in point; and there are also many others. Yet, there is no automaticity about this process of spill-over, nor is there any guarantee of its irreversibility in individual policy areas or sectors. Although there may be good reasons to expect in the future a further transfer of powers to the centre (*pace* Mrs Thatcher and the other diehards of undiluted national sovereignty), the European political system is likely to remain weak, and this weakness will be mainly the result of political and social integration lagging behind economic integration.

The weakness of common institutions at the European level is related to a shortage of legitimacy and administrative resources and also to the hybrid nature of those institutions. This has an important effect in terms of the prevailing economic order and the distribution of power between private and public agents, a subject which has been discussed above. But it also has other important implications. Economic action at the European level tends to be slow and often inefficient, because of cumbersome decision-making procedures and the inflexibility of common institutions. The cost-effectiveness of policy measures leaves much to be desired and so does the transparency of decision-making and the exercise of democratic control. On the other hand, the difficulties experienced in the implementation of jointly agreed measures at the national level have been steadily growing and so has the number of infringements of EC rules, thus raising awkward questions about the credibility of the European political system.

An interesting feature of the emerging political system, strongly reminiscent of practices on the other side of the Atlantic, is the growing importance of lobbies both in terms of numbers and influence with respect to policy outcomes at the European level. This development is probably directly related to the confederal nature of the European political system and its large size. But this is also bound to have a distributional effect, favouring those economic interests which can command better organization and knowledge, and the money which will usually buy the other two.

The latest phase of integration has seen the strengthening of the role of the EC Commission in the initiation, adoption, and execution stages of common policies. Similarly, there has been a reinforcement of its role as a representative of the Community interest *vis-à-vis* the rest of the world. This is at least partly attributable to the personality of its President since 1985; and thus also partly reversible in the future. This increase in power also carries with it considerable risks. The continuous growth of responsibilities undertaken by the Brussels institution has gradually stretched the limits of its own administrative capacities; and this could eventually work as a boomerang, especially when the management of ever increasing amounts of money is concerned. The annual reports of the EC Court of Auditors provide serious cause for concern. On the other hand, the more assertive role of the Commission has already provoked some strong reactions on behalf of national governments.

The so-called 'democratic deficit' of the Community is well known, and it is the product of the inability of national parliaments to control the European executive, while the European Parliament is still severely constrained by the treaties in playing such a role. Although boosted by new provisions contained in the SEA and now again with the Maastricht treaty, the role of the European Parliament is still heavily circumscribed. Furthermore, the lack of effective powers sometimes breeds irresponsibility. On the other hand, the Council of Ministers remains the main legislative body, and this means that ultimate power still resides with national governments represented in the Council; even though the intergovernmental character of decision-making has been progressively diluted, partly through the increasing recourse to majority voting. A more effective functioning of the Council in the future will require a much stronger administrative backing in Brussels.

One very interesting and almost totally unexpected development has been the way in which the system of the rotating six-month Presidency of the Council has worked in recent years as an important catalyst for the acceleration of the integration process. The success of each Presidency has been measured in terms of the number and importance of EC regulations and directives adopted during those six months, and this has acted as a powerful incentive for national governments occupying the chair in search of European acclaim and ways of impressing their public opinion.

Franco-German agreements have continued to provide the political basis for a large number of EC decisions. German economic power has been combined with French political influence, an influence which has been the product of recent European history and the domestic political system. It remains to be seen how this bilateral co-operation, and also the intra-European balance of power, will be affected by the increased political weight of united Germany which is no longer subject to the political constraints imposed on it by the victorious Allies after the end of the Second World War. Relative economic weakness and a certain ambivalence in British attitudes towards European integration, partly explained by historical factors and the more international orientation of the economy, have so far limited Britain's influence in the European decision-making process. On the other hand, Italy and also Spain, as a newcomer to the European club, have situated themselves in the mainstream of European politics, and they have thus exercised significant influence in the integration process.

As for the smaller members, a distinction can be drawn between the Benelux countries, which have been politically and economically better integrated in the core group, and the more recent members of the EC. The ability of countries to influence decisions has been largely a function of their relative economic weight, the efficiency of their administrative structures and their ability to form alliances in a complex multinational negotiation. Virtually all of the smaller countries have gradually discovered through experience that strong and effective EC institutions provide a better guarantee for the defence of their long-term national interests than traditional intergovernmentalism in the context of which those interests are more likely to be trampled down under the feet of the bigger powers. Hence the support of smaller countries for more supranational arrangements.

Asymmetries have always existed in the EC, and they have increased substantially with successive enlargements. One of the big questions in the next few years will be whether those asymmetries will take a more permanent and institutionalized form with the creation of different tiers; or will it be more appropriate to talk about non-uniform rules and a *géométrie variable*? The answer will mainly depend on whether it is the same countries which tend to drop out of different common policies and arrangements. The Schengen agreement for the elimination of all frontier controls is one recent and important example of an agreement which will not, at least for some time, apply to all member countries. It was started by the six founding members of the Community, and subsequently Spain, Portugal, and Greece also joined. On the other hand, in view of the importance of EMU, a long absence of some countries from the final stage could have major implications for the basic structure and the internal balance of power inside the Community, especially since countries like Italy and Belgium risk being left out.

The political game remains predominantly national. It is usually played as if national units were more independent economically than they actually are. It is true that both symbols and public money are still very much in the hands of national governments and this largely determines the direction of loyalties and expectations of their citizens. But it is also a problem of time-lag in the perception of change by political elites and even more so by public opinion. Malinvaud (1989: 374) has argued that 'achieving European unification while maintaining national autonomy' is like a tragedy in the spirit of Corneille; and national governments, together with their citizens, have been the main protagonists in this tragedy. The discrepancy between the growing Europeanization (and internationalization) of economic forces and the persistence of national political realities has also much to do with the predominantly technocratic nature of European integration.

Even when opinion polls show the existence of large majorities in favour of further integration, they essentially point to the existence of some kind of permissive consensus; and this consensus is not entirely solid. This lack of solidity has been manifested in several countries in the aftermath of the signing of the new treaty at Maastricht. Most European citizens do not feel directly involved in the integration process, and the decisions taken by the central European institutions are often seen as coming from outer space. Political parties, still very much national in their outlook and internal organization, have largely failed until now as two-way transmission belts in this respect. And certainly, it is not the harmonization of standards which will send people waving banners down the streets of Newcastle and Cagliari. Europe is being built from the top downwards and integration remains largely an affair of élites; it is the privileged who tend to identify more with Europe. Business interests play a more active part than labour, and this has been repeatedly confirmed; one characteristic example being the informal processes which had prepared the ground for the internal market programme back in the mid-1980s.

Economic integration has never been considered only as an end in itself, and this has been evident at every stage of the process. Wider political considerations have always occupied the most prominent position. Back in the 1950s, when the three Communities were set up, the emphasis on economic instruments was mainly for reasons of political expediency. European integration was seen mainly as a means of dealing with the German problem and the Cold War and, only to a lesser extent, was it an instrument for tackling the collective weakness of European economies. Regional integration in Western Europe is generally acclaimed as a major success, while most similar attempts in other parts of the world have had a very different fate. Western Europe has lived so far through one of the longest periods of peace and growing prosperity, and regional integration has played no small part in attaining those goals. The recent collapse of the Soviet empire has left the Community as the most solid piece of the European architecture; and also the most powerful, together with the old protector, namely the United States, whose role and presence in Europe is likely to be substantially reduced in the near future.

The nature of the problems faced some forty years ago by Monnet, Schuman, Adenauer, and de Gasperi has been substantially transformed in the meantime, although the problems themselves have not altogether disappeared. The defeated and divided Germany was successfully integrated into the EC and the Western alliance. But in the eyes of many Europeans, the German problem now presents itself in a new form: how to integrate a big and powerful Germany inside a strong Community, while removing the risk of domination over the other countries. This has become one of the main catalysts of the integration process; and this is also very much true of the latest constitutional revision.

Despite the creation of the two new pillars of European Union (foreign and security policy, justice and home affairs), the main emphasis is still on economics as a means to a political end. EMU has been chosen, largely for political reasons, as the key instrument for further integration in the post-1992 period. But it is doubtful whether it can have a similar effect on market expectations as the internal market programme had some years earlier. The economic boom of the late 1980s was a key to the successful implementation of the internal market programme, and it was due both to exogenous factors and the supply-side programme adopted at the EC level. A repetition of this positive experience during the first half of the 1990s may not prove that easy. International macroeconomic conditions are now different; and furthermore, a number of European governments will be forced to adopt contractionary policies in the next few years in order to meet the criteria for admission to the final stage of EMU. Modest rates of economic growth and even higher levels of unemployment would not be the best stimulant for 'Euro-euphoria'; as in the past, very much will depend on the state of the economy. On the other hand, popular reactions in the aftermath of the signing of the Maastricht treaty may suggest that the gap between economic and political integration (meaning here popular support for the further transfer of powers to central institutions) risks becoming too wide. And this is not unrelated to growing public disenchantment with governments and traditional political parties which is being witnessed in several European countries. Are we asking our economic feet to run much faster than our political heart can take? And more crucially, are we running in the right direction?

At the same time, the EC is faced with a double challenge as far as its immediate neighbours are concerned. While it embarks on new enlargement negotiations, it also needs to rethink its policy towards the other European countries which are likely to remain outside the Community for quite some time. Further enlargement is linked to deepening which should include important institutional reforms as an essential precondition for the functioning of a Community with more than twelve members. The Maastricht treaty has introduced relatively minor changes in this respect. With the prospect of successive rounds of enlargement in the future, some basic rethinking will have to be done before the next intergovernmental conference.

The EC will also need to develop a new policy *vis-à-vis* the other European countries which are not included in the first round of enlargement. This new policy will require imagination in designing new forms of institutionalized co-operation, painful internal economic restructuring, and more financial resources. It may also require a more assertive political role on behalf of the EC. The withdrawal of the Soviet army from the countries of Eastern Europe and the fall of communist regimes has marked the end of the Cold War. But this does not necessarily mean a more peaceful world. The experience until now demonstrates exactly the opposite. Will intra-European co-operation in the security field take a more concrete form in the foreseeable future and will the EC acquire a new role in this area, thus jettisoning the straightjacket of the 'civilian power'? And if so, what will be the effects on economic integration? The clear dividing line between high and low politics has only existed in some student textbooks, but never in the real world.

The danger of serious crises and prolonged instability in several parts of Eastern Europe looms large, while the old fear of invasion by the Soviet army has been replaced by the fear of invasion of a large number of refugees from the East. In fact, the problem is much wider. With the exception of a few politically stable and economically prosperous countries, most of whom are anyway expected to join the EC sooner rather than later, the whole of the Community's immediate neighbourhood is unlikely to be a model of stability in the foreseeable future; and this is clearly meant as an understatement. This applies not only to Eastern Europe but even more so to most countries in the Mediterranean where demographic growth, economic stagnation, and Islam can produce a very explosive mixture, not to mention the large waves of immigrants which should be expected to swell further.

The fear of being uncomfortably squeezed between the technological prowess of the United States and Japan on the one hand and the economic dynamism of new industrial powers on the other has been one of the primary motives of the latest phase of integration. The source of this fear has not disappeared, while the process of economic internationalization is likely to continue further. External competitiveness is linked to internal adjustment. On the other hand, the construction of 'Fortress Europe' has never been a realistic option. It remains to be seen whether Europe collectively shows any greater interest and readiness, compared with the other two major powers on the world economic scene, in strengthening international institutions which would assume the task of managing global economic interdependence. Intergovernmental co-operation at the multilateral or trilateral level has hardly proved sufficient for this task.

The European Community faces major challenges in a political and economic environment which is undergoing rapid transformation. It has changed and grown a great deal in recent years, and there is likely to be even greater pressure for change and adjustment in the near future. Its aspiration to construct a new European order without hegemony needs to be successfully married with the legitimate desire to play an active role in world affairs. External policy is, of course, closely linked to the functioning of the internal system. The objective of further integration in carefully selected areas needs to be reconciled with larger numbers and the persistence of wide economic and social diversity, while the search for more effective central institutions should be made compatible with the reality of a highly decentralized political system. On the other hand, further deepening of the process of integration and the strengthening of the European political system cannot happen without the existence of a solid popular base. It is basically a question of political integration slowly catching up with economic integration. This will be, indeed, a difficult challenge for the oldest nation-states in the international system. But, after all, national sovereignty has become a relative concept, especially in Europe.

Postscript

These lines are written in October 1992, before the final proofs are returned to the printer. A last chance to catch up with events as they rapidly unfold before our eyes. We now live in a period of crisis, with the Maastricht treaty in some doubt, the EMS badly shaken, and the European economies in deep recession. Closely interrelated, these factors will affect the functioning of the EC and general expectations about the future.

The realignment of exchange rates inside the ERM had been long overdue; it was finally imposed on recalcitrant governments by the market, and there may be more to come. As with Bretton-Woods, politicians were tempted to forget that the system in operation was one of fixed but adjustable exchange rates. Economic convergence, although considerable, was still not sufficient to justify the abandonment of the 'adjustable' element of the system. Large public deficits in Germany, resulting from huge transfers to the new *Länder* and the inability of the political system to match them with corresponding increases in taxation, forced the central bank to rely on high nominal interest rates as the main instrument in the fight against inflation.

When the Italians or the Greeks behave irresponsibly in terms of their fiscal policy, it is basically only they who will eventually suffer the consequences. When the Germans do the same, the other Europeans are also directly affected. High German interest rates forced the other countries of the ERM to follow suit in order to preserve existing intra-ERM exchange rates. But in times of serious economic recession the market seemed to doubt the ability of some governments to sustain a highly deflationary policy for long. On the other hand, the large interest rate differential between Germany and the United States, with the latter resorting to monetary policy as a means of bringing back to its feet an economy still suffering from the hangover caused by the huge debt explosion of the 1980s, produced large international capital inflows into Germany and further added to the tension inside the ERM. This was a repetition of an old, familiar play which several observers had been tempted to consider as increasingly out of fashion, perhaps imagining more symmetry between the European currencies than actually existed.

And then, an important political element was added. Opinion polls in France on the eve of the referendum on the Maastricht treaty suggested that there was a strong chance of a negative vote which would lead to an unceremonious burial of the treaty and the project for EMU. This cast further doubts on the sustainability of intra-ERM exchange rates, which in turn unleashed an unprecedented speculative attack on the weaker currencies. Massive central bank intervention and the further raising of interest rates to support the beleaguered currencies did not provide a strong enough defence of the status quo. The result was the devaluation of the Italian lira, followed by the temporary (?) withdrawal of sterling and the lira from the ERM, the devaluation of the peseta, and the reintroduction of capital controls in several member countries.

There was a close resemblance with events approximately twenty years ago, when the international crisis and the growing policy divergence inside the Community had turned an ambitious exchange rate arrangement (the snake), seen then as the precursor to a complete EMU, to a small DM-zone. The members of the core group had remained the same: Germany, the Benelux countries and, to a lesser extent, Denmark. As for the French franc, it also came under strong speculative attack, even though the economic fundamentals did not justify a devaluation. The continued participation of the French currency in the ERM is absolutely crucial for what has been left of the credibility of the system on which the future EMU is expected to be built. Again a repetition of the 1970s story, although, this time, the end of the story is still unknown.

The EMS has suffered a serious blow; the credibility of the system has badly cracked, and it will take a long time before even the best glue takes its effect. German unification has acted as a strong asymmetric shock on the system, and the solidity of the ERM in the near future (or more precisely, what has been left of it) will depend essentially on the German policy-mix and the general macroeconomic environment. The questions and doubts raised in the book with respect to the long transitional period leading to EMU, the criteria for admission to the final stage, the weakness of the fiscal arm of the future union, and the weakness of the European political system in general have, if anything, been strengthened by recent developments. A continued crisis of the EMS would seriously undermine the EMU project, even with the Maastricht treaty in operation; but it could also lead to an acceleration of the process of monetary integration in an attempt to cut short the transitional period of fixed but adjustable exchange rates, and thus deal effectively with the uncertainty associated with it. Such a strategy would require eventually an adjustment to the Maastricht treaty and it would also, almost inevitably, need to rely on a core group of currencies, at least in the initial stage. The wider political consequences of such an action cannot be underestimated.

It may be paradoxical that the twelve governments are struggling to ratify a treaty, the most important part of which, namely EMU, is very likely to require major changes before it is actually applied. But since the crucial decisions with respect to EMU will be taken only some years later, the ratification of the existing treaty would help to buy precious time in a difficult period. The alternative would be, arguably, much worse: a failure to ratify the treaty could provoke a serious internal crisis in the Community when the latter is expected to act as an important factor of stability on the European continent. As for the option of renegotiation of the treaty, this appears to be highly unrealistic, especially in view of the absolutely minimalist attitude adopted by two countries, namely Britain and Denmark. Some of the old intra-European divisions on the nature of the integration process have survived for a long time.

Although with a marginally positive result, the French referendum has confirmed some of the weaknesses of European integration: the gap between the economic and the political dimension, not only in terms of institutions but also in terms of the political discourse and popular perceptions; the fears of potential losers at a time of rapid economic liberalization and while the safety net of national welfare systems becomes less solid; and the technocratic, and often incomprehensible to the public, nature of European agreements, themselves the product of delicate compromises between member governments. Although, undoubtedly, the scepticism expressed in France and other European countries with respect to the Maastricht treaty has also much to do with factors which are not directly related to the process of regional integration, such as the adverse economic environment prevailing at the time, the anxiety caused by developments on the Com-

munity's eastern frontier, and the crisis of governance manifest-
ing itself in several national political systems, it is, however,
indicative of inherent weaknesses of the European construction.

It is possible that we have already reached the end of another
phase in European integration, although it is still too early to
pronounce a clear judgement. Very different conclusions have
been drawn from the apparent lack of enthusiasm with which a
large section of European public opinion, and most notably
approximately one half of the voters in Denmark and France
(although significantly less in Ireland), has responded to the
Maastricht treaty. The answer to the problem lies, for some
people, in the wider application of the subsidiarity principle and
the return of several powers back to the national level. Thus, an
accusing finger is being pointed at the EC Commission as an
over-zealous producer of unnecessary European legislation, with
social policy being a prominent example. For many others, however, the solution to the crisis points in exactly the opposite
direction, namely the strengthening of European institutions, and
the further acceleration of the process of economic and, even
more so, political integration. This should also involve the
injection of more democracy in the Community decision-making
process. The Maastricht treaty, it is thus argued, is not very
popular largely because it does not go far enough in the construction of a proper federal system.

Although hardly anybody dares to object, at least in public, to
the principle of subsidiarity (still a very unfamiliar term in virtually
all European languages), the debate which has developed around
it often hides important differences in terms of interest and approach. At a rather superficial level, the main division appears to
be between the centralizers (or Euro-fanatics) and the defenders
of national sovereignty and pluralism, or alternatively those who
want to bring decisions closer to the people directly affected
by them. The term 'federalist' is deliberately not used here to
describe the former category, since true federalists should be,
precisely, in favour of a strong decentralization of powers. Interestingly enough, national politicians who have acted as flagbearers
for subsidiarity have often been themselves strongly opposed to
any kind of decentralization of power inside their own country.
The justification, of course, refers to some almost mythical
qualities of the nation-state.

The division has, in fact, more than one dimension. Weak

central institutions and a more intergovernmental kind of co-operation or integration is more akin to the interests of the bigger countries. It is not, therefore, surprising that virtually all small- and medium-sized countries of the EC have not been much impressed by the campaign on subsidiarity nor have they sup-ported the attacks against the Commission, even though they may, at least unofficially, recognize some of the excesses or weaknesses of the Brussels executive. In any event, the latter is most unlikely to escape unharmed from this confrontation. There is yet another, much less obvious, dimension which is related to the kind of economic order for the emerging regional entity. We have argued in the book that the process of economic integration, and especially the latest phase associated with the internal market programme, has weakened the role of the State in the European mixed economy. If market integration continues to proceed much faster than political integration, this will have further effects on the ability of the State to regulate and redistribute. A weak European political system in an increasingly integrated regional economy would also have important consequences for the balance between different economic and social groups and the kind of society we live in. This could become in the future an important issue as the political debate becomes more European and less national.

References

Adams, Heinz, and Rekittke, Karl (1989), 'Ein weites Feld: Standort Europa'. In H. Adams (ed.), *Europa 1992: Strategien, Strukturen, Ressourcen*. Frankfurt: Frankfurter Allgemeine Zeitung.

Aglietta, Michel, Brender, Anton, and Coudert, Virginie (1990), *Globalisation financière: L'Aventure obligée*. Paris: Economica.

Albert, M., and Ball, R. J. (1983), *Towards European Recovery in the 1980s*. Luxembourg: European Parliament, Working Documents 1983–4.

Allsopp, Christopher, and Chrystal, K. Alec (1989), 'Exchange rate policy in the 1990s', *Oxford Review of Economic Policy*, Autumn.

Atkinson, Anthony *et al.* (1992), *La Désinflation Compétitive, Le Mark et les Politiques Budgétaires en Europe*. Paris: Seuil.

Aubry, M. (1989), *Relations Sociales et Emploi*. Paris: La Documentation Française.

Balassa, Bela (1961), *The Theory of Economic Integration*. London: Allen and Unwin.

Baldwin, Richard (1989), 'The growth effects of 1992', *Economic Policy*, 9, October.

Baltensberger, Ernst, and Dermine, Jean (1987), 'The role of public policy in ensuring financial stability: a cross-country, comparative perspective'. In R. Portes and A. K. Swoboda (eds.), *Threats to International Financial Stability*. Cambridge: CEPR/Cambridge University Press.

Bank for International Settlements (1990), *60th Annual Report*. Basle: BIS.

—— (1991), *61st Annual Report*. Basle: BIS.

Bank of England (1989), 'The Single European Market: survey of the UK financial services industry', *Bank of England Quarterly Bulletin*, 3, May.

Barry, Andrew (1990), 'Technical harmonisation as a political project'. In G. Locksley (ed.), *The Single European Market and the Information and Communication Technologies*. London: Belhaven.

Begg, Iain (1989), 'European integration and regional policy', *Oxford Review of Economic Policy*, Summer.

Bieber, Roland, Jacqué, Jean-Paul, and Weiler, Joseph H. H. (eds.) (1985), *An Ever Closer Union*. Brussels: Commission of the EC/The European Perspectives Series.

Biehl, Dieter (1988a), 'On maximal versus optimal tax harmonization'.

In R. Bieber *et al.* (eds.), *1992: One European Market?*. Baden-Baden: Nomos for the European Policy Unit, Florence.

—— (1988*b*), 'Die Reform der EG-Finanzverfassung aus der Sicht einer ökonomischen Theorie der Föderalismus'. In M. Streit (ed.), *Wirtschaftspolitik zwischen ökonomischer und politischer Rationalität*. Wiesbaden.

Bisignano, Joseph (1990), 'Banking in the EEC: structure, competition and public policy'. In G. Kaufman (ed.), *Bank Structure in Major Countries*. New York: Kluwer.

Blanpain, Roger, *et al.* (1983), *The Vredeling Proposal*. Deventer: Kluwer.

Boltho, Andrea (ed.) (1982), *The European Economy*. Oxford: Oxford University Press.

—— (1989), 'European and United States regional differentials: a note', *Oxford Review of Economic Policy*, Summer.

—— and Allsopp, Christopher (1987), 'The assessment: trade and trade policy', *Oxford Review of Economic Policy*, Spring.

Bourguinat, Henri (ed.) (1989), *Conflits et négociations dans le commerce international: L'Uruguay round*. Paris: Economica.

Bressand, Albert (1990), 'Beyond interdependence: 1992 as a global challenge', *International Affairs*, January.

Buckwell, Alan, *et al.* (1982), *The Costs of the Common Agricultural Policy*. London: Croom Helm.

Buigues, Pierre, and Goybet, Philippe (1989), 'The Community's industrial competitiveness and international trade in manufactured goods'. In A. Jacquemin and A. Sapir (eds.), *The European Internal Market*. Oxford: Oxford University Press.

Buigues, Pierre, and Jacquemin, Alexis (1989), 'Strategies of firms and structural environments in the large internal market', *Journal of Common Market Studies*, September.

Calleo, David P. (1982), *The Imperious Economy*. Cambridge, Mass.: Harvard University Press.

Camps, Miriam (1964), *Britain and the European Community 1955–1963*. Oxford: Oxford University Press.

—— (1967), *European Unification in the Sixties*. London: Oxford University Press for The Royal Institute of International Affairs/Council on Foreign Relations.

Cecchini, Paolo (1988), *The European Challenge: 1992*. Aldershot: Wildwood House.

Centre for European Policy Studies (1989), *Indirect Tax Harmonization in the European Community*, Report of a Working Group, July.

Cline, William (ed.) (1983*a*), *Trade Policy in the 1980s*. Washington, DC: Institute for International Economics.

—— (1983*b*), '"Reciprocity": a new approach to world trade policy?'.

In id. (ed.), *Trade Policy in the 1980s*. Washington, DC: Institute for International Economics.

Cnossen, Sijbren, and Shoup, Carl (1987), 'Co-ordination of value-added taxes'. In S. Cnossen (ed.), *Tax Co-ordination in the EC*. Deventer: Kluwer.

Colchester, Nicholas, and Buchan, David (1990), *Europe Relaunched: Truths and Illusions on the Way to 1992*. London: The Economist Books/Hutchinson.

Collins, Doreen (1983), *The Operation of the European Social Fund*. London: Croom Helm.

—— (1985), 'Social policy'. In A. El-Agraa (ed.), *The Economics of the European Community*. Oxford: Philip Allan.

Commissariat du Plan (1980), *L'Europe: Les Vingt Prochaines Années*. Paris: La Documentation Française.

Commission of the EC (1977), *The Role of Public Finance in European Economic Integration* (MacDougall report), i and ii. Brussels.

—— (1985*a*), *Completing the Internal Market. White Paper from the Commission to the European Council*. Luxembourg: Office for Official Publications of the European Communities, June.

—— (1985*b*), 'Annual Economic Report 1985–86', *European Economy*, 26, November.

—— (1988*a*), 'The economics of 1992', *European Economy*, 35, March.

—— (1988*b*), 'Creation of a European financial area', *European Economy*, 36, May.

—— (1988*c*), 'The social dimension of the internal market', *SEC* (88) 1148.

—— (1989*a*), 'Facing the challenges of the early 1990s', *European Economy*, 42, November.

—— (1989*b*), 'Implementation of the legal acts required to build the single market'. Brussels: *COM* (89) 422, 7 September.

—— (1989*c*), *Eighteenth Report on Competition Policy*. Luxembourg: Office for Official Publications of the European Communities.

—— (1990*a*), 'Completing the internal market: an area without internal frontiers'. Brussels: *COM* (90) 552, 23 November.

—— (1990*b*), *Second Survey on State Aids in the European Community in the Manufacturing and Certain Other Sectors*. Luxembourg: Office for Official Publications of the European Communities.

—— (1990*c*), 'The impact of the internal market by industrial sector: the challenge for the Member States', *European Economy*, Special Edition.

—— (1990*d*), *Employment in Europe*. Luxembourg: Office for Official Publications of the European Communities.

—— (1990*e*), 'One market, one money', *European Economy*, 44, October.

—— (1990*f*), *European Regional Development Fund. Fourteenth*

Annual Report (1988). Luxembourg: Office for Official Publications of the European Communities.

—— (1991*a*), 'Annual economic report 1991–92', *European Economy*, 50, December.

—— (1991*b*), *Employment in Europe*. Luxembourg: Office for Official Publications of the European Communities.

—— (1991*c*), *Panorama of EC Industries 1991–92*. Luxembourg: Office for Official Publications of the European Communities.

—— (1991*c*), *The Regions in the 1990s*. Luxembourg: Office for Official Publications of the European Communities.

Committee for the Study of Economic and Monetary Union (Delors report) (1989), *Report on Economic and Monetary Union in the European Community*. Luxembourg: Office for Official Publications of the European Communities.

Crandall, Robert (1986), 'The EC-US steel trade crisis'. In L. Tsoukalis (ed.), *Europe, America and the World Economy*. Oxford: Blackwell for the College of Europe.

Dahrendorf, Ralf (1988), *The Modern Social Conflict*. London: Weidenfeld and Nicholson.

Dashwood, Alan (1983), 'Hastening slowly: the Community's path towards harmonisation'. In H. Wallace, W. Wallace, and C. Webb (eds.), *Policy Making in the European Community*, 2nd edn. Chichester: Wiley.

Davenport, Michael (1982), 'The economic impact of the EEC'. In A. Boltho (ed.), *The European Economy*. Oxford: Oxford University Press.

—— (1989), *The Charybdis of Anti-Dumping. A New Form of EC Industrial Policy?* London: The Royal Institute of International Affairs, Discussion Paper No. 22.

de Cecco, Marcello (1989), 'The European Monetary System and National Interests'. In P. Guerrieri and P. C. Padoan (eds.), *The Political Economy of European Integration*. Hemel Hempstead: Harvester Wheatsheaf.

de Grauwe, Paul (1987), 'International trade and economic growth in the European Monetary System', *European Economic Review*, 31.

—— and Papademos, Lucas (eds.) (1990), *The European Monetary System in the 1990s*. London: Longman for CEPS and the Bank of Greece.

—— and Gros, Daniel (1991), 'Convergence and divergence in the Community's economy on the eve of Economic and Monetary Union'. In P. Ludlow (ed.), *Setting European Community Priorities 1991–92*. London: Brassey's/Centre for European Policy Studies.

Dekker, W. (1985), *Europe-1990*. Eindhoven: Philips.

de la Serre, Françoise (1991), 'La Communauté européenne et l'Europe central, et orientale', *Revue du Marché Commun et de l'Union Européenne*, August.

Demaret, Paul (1986), 'The extraterritoriality issue in the transatlantic context: a question of law or diplomacy?'. In L. Tsoukalis (ed.), *Europe, America and the World Economy*. Oxford: Blackwell.

Denton, Geoffrey (1984), 'Restructuring the EC budget: implications of the Fontainebleau agreement', *Journal of Common Market Studies*, December.

Dermine, Jean (ed.) (1990), *European Banking in the 1990s*. Oxford: Blackwell.

Diamantouros, Nikiforos (1986), 'The southern European NICs', *International Organization*, Spring.

Dicke, Hugo (1989), 'Die Vollendung des Binnenmarktes — der Versuch einer Zwischenbilanz', *Die Weltwirtschaft*, 1.

Donges, Jurgen (1989), *Die EG auf dem Weg zum Binnenmarkt: Erwartungen, Konsequenzen, Probleme*. Kiel Working Paper. No. 360.

Dornbusch, Rudiger (1988), 'The European Monetary System, the dollar and the yen'. In F. Giavazzi, S. Micossi, and M. Miller (eds.), *The European Monetary System*. Cambridge: Cambridge University Press.

—— (1989), 'Credibility, debt and unemployment: Ireland's failed stabilization', *Economic Policy*, 8, April.

Drèze, Jacques, *et al.* (1987), *The Two-Handed Growth Strategy for Europe: Autonomy Through Flexible Cooperation*. Brussels: Centre for European Policy Studies Paper No. 34.

Duchêne, François (1973), 'The European Community and uncertainties of interdependence'. In M. Kohnstamm and W. Hager (eds.), *Nation Writ Large*. London: Macmillan.

Eichengreen, Barry (1990), 'One money for Europe: lessons from the US currency union', *Economic Policy*, 10, April.

Emerson, Michael (ed.) (1984), *Europe's Stagflation*. Oxford: Oxford University Press.

—— (1988), *What Model for Europe?* Cambridge, Mass.: MIT Press.

European Investment Bank (1991), *Annual Report 1990*. Luxembourg.

Fels, Gerhard (1988), 'Der Standort Bundesrepublik im internationalen Wettbewerb', *Hamburger Jahrbuch für Wirtschafts- und Gesellschaftspolitik*.

Flanagan, Robert (1987), 'Labor market behavior and European economic growth'. In R. Lawrence and C. Schultz (eds.), *Barriers to European Growth*. Washington, DC: Brookings.

Franks, Julian, and Mayer, Colin (1990), 'Capital markets and corporate control: a study of France, Germany and the UK', *Economic Policy*, 8, April.

Fratianni, Michele, and Peeters, Theo (eds.) (1978), *One Money for Europe*. London: Macmillan.

Funabashi, Yoichi (1988), *Managing the Dollar: From the Plaza to the Louvre*. Washington DC: Institute for International Economics.

Fursdon, Edward (1980), *The European Defence Community: A History*. London: Macmillan.

George, Ken, and Jacquemin, Alexis (1992), 'Dominant firms and mergers', *The Economic Journal*, January.

Geroski, P. A. (1989), 'The choice between diversity and scale'. In E. Davis *et al., 1992: Myths and Realities*. London: London Business School.

Giannitsis, Tassos (1988), *I Entaxi stin Evropaiki Koinotita kai Epiptoseis sti Viomihania kai sto Exoteriko Emporio* [Accession to the European Community and Effects on Industry and External Trade]. Athens: Institute for Mediterranean Studies.

Giavazzi, Francesco, and Giovannini, Alberto (1989), *Limiting Exchange Rate Flexibility: The European Monetary System*. Cambridge, Mass.: MIT Press.

Giavazzi, F., and Pagano, M. (1988), 'The advantage of tying one's hand: EMS discipline and central bank credibility', *European Economic Review*, 32.

Giavazzi, Francesco, and Spaventa, Luigi (1990), *The 'New' EMS*. CEPR Discussion Paper No. 369.

Giersch, Herbert (ed.) (1983), *Reassessing the Role of Government in the Mixed Economy: Symposium 1982*. Tübingen: Mohr.

—— (1985), *Eurosclerosis*. Kiel: Kiel Discussion Paper No. 112, October.

Gilibert, P. L., and Steinherr, A. (1989), *The Impact of Financial Market Integration on the European Banking Industry*. EIB Paper No. 8, March. Luxembourg: European Investment Bank.

Gilpin, Robert (1987), *The Political Economy of International Relations*. Princeton, N. J.: Princeton University Press.

Giovannini, Alberto (1989), 'National tax systems versus the European capital market', *Economic Policy*, 9, October.

Golembe, Carter, and Holland, David (1990), 'Banking and securities'. In G. C. Hufbauer (ed.), *Europe 1992: An American Perspective*. Washington, DC: Brookings.

Goodhart, Charles (1993), 'The external dimension of EMU'. In L. Tsoukalis (ed.), *Europe and Global Economic Interdependence*. Brussels: Presses Interuniversitaires Européennes for the College of Europe and the Hellenic Centre for European Studies.

Goyens, M. (1992), 'Consumer protection in a Single European Market: What challenge for the EC agenda?', *Common Market Law Review*, February.

Grabitz, Eberhard (ed.) (1984), *Abgestufte Integration: Eine Alternative zum herkömmlichen Integrationskonzept?* Kehl am Rhein: Engel.

Grilli, Vittorio (1989), 'Europe 1992: issues and prospects for the financial markets', *Economic Policy*, 9, October.

Grimm, Doris, Schatz, Werner, and Trapp, Peter (1989), *EG 1992: Strategien, Hindernisse, Erfolgaussichten.* Kiel: Kiel Discussion Paper, No. 151.

Gros, Daniel (1990), *Capital Market Liberalization and the Fiscal Treatment of Savings in the European Community.* CEPS Working Party Report. Brussels: Centre for European Policy Studies.

—— and Thygesen, Niels (1988), *The EMS: Achievements, Current Issues and Directions for the Future.* Brussels: Centre for European Policy Studies Paper No. 35.

—— (1992), *European Monetary Integration: From the European Monetary System Towards Monetary Union.* London: Longmans.

Haas, Ernst (1958), *The Uniting of Europe.* London: Stevens.

Hager, Wolfgang (1982), 'Protectionism and autonomy: how to preserve free trade in Europe', *International Affairs*, Summer.

Helpman, E., and Krugman, P. (1985), *Market Structure and Foreign Trade: Increasing Returns, Imperfect Competition and the International Economy.* Cambridge, Mass.: MIT Press.

Hine, R. C. (1985), *The Political Economy of International Trade*, Brighton: Harvester.

—— (1989), 'Customs union and enlargement: Spain's accession to the European Community', *Journal of Common Market Studies*, September.

Hufbauer, Gary C. (1990), 'An overview'. In id. (ed.), *Europe 1992: An American Perspective.* Washington, DC: Brookings.

Ishikawa, Kenjiro (1990), *Japan and the Challenge of Europe 1992.* London: Pinter/The Royal Institute of International Affairs.

Jacquemin, Alexis, Buigues, Pierre, and Ilzkovitz, Fabienne (1989) 'Horizontal mergers and competition policy in the European Community, *European Economy*, 40, May.

Jacquemin, Alexis, and Sapir, André (1988), 'European integration or world integration?', *Weltwirtschaftliches Archiv*, 124/1.

—— (1990), 'La perspective 1992 et l'après Uruguay round', *Economie prospective internationale*, 44.

Joerges, Christian (1988), 'The new approach to technical harmonization and the interests of consumers: reflections on the requirements and difficulties of a Europeanization of product safety policy'. In R. Bieber *et al.* (eds.), *1992: One European Market?* Baden-Baden: Nomos for the European Policy Unit, Florence.

Julius, DeAnne (1990), *Global Companies and Public Policy: The Growing Challenge of Foreign Direct Investment.* London: Pinter/The

Royal Institute of International Affairs.

Kay, J. A. (1989), 'Myths and realities'. In E. Davis *et al., 1992: Myths and Realities*. London: London Business School.

Kelly, Margaret, *et al.*, (1988), *Issues and developments in international trade policy*, IMF Occasional Paper No. 63.

Kennedy, Paul (1988), *The Rise and Fall of the Great Powers*. London: Unwin Hyman.

Keohane, Robert O. (1984), *After Hegemony*. Princeton, NJ: Princeton University Press.

—— and Hoffmann, Stanley (1991), 'Institutional change in Europe in the 1980s'. In R. Keohane, and S. Hoffmann (eds.), *The New European Community: Decisionmaking and Institutional Change*. Boulder – San Francisco-Oxford: Westview Press.

Krasner, Stephen D. (1982), 'Structural causes and regime consequences: regimes as intervening variables', *International Organization*, 36/2.

Krugman, Paul (1986), 'Introduction: new thinking about trade policy'. In id. (ed.), *Strategic Trade Policy and the New International Economics*. Cambridge, Mass.: MIT Press.

—— (1987), 'Economic integration in Europe: some conceptual issues'. In T. Padoa-Schioppa *et al., Efficiency, Stability and Equity*. Oxford: Oxford University Press.

—— (1988), *EFTA and 1992*. Geneva: EFTA Occasional Paper No. 23.

—— (1989), *Exchange Rate Instability*. Cambridge, Mass.: MIT Press.

—— (1990), 'Policy problems of a monetary union'. In P. de Grauwe, and L. Papademos (eds.), *The European Monetary System in the 1990's*. London: Longman for CEPS and the Bank of Greece.

Kruse, D. C. (1980), *Monetary Integration in Western Europe: EMU, EMS and Beyond*. London: Butterworths.

Lafay, Gerard *et al.*, (1989), *Commerce international: La Fin des avantages acquis*. Paris: Economica.

Lagrange, Maurice (1971), 'L'Europe institutionelle: réflexions d'un témoin', *Revue du Marché Commun*, June.

Lamfalussy, A. (1989), 'Macro-coordination of fiscal policies in an economic and monetary union in Europe'. In Committee for the Study of Economic and Monetary Union, *Report on Economic and Monetary Union in the European Community*. Luxembourg: Office for Official Publications of the European Communities.

Lawrence, Robert Z., and Schultze, Charles L. (eds.), (1987), *Barriers to European Growth*. Washington DC: Brookings.

Lawrence, Robert Z. (1991), 'Emerging regional arrangements: building blocks or stumbling blocks?'. In R. O'Brien (ed.), *Finance and the International Economy: 5 — The Amex Bank Review Prize Essays*. Oxford: Oxford University Press for Amex Bank Review.

Lebègue, Daniel (1988), *La Fiscalité de l'épargne dans le cadre du marché intérieur européen*. Rapport du groupe de travail du Conseil National du Crédit, June.

Leskelä, Jukka and Parviainen, Seija, (1990), *EFTA Countries Foreign Direct Investments*. Geneva: EFTA Occasional Paper No. 34.

Lindberg, Leon N. (1963), *The Political Dynamics of European Economic Integration*. Stanford: Stanford University Press.

Lister, Marjorie (1988), *The European Community and the Developing World*. Aldershot: Avebury.

Ludlow, Peter (1982), *The Making of the European Monetary System*. London: Butterworths.

McAleese, Dermot, and Matthews, Alan (1987), 'The Single European Act and Ireland: implications for a small member state', *Journal of Common Market Studies*, September.

MacBean, A. I., and Snowden, P. N. (1981), *International Institutions in Trade and Finance*. London: Allen & Unwin.

McGowan, Francis, and Seabright, Paul (1989), 'Deregulating European airlines', *Economic Policy*, 9, October.

McKinnon, Ronald (1963), 'Optimum currency areas', *American Economic Review*, 53.

Maddison, Angus (1982), *Phases of Capitalist Development*. Oxford: Oxford University Press.

Malinvaud, Edmond (1989), Comment on A. Giovannini, 'National tax systems versus the European capital market', *Economic Policy*, 9, October.

Markovits, Andrei (ed.) (1982), *The Political Economy of West Germany*. New York: Praeger.

Mattera, Adolfo (1988), *Marché unique européen: Ses règles, son fonctionnement*. Paris: Jupiter.

Mayer, Otto (1989), 'Zur sozialen Dimension des europäischen Binnenmarktes'. In O. Mayer *et al.* (eds.), *Der europäische Binnenmarkt: Perspektiven und Probleme*. Hamburg: Weltarchiv.

Mayes, David (1989), 'The effects of economic integration on trade'. In A. Jacquemin and A. Sapir (eds.), *The European Internal Market*. Oxford: Oxford University Press.

Melitz, Jacques (1988), 'Monetary discipline and co-operation in the European Monetary System: a synthesis'. In F. Giavazzi, S. Micossi, and M. Miller (eds.), *The European Monetary System*. Cambridge: Cambridge University Press.

Messerlin, Patrick (1987), 'The European iron and steel industry and the world crisis'. In Y. Mény and V. Wright (eds.), *The Politics of Steel: Western Europe and the Steel Industry in the Crisis Years (1979–1984)*. Berlin: Walter de Gruyter.

—— (1989), 'The EC anti-dumping regulations: a first economic appraisal, 1980–1985', *Weltwirtschaftliches Archiv*, 125/3.

Michalski, Anna, and Wallace, Helen (1992), *The European Community:*

The Challenge of Enlargement. London: Chatham House Discussion Paper.

Milward, Alan S. (1984), *The Reconstruction of Western Europe 1945–51*. London: Methuen.

Minford, Patrick (1985), *Unemployment: Cause and Cure*, 2nd edn. Oxford: Blackwell.

Mitsos, Achilleas (1989), *I Elliniki Viomichania sti Diethni Agora* [Greek Industry in the International Market]. Athens: Themelio.

Molle, Willem (1990), *The Economics of European Integration*. Aldershot: Dartmouth.

Monnet, Jean (1976), *Mémoires*. Paris: Arthème Fayard.

Montagnier, Gabriel (1991), 'Harmonisation fiscale communautaire', *Revue trimestrielle de droit européen*, Jan.-Mar.

Moravcik, Andrew (1991), 'Negotiating the Single European Act'. In R. Keohane, and S. Hoffmann (eds.), *The New European Community: Decisionmaking and Institutional Change*. Boulder-San Francisco – Oxford: Westview Press.

Mortensen, Jorgen (1990), *Federalism vs. Co-ordination: Macroeconomic Policy in the European Community*. Brussels: Centre for European Policy Studies, CEPS Paper No. 47.

Moyer, Wayne, and Josling, Timothy (1990), *Agricultural Policy Reform: Politics and Process in the EC and USA*. London: Harvester Wheatsheaf.

Müller, Jurgen, and Owen, Nicholas (1989), 'The effect of trade on plant size'. In A. Jacquemin and A. Sapir (eds.), *The European Internal Market*. Oxford: Oxford University Press.

Mundell, Robert (1961), 'A theory of optimum currency areas', *American Economic Review*, 51.

Murphy, Anna (1990), *The European Community and the International Trading System*, i and ii. Brussels: Centre for European Policy Studies, Nos. 43 and 48.

Myrdal, Gunnar (1956), *An International Economy*. London: Routledge and Kegan Paul.

—— (1957), *Economic Theory and Underdeveloped Regions*. London: Duckworth.

Mytelka, Lynn Krieger (1989), 'Les alliances stratégiques au sein du programme européen ESPRIT', *Economie prospective internationale*, 37.

National Economic and Social Council (1989), *Ireland in the European Community: Performance, Prospects and Strategy*. Dublin: NESC.

Nell, Philippe (1990), 'EFTA in the 1990s: the search for a new identity', *Journal of Common Market Studies*, June.

Nello, Susan S. (1991), *The New Europe: Changing Economic Relations Between East and West*. Hemel Hempstead: Harvester Wheatsheaf.

Neven, Damien (1990), 'EEC integration towards 1992: some distributional aspects', *Economic Policy*, 10, April.

Nicol, William, and Yuill, Douglas (1982), 'Regional problems and policy'. In A. Boltho (ed.), *The European Economy*. Oxford: Oxford University Press.

Noelke, M., and Taylor, R. (1981), *EEC Protectionism: Present Practice and Future Trends*, i. Brussels: European Research Associates.

Nye, Joseph (1990), *Bound to Lead: The Changing Nature of American Power*. New York: Basic Books.

Odell, Peter (1986), *Oil and World Power*, 8th edn. Harmondsworth: Penguin.

OECD, *Main Economic Indicators*, various issues. Paris.

—— (1987), *The Future of Migration*. Paris: OECD.

—— (1989), *OECD Economic Surveys: Spain*. Paris: OECD.

—— (1990), *OECD Economic Surveys: Greece*. Paris: OECD.

—— (1991), *OECD Economic Surveys: Ireland*. Paris: OECD.

Olson, Mancur (1982), *Rise and Decline of Nations*. New Haven, Conn.: Yale University Press.

Owen, Nicholas (1983), *Economies of Scale, Competitiveness and Trade Patterns in the European Community*. Oxford: Clarendon Press.

Padoa-Schioppa, T. *et al.* (1987), *Efficiency, Stability and Equity*. Oxford: Oxford University Press.

—— (1988), 'The European Monetary System: a long-term view'. In F. Giavazzi, S. Micossi, and M. Miller, (eds.), *The European Monetary System*. Cambridge: Cambridge University Press.

Page, Sheila (1981), 'The revival of protectionism and its consequences for Europe', *Journal of Common Market Studies*, September.

Pearce, Joan, and Sutton, John (1986), *Protection and Industrial Policy in Europe*. London: Routledge and Kegan Paul for The Royal Institute of International Affairs.

Pelkmans, Jacques (1980), 'Economic theories of integration revisited', *Journal of Common Market Studies*, June.

—— (1984), *Market Integration in the European Community*. The Hague: Martinus Nijhoff.

—— (1987), 'The new approach to technical harmonization and standardization', *Journal of Common Market Studies*, March.

—— (1988), 'A grand design by the piece? An appraisal of the internal market strategy'. In R. Bieber *et al.* (eds.), *1992: One European Market?* Baden-Baden: Nomos for the European Policy Unit, Florence.

—— (1990), 'Regulation and the single market: an economic perspective'. In H. Siebert (ed.), *The Completion of the Internal Market*. Tübingen: Mohr/Institut für Weltwirtschaft an der Universität Kiel.

—— (1993), 'Regionalism in world trade: vice or virtue?'. In L. Tsoukalis (ed.), *Europe and Global Economic Interdependence*.

Brussels: Presses Interuniversitaires Européennes for the College of Europe and the Hellenic Centre for European Studies.

—— and Sutherland, Peter (1990), 'Unfinished business: the credibility of 1992'. In *Governing Europe*, i. Brussels: Centre for European Policy Studies.

—— and Winters, Alan (1988), *Europe's Domestic Market*. London: Routledge and Kegan Paul for The Royal Institute of International Affairs, Chatham House Papers, 43.

Perez Amorós, Francisco and Rojo, Eduardo (1991), 'Implications of the single European market for labour and social policy in Spain', *International Labour Review*, vol. 130, No. 3.

Perroux, François (1959), 'Les formes de concurrence dans le marché commun', *Revue d'économie politique*, 1.

Pijpers, A., *et al.* (1988), *European Political Cooperation in the 1980s: A Common Foreign Policy for Western Europe?* Dordrecht: Nijhoff.

Pinder, David (1986), 'Small firms, regional development and the European Investment Bank', *Journal of Common Market Studies*, March.

Pipkorn, J. (1986), 'Die Mitbestimmung der Arbeitnehmer als Gegenstand gemeinschaftlicher Rechtsentwicklung'. In H. Von Lichtenberg (ed.), *Sozialpolitik in der EG*. Baden-Baden: Nomos.

Pomfret, Richard (1986), *Mediterranean Policy of the European Community: A Study in Discrimination in Trade*. London: Macmillan.

Prest, Alan (1983), 'Fiscal policy'. In P. Coffey (ed.), *Main Economic Policy Areas of the EEC*. The Hague: Martinus Nijhoff.

Puchala, Donald (1972), 'Of blind men, elephants and international integration', *Journal of Common Market Studies*, March.

Rhodes, Martin (1992), 'The future of the "social dimension": labour market regulation in post-1992 Europe', *Journal of Common Market Studies*, March.

Robson, Peter (1987), *The Economics of International Integration*, 3rd edn. London: Allen and Unwin.

Romero, Federico (1990), 'Cross-border population movements'. In W. Wallace (ed.), *The Dynamics of European Integration*. London: Pinter for The Royal Institute of International Affairs.

Rosenblatt, Julius, *et al.* (1988), *The Common Agricultural Policy of the European Community*, International Monetary Fund, Occasional Paper No. 62, November.

Rosenthal, Douglas (1990), 'Competition policy'. In G. C. Hufbauer (ed.), *Europe 1992: An American Perspective*. Washington, DC: Brookings.

Sachs, J., and Wyplosz, C. (1986), 'The economic consequences of President Mitterrand', *Economic Policy*, 2.

Sandholz, Wayne, and Zysman, John (1989), '1992: recasting the European bargain', *World Politics*, October.

Sapir, André (1989), 'Does 1992 come before or after 1990? On regional versus multilateral integration'. In R. Jones and A. Krueger (eds.), *The Political Economy of International Trade*. Oxford: Blackwell.

Schmitt von Sydow, Helmut (1988), 'The basic strategies of the Commission's White Paper'. In R. Bieber *et al.* (eds.), *1992: One European Market?* Baden-Baden: Nomos for the European Policy Unit, Florence.

Sellekaerts, W. (1973), 'How meaningful are empirical studies on trade creation and diversion?', *Weltwirtschaftliches Archiv*, 109.

Servan-Schreiber, Jean-Jacques (1967), *Le Défi américain*. Paris: Denoël.

Shackleton, Michael (1990), *Financing the European Community*. London: Pinter for The Royal Institute of International Affairs.

Sharp, Margaret (1990), 'Technology and the dynamics of integration'. In W. Wallace (ed.), *The Dynamics of European Integration*. London: Pinter for The Royal Institute of International Affairs.

—— and Shearman, Claire (1987), *European Technological Collaboration*. London: Routledge and Kegan Paul for The Royal Institute of International Affairs, Chatham House Papers 36.

Shonfield, Andrew (1976) 'International economic relations of the Western world: an overall view'. In A. Shonfield (ed.), *International Economic Relations of the Western World 1959–1971*. London: Oxford University Press for The Royal Institute of International Affairs.

Siebert, Horst (1989), *The Harmonization Issue in Europe: Prior Agreement or a Competitive Process?* Kiel Working Papers, 377, June.

—— (1991), 'German unification: the economics of transition', *Economic Policy*, October.

Siedentopf, Heinrich, and Hauschild, Christoph (1988), 'The implementation of Community legislation by the member states: a comparative analysis'. In H. Siedentopf and J. Ziller (eds.), *Making European Policies Work: The Implementation of Community Legislation in the Member States*, i. London: Sage/European Institute of Public Administration.

Smith, Alasdair, and Venables, Anthony (1990), 'Automobiles'. In G. C. Hufbauer (ed.), *Europe 1992: An American Perspective*. Washington, DC: Brookings.

Strange, Susan (1976), 'International monetary relations'. In A. Shonfield (ed.), *International Economic Relations of the Western World 1959–1971*. London: Oxford University Press for The Royal Institute of International Affairs.

—— (1985), 'Interpretations of a decade'. In L. Tsoukalis (ed.), *The Political Economy of International Money*. London: Sage for The Royal Institute of International Affairs.

—— (1987), *Casino Capitalism*. Oxford: Blackwell.

—— (1988), 'A "dissident" view'. In R. Bieber *et al.* (eds.), *1992: One*

European Market? Baden-Baden: Nomos for the European Policy Institute, Florence.

Strasser, Daniel (1990), *Les Finances de l'Europe: Le Droit budgetaire et financier des Communautés Européennes*, 6th edn. Paris.

Straubhaar, Thomas (1988), 'International labour migration within a Common Market: some aspects of EC experience', *Journal of Common Market Studies*, September.

Tarditi, Secondo, *et al.* (1989), *Agricultural Trade Liberalization and the European Community*. Oxford: Oxford University Press.

Taylor, Paul (1983), *The Limits of European Integration*. London: Croom Helm.

—— (1989), 'The new dynamics of EC integration in the 1980s'. In J. Lodge (ed.), *The European Community and the Challenge of the Future*. London: Pinter.

Thygesen, Niels (1979), 'The emerging European Monetary System: precursors, first steps and policy options', *Bulletin of the National Bank of Belgium*, April.

Tinbergen, Jan (1954), *International Economic Integration*. Amsterdam: Elsevier.

Tindemans, Leo (1975), Report on European Union, Supplement to *Bulletin of the EC*, 1.

Tracy, Michael (1989), *Government and Agriculture in Western Europe 1880–1988*, 3rd edn. London: Harvester Wheatsheaf.

Tsoukalis, Loukas (1977a), *The Politics and Economics of European Monetary Integration*. London: Allen and Unwin.

—— (1977b), 'The EEC and the Mediterranean: is "global" policy a misnomer?' *International Affairs*, July.

—— (1981), *The European Community and its Mediterranean Enlargement*. London: Allen and Unwin.

—— (ed.) (1992), *I Ellada kai i Evropaiki Koinotita: I Proklisi tis Prosarmogis* [Greece and the European Community: The Challenge of Adjustment]. Athens: Papazissis for the Hellenic Centre for European Studies.

—— and Strauss, Robert (1985), 'Crisis and adjustment in European steel: beyond laisser-faire', *Journal of Common Market Studies*, March.

Ungerer, Horst *et al.* (1990), *The European Monetary System: Developments and Perspectives*. Washington DC: International Monetary Fund, Occasional Paper, No. 73.

United Nations Centre on Transnational Corporations (1991), *Report 1991: The Triad in Foreign Direct Investment*. New York.

Van Ginderachter, J. (1989), 'La réforme des fonds structurels', *Revue du Marché Commun*, May.

Vanhove, Norbert, and Klaassen, Leo (1987), *Regional Policy: A European Approach*, 2nd edn. Aldershot: Avebury.

VerLoren van Themaat, Pieter (1988), 'The contributions to the establishment of the internal market by the case-law of the Court of Justice of the European Communities'. In R. Bieber *et al.* (eds.), *1992: One European Market?* Baden-Baden: Nomos for the European Policy Unit, Florence.

Vernon, Raymond (1971), *Sovereignty at Bay*. New York: Basic Books.

Viñals, José, *et al.* (1990), 'Spain and the "EEC cum 1992" shock'. In C. Bliss and J. Braga de Macedo (eds.), *Unity With Diversity in the European Economy*. Cambridge: Cambridge University Press/Centre for Economic Policy Research.

Vogel-Polsky, Eliane (1989), 'L'acte unique ouvre-t-il l'espace social européen?', *Droit social*, February.

Wallace, Helen (1983), 'Distributional politics: dividing up the Community cake'. In H. Wallace, W. Wallace, and C. Webb (eds.), *Policy Making in the European Community*, 2nd edn. Chichester: John Wiley.

Wallace, William (1990), *The Transformation of Western Europe*. London: Pinter for The Royal Institute of International Affairs.

Walters, Alan (1990), *Sterling in Danger*. London: Fontana.

Wegner, Manfred (1985), 'External adjustment in a world of floating: different national experiences in Europe'. In L. Tsoukalis (ed.), *The Political Economy of International Money*. London: Sage for The Royal Institute of International Affairs.

Weiler, Joseph (1983), 'Community member states and European integration: is the law relevant?' In L. Tsoukalis (ed.), *The European Community: Past, Present and Future*. Oxford: Blackwell.

Werner Report (1970), 'Report to the Council and the Commission on the realization by stages of Economic and Monetary Union in the Community', Supplement to *Bulletin of the EC*, 3.

Winham, Gilbert R. (1986), *International Trade and the Tokyo Round Negotiation*. Princeton, NJ: Princeton University Press.

Woolcock, Stephen (1989), *European Mergers: National or Community Controls?* The Royal Institute of International Affairs, RIIA Discussion Paper No. 15.

Woolcock, Stephen, Hodges, Michael, and Schreiber, Kristin (1991), *Britain, Germany and 1992: The Limits of Deregulation*. London: Pinter for The Royal Institute of International Affairs.

Wyplosz, Charles (1990), 'Macro-economic implications of 1992'. In J. Dermine (ed.), *European Banking in the 1990s*. Oxford: Blackwell.

Yannopoulos, George (1989), 'The management of trade-induced structural adjustment: an evaluation of the EC's Integrated Mediterranean Programmes', *Journal of Common Market Studies*, June.

Zavvos, George (1988), 'The EEC banking policy for 1992', *Revue de la banque*, March/April.

Index